Dr. Bryen Lorenz

Microelectronics

Microelectronics

CHARLES L. ALLEY
KENNETH W. ATWOOD

A Reston Book
Prentice-Hall
Englewood Cliffs, New Jersey 07632

Library of Congress Cataloging in Publication Data

Alley, Charles L.
  Microelectronics.

  1. Microelectronics.   I. Atwood, Kenneth W.
II. Title.
  TK7874.A429   1986   621.381′7   85-11933
  ISBN 0-8359-4587-1

A Reston Book
Published by Prentice-Hall
A Division of Simon & Schuster, Inc.
Englewood Cliffs, NJ 07632

10   9   8   7   6   5   4   3   2   1

Contents

PREFACE xi

1   INTRODUCTION   1
   1.1   Typical Electronic Systems, 1
   1.2   The Operational Amplifier, 3
   1.3   The Cathode-Ray Tube, 9
   1.4   The Oscilloscope, 11
          Problems, 12

2   SEMICONDUCTORS   17
   2.1   Conductors, Insulators, and Semiconductors, 17
   2.2   Conduction in a Semiconductor, 20
   2.3   Charge Carrier Density in a Semiconductor, 23
   2.4   Doped Semiconductors, 27
   2.5   Gallium Arsenide Semiconductors, 32
          Problems, 35

3   DIODES   39
   3.1   The p-n Junction, 39
   3.2   Diffusion and Recombination, 41
   3.3   Diffusion Currents and Drift Currents, 42
   3.4   Minority Carrier Diffusion, 46
   3.5   The Diode Equation, 49
   3.6   The Forward-Biased Diode, 51
   3.7   The Reverse-Biased Diode, 56
   3.8   Rectifier-Filter Circuits, 57
   3.9   Reference Diodes, 62
   3.10   Junction Capacitance, 68
   3.11   Diffusion Capacitance, 72
          Problems, 73

4   JUNCTION TRANSISTORS   77
   4.1   Bipolar Junction Transistors, 77
   4.2   The Ebers-Moll Model, 83
   4.3   Characteristic Curves, 85
   4.4   The Common-Emitter Configuration, 89
   4.5   Regions of Operation, 91
   4.6   Base Width Modulation, 93
   4.7   Transistor Ratings, 95
   4.8   Basic Amplifiers, 99
   4.9   Transistor Capacitances, 101
          Problems, 104

5   TRANSISTOR CHARACTERISTICS AND MODELS   107
     5.1   A Small-Signal Model, 107
     5.2   A Low-Frequency Model, 111
     5.3   The Hybrid-π Model, 113
     5.4   Two-Port Network Theory, 115
     5.5   The H-Parameters, 118
     5.6   Determination of Hybrid-π Parameters, 121
     5.7   The Graded-Base Transistor, 123
     5.8   Transistor High-Frequency Characteristics, 126
            Problems, 129

6   TRANSISTOR AMPLIFIER CONFIGURATIONS   131
     6.1   Common-Emitter Amplifier Analysis, 132
     6.2   The Common-Base Amplifier, 139
     6.3   The Common-Collector Amplifier, 145
     6.4   Comparison of CE, CB, and CC Amplifiers, 150
     6.5   Transducer Gain, 150
            Problems, 154

7   TRANSISTOR AMPLIFIER DESIGN   159
     7.1   Transistor Selection, 159
     7.2   Variations of Transistor Parameters, 161
     7.3   Stabilizing-Bias Circuits, 163
     7.4   Stabilized Bias Design, 167
     7.5   Boot-Strapping for the CC Amplifier, 172
     7.6   The Determination of Coupling and
            Bypass Capacitors, 173
     7.7   Time Response, 176
            Problems, 183

8   FIELD-EFFECT TRANSISTORS   187
     8.1   The Junction FET, 187
     8.2   An FET Circuit Model, 192
     8.3   The Depletion-Mode MOSFET, 197
     8.4   The Enhancement-Mode MOSFET, 200
     8.5   Dual Insulated-Gate Field-Effect Transistors, 206
            Problems, 214

9   FET AMPLIFIERS   217
     9.1   Bias Circuit Design for the FET, 217
     9.2   Common Source Frequency Characteristics, 224
     9.3   The Source Follower, 226
     9.4   The Common-Gate Amplifier, 229
     9.5   Combination FET-BJT Amplifier, 232

9.6   Voltage-Controlled Resistance, 233
Problems, 235

10   DIRECT-COUPLED AMPLIFIERS 239
10.1   The Darlington Connection, 240
10.2   n-p-n–p-n-p Combinations, 243
10.3   Differential Amplifiers, 245
10.4   The Emitter-Coupled Amplifier, 249
10.5   The Cascode Amplifier, 252
10.6   Complementary-Symmetry Amplifiers, 258
Problems, 263

11   MULTISTAGE AMPLIFIERS 267
11.1   Gain and Bandwidth Considerations in Cascaded
Amplifiers, 267
11.2   dB Gain, 271
11.3   The S-Domain and Frequency Domain, 272
11.4   Straight-Line Approximations of Gain and Phase
Characteristics (Bode Plots), 273
11.5   Amplifier Noise, 279
11.6   Resistor Noise, 281
11.7   Diode Noise, 283
11.8   Transistor Noise, 284
11.9   Noise Optimization for Low-Frequency Transistor
Amplifiers, 289
11.10 Noise in FET Amplifiers, 292
11.11 Signal-to-Noise Ratio Calculations, 293
Problems, 293

12   NEGATIVE FEEDBACK 297
12.1   The Effect of Feedback on Gain, Distortion, and
Bandwidth, 297
12.2   The Effect of Negative Feedback on Input Impedance
and Output Impedance, 303
12.3   Feedback Circuits, 307
12.4   Stability of Feedback Circuits, 312
12.5   Phase-Lead Compensation, 318
12.6   Log Gain Plots, 323
12.7   Peaking of the Frequency Response, 325
12.8   Phase-Lag Compensation, 330
Problems, 332

13   LINEAR INTEGRATED CIRCUITS 337
13.1   Monolithic Circuits, 337
13.2   Basic Linear Integrated Circuits, 342

13.3   Operational Amplifiers, 345
13.4   Input Offset and Output Offset Voltages, 348
13.5   The Summing and Integrator Circuit, 350
13.6   Amplifier Compensation, 351
13.7   Slew Rate, 353
13.8   Op Amps Having Fast Rise Time, 356
13.9   Amplifier Input Compensation, 362
13.10   The Voltage Follower, 364
13.11   Voltage Comparators, 367
        Problems, 373

14   ACTIVE FILTERS   377
14.1   Filters Utilizing the Inverting Op Amp
        Configuration, 378
14.2   Active Filter Design, 379
14.3   Stability of Active Filters, 384
14.4   High-Order Filters, 386
14.5   A Double-Pole Low-Pass Filter Utilizing a Voltage
        Follower, 388
14.6   A Double-Pole High-Pass Filter, 391
14.7   Multiple-Feedback Filters, 392
14.8   Twin-T Band-Stop Filter, 398
14.9   The Universal Active Filter, 400
        Problems, 404

15   POWER AMPLIFIERS   407
15.1   Power Output and Efficiency, 407
15.2   Push-Pull Amplifiers, 410
15.3   Power Output and Efficiency of Class B
        Amplifiers, 413
15.4   Complementary-Symmetry Amplifiers, 416
15.5   Quasi-Complementary Symmetry, 420
15.6   IC Power Amplifiers, 425
15.7   VMOS Power Amplifiers, 428
15.8   Thermal Conduction and Thermal Runaway, 435
15.9   Control Circuits for Power Amplifiers, 441
        Problems, 443

16   POWER SUPPLIES   447
16.1   Power Supply Characteristics, 447
16.2   Capacitor Input Filters, 448
16.3   Circuit and Voltages for the Full-Wave Rectifier, 450
16.4   Emitter-Follower Regulators, 455
16.5   Closed-Loop Regulators, 461
16.6   Integrated Voltage Regulators, 466

16.7   Switching Regulators, 473
16.8   Silicon-Controlled Rectifiers and Triacs, 478
16.9   Applications of SCRs, 480
          Problems, 485

17   OSCILLATOR CIRCUITS   489
17.1   RC Oscillators, 490
17.2   Function Generators, 493
17.3   Classes of Operation, 497
17.4   Tuned Amplifiers, 501
17.5   Tuned-Amplifier Bandwidth, 505
17.6   Tuned Coupling Circuits, 507
17.7   Tapped-Tuned Circuits, 511
17.8   LC Oscillators, 515
17.9   Crystal-Controlled Oscillators, 522
          Problems, 527

Appendix   TRANSISTOR CHARACTERISTICS   531

          INDEX   535

Preface

This book is intended as a basic introduction to electronic devices and their application to a variety of electronic circuits and systems. The reader should have a background in electrical circuits equivalent to that obtained in an introductory engineering-circuits course.

The goal of the authors is to introduce the subject in the most exciting, understandable, up-to-date manner. This goal is accomplished by proceeding from specific to general concepts in a logical way, with practical examples and student problems to illustrate and solidify the concepts. An effort has been made to include the newest devices and techniques.

The content of the book is adequate for a three-quarter or two-semester course. The first nine chapters deal primarily with device characteristics and basic circuit applications. The remaining eight chapters deal with applications in more complex but frequently used circuits and systems.

Chapter 1 may be omitted without loss of continuity or understanding of the chapters that follow. Chapter 2 may also be omitted by students who have studied solid-state physics or semiconductor theory. Chapter 2 is intended to be neither a rigorous nor a comprehensive study of solid-state theory, but it is adequate for an understanding of the devices treated in this book.

Microelectronics

1

Introduction

During the twentieth century, technological developments have occurred at a phenomenal rate. The electronic technology has not only contributed to this development but has itself experienced fantastic growth. Thus we find that radio, television, radar, stereo, automatic control systems, and computers are familiar items to most members of our society. While these devices are widely used, the general public considers them too complicated to be understood by anyone but an expert. Actually, most electronic devices can be reduced to a group of interconnected basic electronic building blocks or circuits. It is not too difficult to design electronic equipment if one is well acquainted with the characteristics of these basic blocks or units. In fact, many of these basic electrical circuits are now available in compact packages known as *integrated circuits*. Much of this text is devoted to the analysis and design of these basic blocks or circuits; however, we also consider the problems encountered when the discrete circuits or the integrated circuits are interconnected.

**1.1
TYPICAL
ELECTRONIC
SYSTEMS**

The engineer usually begins the design of an electronic device by laying out a *block diagram*. This block diagram is then used as a guide while he designs the circuits represented by the various blocks in the system.

To illustrate this concept, let us consider the block diagram of an audio amplifier. Assume the input signals will be from a microphone or from a phonograph pickup. The output signal must be large enough to drive (or activate) a loudspeaker. A typical block diagram for this device might be sketched as shown in Fig. 1.1.

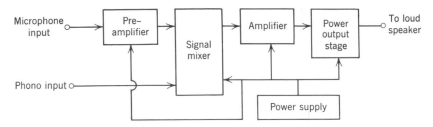

Fig. 1.1. Block diagram of an amplifier.

The electrical signal from a microphone is usually much smaller than the electrical signal from a phonograph pickup. Therefore, the preamplifier is used to amplify the signal from the microphone until it is about as large as the signal from the phonograph pickup. The signal mixer combines the two signals (microphone and phonograph) and is usually designed so an adjustment of one signal will not affect the other signal. The combined signals are then sent through an amplifier to increase the signal amplitude. This amplified signal is used to drive the power output stage. This power output stage must produce enough electrical signal power to drive the loudspeaker at the required signal level. The power supply furnishes enough dc power to operate all of the amplifier stages. In a simple portable circuit, this power supply may be a single battery. However, in a high-quality, high-power amplifier, this power supply may be quite a sophisticated and complex circuit.

The block diagram for a simple radio receiver is given in Fig. 1.2. The radio signal is received by the antenna and amplified in the radio frequency (RF) amplifier. This RF amplifier is tuned to amplify only one radio signal and to reject all the other radio signals which may be present. The detector converts the radio signal to an electrical audio frequency signal. This audio signal is then amplified in the audio amplifier. The amplified audio signal is used to drive the power amplifier stage which furnishes enough power to drive the loudspeaker. Again, a power supply is required to furnish dc power to the various amplifiers in the radio. While this receiver is very simple, it is typical of the early radio receivers. More complex circuits are currently used.

When the engineer has drawn the block diagram of the system and has determined the required specifications for this system, he is then ready to design a circuit for each of the blocks. Some of these circuits may be available as integrated circuit modules. In these cases, the entire block (or in some cases, several blocks) may be available in a single, tiny *integrated circuit* (*IC*) or *chip*.

In order to use integrated circuits or to design the various stages of a system, one must understand how these circuits operate. Since the actual circuits may be quite complex, engineers reduce this complexity

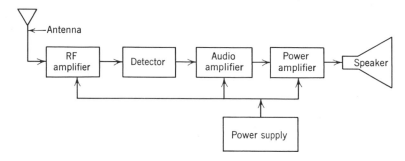

Fig. 1.2. Block diagram of a radio receiver.

by the *modeling* technique. A model of an electrical circuit or device is a simpler configuration which has electrical characteristics which *approximate* the characteristics of the actual circuit. Some of these models may be rather crude and only yield "ball-park" results. For example, if a simple model provides analytical results that agree within 10 percent of the measured parameters of the actual device, the model is often considered to be adequately accurate. In general, models tend to become more complex as we demand a closer approximation to the characteristics of the actual circuit. Thus, a compromise is usually required so the model can be as simple as possible and still yield results with the required accuracy. In order to illustrate this modeling technique, let us consider the operational amplifier.

**1.2
THE
OPERATIONAL
AMPLIFIER**

The concept of the "operational amplifier" was first described in a paper in 1947.[1] These *operational amplifiers* (or *op amps* as they are often called) have the following characteristics:

1. Very high input impedance.

2. Very high voltage gain.

3. Very low output impedance.

In constructing a model for the operational amplifier, *Thévenin's theorem* can be applied at the input terminals. Thévenin's theorem states that a linear two-terminal network, regardless of its complexity, can be replaced by a series circuit composed of an impedance and a voltage source. (If an ac model is required, the circuit may only be valid at one frequency when the impedance is complex.) For the typical op

---

[1] J. R. Ragazzini, R. H. Randall, and F. A. Russell, "Analysis of Problems in Dynamics by Electronic Circuits," *Proceedings of the IRE*, May 1947.

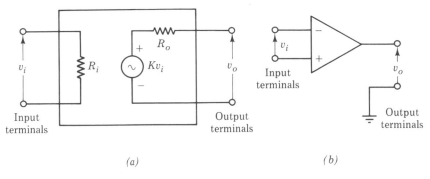

Fig. 1.3. Model of an operational amplifier:
(a) actual model; (b) IEEE symbol.

amp, the equivalent voltage source at the input terminals is essentially zero so the input circuit can be replaced by a single input resistance, R_i, as shown in Fig. 1.3a. This resistance R_i is known as the *input resistance* of the amplifier. Next, Thévenin's theorem is applied to the output terminals. The resulting circuit is a voltage source that is K times as large as the input signal. Note that K is equal to the *voltage gain* of the op amp. The equivalent resistance, R_o, which results when Thévenin's theorem is applied to the output terminals, is known as the *output resistance* of the amplifier. From this development, a model for an operational amplifier may be drawn as shown in Fig. 1.3a.

In order to promote uniformity in the electrical engineering literature, the IEEE (Institute of Electrical and Electronics Engineers) has developed a set of standards for symbols and abbreviations of electrical terms. The standard IEEE symbol for an op amp is shown in Fig. 1.3b. Normally, two input terminals are provided, and the input signal is applied between these terminals. If the − terminal is grounded, the output signal will be identical in shape but much larger in amplitude than the signal on the + terminal. In contrast, if the + terminal is grounded and the input signal is applied to the − terminal, the output signal will not only be much larger in amplitude but will also be inverted with respect to the input signal. Thus, a positive pulse on the − terminal will produce a negative pulse at the output terminal. For this reason, the − terminal is called the *inverting input* and the + terminal is known as the *noninverting input*. Normally, all signal voltages are referenced to a common ground system.

The op amp requires more than just an input signal for proper operation. As noted in Figs. 1.1 and 1.2, a dc power supply is required to operate an amplifier. The amplifier actually converts the dc energy to signal energy. Thus, while the signal does become larger when it passes through an amplifier, there is no net gain of power in the device. In fact, the heat generated by an op amp is due to the power loss in converting dc energy to signal energy. Thus, any dream of perpetual motion through the use of amplifiers is quickly laid to rest!

While many types of operational amplifiers exist, let us consider the 741 type of op amp in some detail. The 741 is manufactured by several different companies (Fairchild makes the μ741, Raytheon makes the RM741 and RC741, Texas Instruments makes the SN52741, etc.) and is quite inexpensive. The 741 is also *internally compensated*. If the op amp is *not* compensated, it will usually oscillate and be useless as an amplifier. In Chapter 13, we will discuss proper compensation techniques for amplifiers that are not internally compensated.

The 741 amplifier is connected as shown in Fig. 1.4. Since the 741 is packaged in over five different configurations, the actual pin connections will vary with the particular package used. The pin connections given in Fig. 1.4 match the metal TO-99 case and also the plastic eight-pin package. The 741 requires two power supplies with equal voltages. One supply must be positive with respect to ground, and the other supply must be negative with respect to ground, as shown in Fig. 1.4. Two batteries or special balanced power supplies are usually used to furnish this dc power. The 10-kΩ potentiometer between pins 1 and 5 is the dc *offset control*. This control is adjusted by shorting both input circuits (pins 2 and 3) to ground. Then the dc power is applied and the 10-kΩ potentiometer is adjusted until the output voltage (pin 6) is equal to zero. The resistor R_L is known as the *load resistance*. This load resistance may be the input resistance of another electrical device

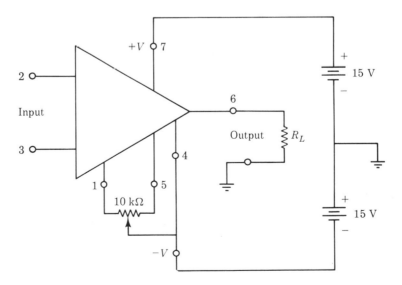

Fig. 1.4. Pin connections for a 741 operational amplifier.

such as a set of earphones or the input resistance of another piece of electronic equipment. In any case, the op amp will be required to furnish signal to some load. R_L represents this load. The 741 has a maximum power dissipation of 85 milliwatts. In order to protect the 741 op amp, the circuit automatically limits the load current to a safe value. Therefore, as the value of R_L decreases below 700 ohms, the output voltage magnitude must also decrease so current limiting will not occur. When current limiting occurs, the output voltage will not have the same shape as the input voltage.

While the parameters of 741 amplifiers vary from unit to unit, typical characteristics are listed as follows:

Voltage gain = 100,000

Output impedance = 75 Ω

Input impedance = 1 MΩ

Output voltages swing from -14 V to $+14$ V (if V+ = +15 V and V− = −15 V)

Many amplifier applications do not require voltage gains as high as 10^5. In these cases, the gain can be adjusted by using *negative feedback*. In the negative feedback configuration, part of the output signal is fed back into the input circuit, so some of the input signal is canceled. The *inverting configuration* is shown in Fig. 1.5. Note that while the offset circuit and the power supply are not shown in Fig. 1.5, these circuits must also be provided. As noted in the 741 characteristics, the peak output signal of the op amp is 14 volts. Thus, the maximum rms value of a sinusoidal output signal, v_o, is about 10 volts. If the gain of the amplifier is 10^5, the input signal, v_1, will have a value of 10 V / 10^5 =

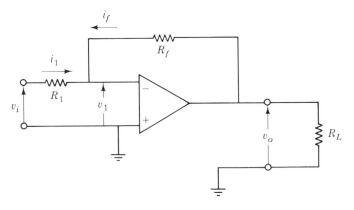

Fig. 1.5. Op amp in the inverting configuration.

10^{-4} volts or 0.1 millivolt. This signal, v_1, is normally so much smaller than v_o or v_i that it is considered to be essentially zero. Then,

$$i_1 \simeq \frac{v_i}{R_1} \tag{1.1}$$

and

$$i_f \simeq \frac{v_o}{R_f} \tag{1.2}$$

Since the input impedance of the op amp is very high and v_1 is almost zero, the current into the op amp is also essentially zero. Then $-i_1 \simeq i_f$, and from Eq. 1.1 and Eq. 1.2,

$$\frac{v_o}{R_f} = -\frac{v_i}{R_1} \tag{1.3}$$

or

$$K_v = \frac{v_o}{v_i} = -\frac{R_f}{R_1} \tag{1.4}$$

The negative sign in Eq. 1.4 means the output signal is inverted with respect to the input signal. Since v_1 is essentially zero, $i_1 \simeq v_i/R_1$ and the input impedance to the circuit in Fig. 1.5 is equal to R_1. The output impedance of the total configuration is less than the output impedance of the op amp alone. The reason for this decrease of output impedance will become clear when feedback is considered in more detail later in the text.

Example 1.1 | Use a 741 op amp to design an inverting amplifier with a gain of -100 and an input impedance of 10 kΩ. This amplifier must furnish the signal to a 1,000-Ω load. The configuration for the amplifier is shown in Fig. 1.6. Since the input impedance is 10 kΩ, R_1 is made equal to 10 kΩ. From Eq. 1.4, $R_f = -K_v R_1$, so $R_f = +100 \times 10 \ k = 1$ MΩ. The load resistance of 1,000 Ω is connected from the op amp output terminal to ground. This load resistance could represent the imped-ance of a pair of earphones.

If it is desirable to avoid inverting the signal, the noninverting config-uration shown in Fig. 1.7 is used. Since v_1 is so very small,

$$v_i \simeq v_f = \frac{R_1}{R_1 + R_f} v_o$$

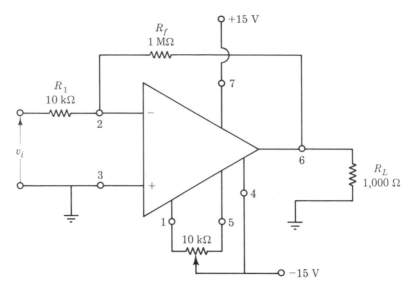

Fig. 1.6. Amplifier with a gain of 100.

so the voltage gain of this configuration is:

$$K_v = \frac{R_1 + R_f}{R_1} \tag{1.5}$$

The input impedance of this configuration is much greater than the input impedance of the op amp alone and is, therefore, many megohms. The output impedance of the total circuit is less than the output impedance of the op amp.

The contents of this section indicate that the engineer can handle op amps as "black boxes" and use these amplifiers without knowing what

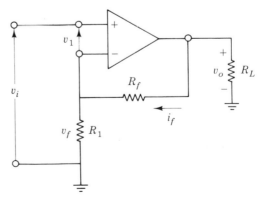

Fig. 1.7. Noninverting amplifier configuration.

is actually inside the "black boxes." For many applications, this superficial treatment may be sufficient. However, in order to really know how to use operational amplifiers under all conditions, the engineer must know what is inside the "black boxes" and also understand how these circuits operate. Consequently, operational amplifiers will be treated in much greater detail in later chapters.

**1.3
THE
CATHODE-RAY
TUBE**

Since we cannot actually see electricity in a circuit, many different types of electrical test instruments have been developed. In general, these instruments usually produce a visual display to indicate the values of the electrical parameters in a circuit. The most useful electrical test instrument is the cathode-ray oscilloscope. With an oscilloscope and some ingenuity, it is possible to measure ac and dc voltages and currents, frequencies, pulse durations, harmonic content, etc.

The heart of the oscilloscope is a vacuum tube known as a cathode-ray tube. The cathode-ray tube is constructed as shown in Fig. 1.8. The *electron gun* produces a narrow beam of high-speed electrons. These electrons are directed along the axis of the evacuated tube. If there are no potential differences between the *deflection plates*, the electrons will travel along the axis of the tube and strike the center of the *fluorescent screen*. The material on the fluorescent screen emits light when it is struck by a fast-moving electron. Thus, a tiny spot of light indicates where the electron beam is striking the fluorescent screen. If the upper deflection plate is made positive and the lower deflection plate is made negative, the electrons (which have a negative charge) will be attracted toward the upper plate and repelled by the lower plate. Consequently, the electron beam will be deflected upward toward the top of the fluorescent screen, as shown in Fig. 1.8. Of course, a negative upper plate

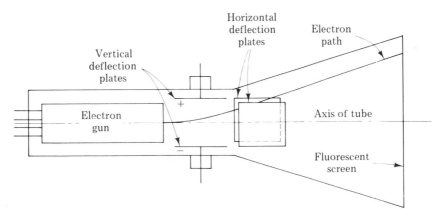

Fig. 1.8. Cathode-ray tube.

and a positive lower plate would deflect the electron beam toward the bottom of the fluorescent screen. The vertical displacement of the electron beam is proportional to the voltage applied to the vertical deflection plates. Thus, the cathode-ray tube produces a visual display (a small spot of light) on the fluorescent screen with a vertical displacement which is proportional to the voltage on the vertical deflection plates.

A second set of deflection plates is fastened at right angles to the vertical deflection plates. These plates are known as the *horizontal deflection plates* and deflect the electron beam to the left or right side of the fluorescent screen when voltages are applied to these plates. By the proper combination of voltages on the deflection plates, the electron beam can be directed to any given position on the fluorescent screen.

The cathode-ray tube can produce such an endless combination of patterns that it is used for TV receivers, radar displays, alpha-numeric displays for computers, and even as a heart-beat monitor in hospital stations. The TV picture tubes use magnetic deflection rather than the electrostatic deflection we have just discussed. Otherwise, the action of the TV picture tubes are the same as the tube in Fig. 1.8.

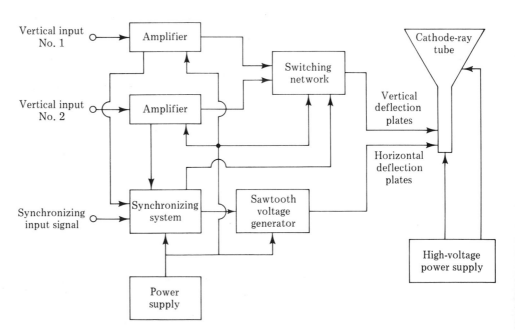

Fig. 1.9. Block diagram of an oscilloscope.

1.4
THE
OSCILLOSCOPE
The oscilloscope is an electronic test instrument that is built around the cathode-ray tube, as previously mentioned. The block diagram for a typical oscilloscope is shown in Fig. 1.9.

The sawtooth voltage generator produces a voltage waveform which varies in time, as shown in Fig. 1.10. Since this voltage increases linearly with time and is applied to the horizontal deflection plates, the electron beam sweeps linearly across the face of the cathode-ray tube (cathode-ray tube is usually abbreviated as CRT) in a horizontal direction. The signal to be analyzed is applied to the vertical deflection plates of the CRT. Then, the electron beam is deflected vertically in accordance with the vertical input signal and simultaneously moves linearly in the horizontal direction. Thus, a plot of the vertical signal versus time is traced on the face of the CRT. As soon as one horizontal trace is completed, the beam returns to the left edge of the CRT to begin the next trace.

The oscilloscope shown in Fig. 1.9 accommodates two vertical input signals. These signals pass through the switching network before they are applied to the vertical deflection plates. The switching network can be controlled so the following signals are applied to the vertical deflection plates of the CRT:

1. Only the signal at input No. 1 appears on the face of the CRT.

2. Only the signal at input No. 2 appears on the face of the CRT.

3. In the *alternate* mode, the signal on inut No. 1 appears on the CRT for one cycle of the sawtooth waveform, and the signal on input No. 2 appears on the CRT for the next cycle of the sawtooth waveform. Since this action keeps repeating, two signals can be displayed almost simultaneously for comparison.

4. In the *chopped* mode, the display is switched very rapidly back and forth between the two vertical inputs. Again, the two different signals appear on the CRT, but close inspection reveals a

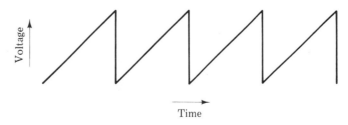

Fig. 1.10. Sawtooth voltage waveform.

dashed-line display as the switching circuit quickly samples the two signals.

5. It is also possible to sum the two input voltages so the display shows the sum of the two vertical input signals. Since it is possible to invert one of the vertical signals, the difference between the two vertical input signals can also be obtained by inverting one signal and summing the resultant two signals.

The synchronizing system is used to start each trace at the same voltage level of the vertical signal. Thus, if the vertical signal is a repeating function, each trace will follow the same path as the preceding trace. Then, a fixed pattern appears on the CRT for analysis and measurement. If the switching circuit is in the alternate mode, each input signal will trigger the sweep when the appropriate voltage level is present. In this instance, any phase difference between the two signals may be lost. To alleviate this problem, most dual-trace oscilloscopes provide for the sweep to be triggered by only one input signal. Then the true phase relationships between the two signals will be displayed. Since each model of oscilloscope is somewhat different, the instruction book should be studied so the features unique to each model are well understood by the student. Remember: *When all else fails, consult the instruction manual.*

Problems   *Section 1.1*

1.1 Since a signal generator does not have signal input terminals, the model or equivalent circuit for a signal generator may be drawn as a voltage source in series with an internal impedance. A given signal generator with an ac open-circuit output voltage of 1-volt peak has an internal resistance of 5,000 Ω. An ac voltmeter with an internal resistance of 10,000 Ω is connected across the output terminals of the signal generator.

   a. Using models, draw an equivalent circuit for this configuration.
   b. What will be the reading of the voltmeter?

   Answer: 0.47 V (rms).

1.2 A signal generator produces an ac open-circuit signal voltage of 10-V peak at a frequency of 100 kHz. The internal resistance of this signal generator is 5,000 Ω. An ac voltmeter with an internal resistance of 10,000 Ω in parallel with a capacitance of 159 pF is connected across the output terminals of this signal generator.

   a. Draw an equivalent circuit for this configuration.
   b. What will be the reading of the voltmeter?

   Answer: 4.47 V (rms).

Section 1.2

1.3 A given signal generator produces an ac signal with an open-circuit voltage of 5-V peak. The internal resistance of this signal generator is 500 Ω. A typical 741 op amp is connected in the inverting mode with a gain of -10. The input resistance of the total op amp configuration is 10 kΩ. The output resistance of the op amp (with feedback) is 7.5 Ω. A load of 1,000 Ω is connected to the output of the amplifier.

   a. Draw the circuit diagram for this configuration.
   b. Find the values of the resistors in the circuit.
   c. Find the ac voltage on the terminals of the signal generator.
   d. Find the ac voltage on the output terminals of the op amp.
   e. Find the ac voltage on the input terminals of the op amp.

1.4 A sinusoidal signal source has an open-circuit output voltage of 1.414-volts peak. The (internal) output resistance of this generator is 600 Ω.

   a. What is the rms value of the open-circuit output voltage?
   b. How much ac power will be delivered to a set of earphones which is connected to this signal generator? The earphones have 600-Ω internal resistance.
   c. How much ac power will the signal generator deliver to a 16-Ω loudspeaker? Assume that the 16 Ω is a pure resistance.
   d. The signal generator is connected to the input of an operational amplifier. The op amp is connected in the inverting mode so the configuration has an input resistance of 10 kΩ, an output resistance of 1 Ω, and a voltage gain of minus one. The output of the op amp is used to drive the 600-Ω earphones. Draw the circuit diagram for this configuration and indicate the magnitude of each resistor in your circuit.
   e. How much ac power is delivered to the 600-Ω earphones in part (d)?
   f. The circuit is connected as in part (d) but the 16-Ω loud-

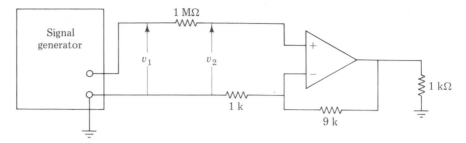

Fig. 1.11. Configuration for Prob. 1.6.

speaker replaces the 600-Ω earphones. How much power does the op amp deliver to the loudspeaker?

g.   Compare the results of part (c) and part (f). Explain why these results are different.

Note: The 741 op amp is not capable of supplying the power determined in part (f) to a 16-Ω load. The load resistance for the 741 should not be less than 100 Ω. Integrated circuits with much higher power output capability are available, however.

1.5   A signal generator produces a sinusoidal signal. A voltmeter with an internal impedance of 1 MΩ reads 5 V when it is connected across the signal generator terminals. When 100 Ω is connected across the signal generator terminals in parallel with the voltmeter, the voltmeter reads 3.33 V. What is the internal resistance of the signal generator?

1.6   A circuit is connected as shown in Fig. 1.11. The signal generator produces an ac signal. A voltmeter is used to measure the voltage v_1 and v_2. Find the input resistance to the op amp if $v_1 = 6$ V and $v_2 = 4$ V. What is the gain of the op amp?

Section 1.4

1.7   A given dual-trace oscilloscope having the block diagram shown in Fig. 1.9 is capable of displaying input signals having frequencies up to 100 MHz from either vertical input.

a.   What must be the frequency of the square wave that turns the switch on and off if the input No. 1 frequency is 100 MHz, the mode is *alternate*, and two full cycles of the input signal are seen on No. 1 trace?

b. What must be the frequency of the square wave switching signal if the same conditions of part (a) are desired but the mode is *chopped* and the two-cycle trace consists of 20 line segments?

c. What is the frequency of the sawtooth sweep voltage of Fig. 1.10 for the conditions of (a) and (b)?

2

Semiconductors

As the name implies, *semiconductors* are those materials which are neither good conductors nor good insulators. In order to understand the operation of most "solid-state" electronic devices, a knowledge of semiconductor physics is required. This chapter will provide an insight into the behavior of semiconductor materials, but will not treat semiconductor physics in depth. The in-depth study of semiconductor materials and junction theory is properly the content of an entire course. Such a course usually either precedes or follows an *electronic circuits course*, which is the main content of this book. If your background includes a semiconductor theory course, you may wish to proceed immediately to Chapter 3, or Chapter 4, depending upon your background.

**2.1
CONDUCTORS,
INSULATORS,
AND SEMI-
CONDUCTORS**

The chemical properties of an atom depend almost entirely on the number of electrons in the outer shell, which is known as the valence shell. The atoms which have four valance electrons, such as carbon, silicon, and germanium, are known as tetravalent atoms and are of special interest to us because they include the semiconductors. Most elements combine chemically with other elements so that the outer electron shells contain eight electrons. The total energy of the atom, or molecule, is minimum in this arrangement. It is possible for the tetravalent atoms to achieve this desired configuration by each atom sharing its valence electrons with its four adjacent neighbors. In this configuration, the pure tetravalent material forms a crystal. While the actual crystal is, of course, a three-dimensional structure, we can represent the structure schematically, as shown in Fig. 2.1. The nucleus and the filled

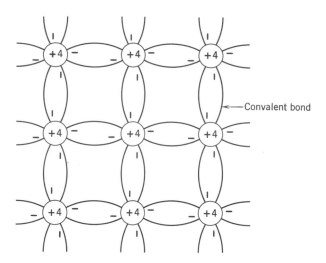

*Fig. 2.1. Two-dimensional representation of covalent bonds
in a tetravalent crystal.*

inner shells are represented by the circle with the +4, which is the net
positive charge of this group. The valence electrons are represented by
the negative signs, and the bonds which result from the reduced energy
and which hold the crystal together are represented by the curved lines.
These bonds are known as *covalent bonds*. In this crystalline form,
carbon is known as diamond.

An electrical conductor is a material which will permit electric
charge to flow (current) when a potential difference is applied to the
material. Therefore, conduction relies on the availability of charge carriers, such as electrons, which can move freely about in the material. A
good conductor has many free electrons, but a semiconductor has comparatively few electrons at normal temperatures. Of course, an insulator
has essentially no free electrons. At 0°K (absolute zero), all the electrons
in a semiconductor crystal are bound by the covalent bonds, as shown
in Fig. 2.1. Therefore, a semiconductor is actually an insulator at
$T = 0°K$.

A semiconductor can become a fairly good conductor, however, if its
temperature is raised high enough to give a sufficient number of electrons enough energy to break their covalent bonds and drift freely
through the crystal. This electron liberation process compares roughly
with the liberation of a rocket from the earth by transferring the energy
of the rocket fuel, first primarily to kinetic energy and then, as the rocket
gains altitude, to potential energy until the rocket gains an altitude
equal to many diameters of the earth. The gravitational pull of the earth
is then negligible and the rocket is free to drift through the solar system.

The tetravalent atoms such as carbon, silicon, germanium, and tin do not form crystals of equal conductivity at a given temperature, because the energy required to break a covalent bond depends upon the closeness of the atomic spacing in the crystal. The smaller the atom, the closer the spacing, and the greater the energy required to break the covalent bonds. Figure 2.2 shows the relative atomic spacing and the relative energy required to break a covalent bond for carbon, silicon, germanium, and tin. Compared with carbon, silicon has one more filled shell, germanium has two more, and tin has three more filled shells between the valence shell and the nucleus; therefore, the atom size and spacing between atoms increases in that order. The energy represented by the forbidden band is the minimum energy required by a valance (band) electron to break a covalent bond and become a conduction (band) electron. Smaller amounts of energy cannot free the electron from the valence band. Therefore, electrons can pass through the forbidden band but cannot remain there.

At room temperature, germanium has some free electrons, but not nearly as many as tin. Consequently, germanium is neither a good insulator nor a good conductor; hence, it is a semiconductor.

Diamond is an insulator, and tin is a fairly good conductor at normal room temperatures. At very high temperatures, assuming the crystal to remain intact, they all become conductors, while at 0°K, they all become insulators. Photons, or light energy, and strong electric fields may also break covalent bonds and elevate electrons to the conduction band, thus increasing the conductivity of the material.

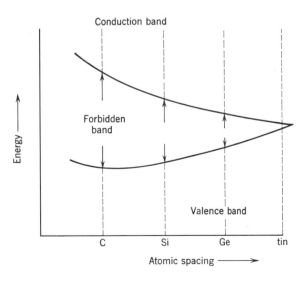

Fig. 2.2. Energy bands of tetravalent crystals as a function of lattice spacing.

**2.2
CONDUCTION
IN A SEMI-
CONDUCTOR**

As mentioned, heat or other sources of energy, such as light, cause valence band electrons to break their covalent bonds and become free electrons in the conduction band, thus producing modest conductivity in a semiconductor at normal temperatures. As the electron leaves the valence band, it creates a missing covalent bond which is known as a *hole*. This hole permits charge movement and, hence, conduction by the valence electrons, as illustrated by Fig. 2.3. A nearby valence band electron can fill a hole and thus create another hole with practically no exchange of energy. If an observer could see electrons and holes, he would notice that the holes move as well as the electrons. The hole acts as a positive charge, since the atom where the hole is located is missing an electron and has a net positive charge of 1.6×10^{-19} C. Thus, electric conduction is caused by both free electrons and free holes, which are known as *charge carriers*. Unfortunately, the detailed mechanism by which hole conduction occurs is not as simple as indicated by the above discussion. However, this concept is sufficiently accurate for our purposes.

The conduction due to the electrons in the conduction band is a different process than the conduction due to the holes left in the valence band. *In the intrinsic or pure semiconductor material, there are as many holes as there are free electrons* because the free electron leaves a missing covalent bond or hole. An analogy (which is attributed to Shockley) might be used to illustrate the conduction processes. A parking garage with two floors has the lower floor completely filled with identical automobiles and the upper floor completely empty. Under

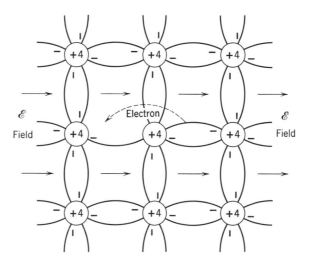

Fig. 2.3. Illustration of conduction resulting from broken covalent bonds or holes.

these conditions, there can be no movement of automobiles on either floor. However, if one automobile is elevated from the lower to the upper floor, there can be motion of automobiles on each floor. The auto on the upper floor may move freely over comparatively large distances. In contrast, the motion on the lower floor is accomplished by moving one car at a time into the available space. Hence, an observer near the ceiling of the first floor would see the "hole" move on the lower floor. The hole would have less mobility than the auto on the upper floor. Nevertheless, both would contribute to the total motion. Similarly, a free electron in a semiconductor has greater mobility, or ease of movement, than a hole.

We now know that electrical conductivity depends upon free charge carriers in a material, and that semiconductors have two different charge carrier mechanisms known as electrons and holes. Our next goal is to determine the conductance (or resistance) of a piece of semiconductor material. But before that can be done, we need to know how many of each type of carrier there are (or the carrier densities) in the material, and how the conductivity is related to the carrier densities. Since the latter problem is the easiest, it will be considered first.

As you probably know,

$$\text{Resistance } R = \frac{V}{I} = \rho \frac{\ell}{A} \tag{2.1}$$

where ρ is the resistivity of the material, ℓ is the length, and A is the cross-sectional area of the conductor.

Similarly,

$$\text{Conductance } G = \frac{I}{V} = \sigma \frac{A}{\ell} \tag{2.2}$$

where $\sigma = 1/\rho$ is the conductivity of the material.

Then, from Eq. 2.2,

$$\sigma = \frac{I/A}{V/\ell} = \frac{J}{\mathscr{E}} \tag{2.3}$$

where $I/A = J$ is the current per unit cross-sectional area, or *current density*. The term $V/\ell = \mathscr{E}$ is known as the electric field intensity. The electric field intensity $\mathscr{E}$ is constant in a uniform conductor of constant cross-section (area A). Equation 2.3 is therefore Ohm's law on a *per unit* basis.

Since current density is the rate of flow of charge through a unit cross-sectional area,

$$J = qn\bar{v} \tag{2.4}$$

where n is the number of carriers (electrons, for example) per cubic meter, q is the charge per carrier (1.6×10^{-19} C for electrons or holes), and $\bar{v}$ is the average drift velocity of the carriers due to the electric field. Substituting the expression for J in Eq. 2.4 into Eq. 2.3,

$$\sigma = \frac{qn\bar{v}}{\mathscr{E}} \tag{2.5}$$

One problem remains. Neither the drift velocity $\bar{v}$ nor the electric field are usually known. Fortunately, however, their ratio $\bar{v}/\mathscr{E}$ is known and is constant at a given temperature for a given material. This constant is known as the carrier mobility and has been given the symbol μ. Thus,

$$\mu = \frac{\bar{v}}{\mathscr{E}} \tag{2.6}$$

As previously mentioned, the mobility of electrons is greater than the mobility of holes.

Since the total conductivity is the sum of the conductivities due to electrons and holes,

$$\sigma = qn\mu_n + qp\mu_p = q(n\mu_n + p\mu_p) \tag{2.7}$$

where

n is free electron density

p is free hole density

μ_n is electron mobility

μ_P is hole mobility

The mobilities of electrons and holes in both silicon and germanium at 300°K are given in Table 2.1. As the reader may suspect, these mobilities are temperature dependent. The approximate temperature dependence for each mobility is also given in Table 2.1.

Example 2.1

The density of free electrons in pure, or *intrinsic*, germanium at 300°K (room temperature) is about 2.4×10^{19} electrons per cubic meter. Let us determine the resistance of a bar of germanium which is 1 millimeter by 2 millimeters and 1 centimeter long.

The conductivity (from Eq. 2.7) is $\sigma = q(n\mu_n + p\mu_p) = 1.6 \times 10^{-19}$ $(2.4 \times 10^{19} \times .39 + 2.4 \times 10^{19} \times 0.19) = 2.22$ siemans / meter. The resistivity is $\rho = 1/\sigma = 1/2.22 = 0.45$ ohm-meter at 300°K. The resistance of the germanium bar is $R = \rho\ell/A = 0.45 \times 0.01/(0.001 \times 0.002) = 2{,}250 \ \Omega$.

Table 2.1.   Carrier mobilities.[1]

Material	Mobility at 300°K (m²/volts-s)	Approximate Mobility at a Given Temperature T
Ge		
Free electrons	.3900	$4.9 \times 10^7 \, T^{-1.66}$ (100–300°K)
Holes	.1900	$1.05 \times 10^9 \, T^{-2.33}$ (125–300°K)
Si		
Free electrons	.1350	$2.1 \times 10^9 \, T^{-2.5}$ (160–400°K)
Holes	.0480	$2.3 \times 10^9 \, T^{-2.7}$ (150–400°K)

2.3 CHARGE CARRIER DENSITY IN A SEMI-CONDUCTOR

The next problem to be treated is that of determining the free carrier density as a function of temperature and forbidden band energy in a semiconductor. A formal derivation, or even an informal one, is beyond the scope of this work. However, an analogy will be used which will make plausible the carrier density formula to be presented later.

Gas molecules in the atmosphere exist in a gravitational force field and have kinetic energy which is related to the thermal energy of the gas molecules. Some of the thermal energy can be converted to potential energy as the gas molecules rise to higher elevations above the earth. Similarly, the valence electrons in a semiconductor exist in an electrical force field and have kinetic energy which is related to the thermal energy of the electrons. Some of this kinetic energy can be converted to potential energy as electrons break the covalent bonds and become free electrons. This similarity of the gas atoms in a gravitational field and the electrons in an electric field will be used to develop a model to predict the density of electrons in the conduction state of a semiconductor. Figure 2.4 is a sketch of an instrumented balloon which rises vertically and measures the density of air molecules in the atmosphere.

As a beginning point, we should recall that the atmosphere exists only because of its temperature, which provides kinetic energy and mobility to the air molecules. At 0°K, there would be no atmosphere, only a thin film of solid oxygen and nitrogen on the surface of the earth. As the temperature is increased, the air molecules gain sufficient kinetic energy, which can be exchanged for potential energy, to place them high above the earth. If the temperature becomes sufficiently high, many molecules can reach escape velocity and go into outer space. This escape compares with thermionic emission where the electron is ejected from the heated surface of the emitting material.

Boltzmann has shown mathematically that if the temperature of a gas, such as the atmosphere, is constant, the density of the gas is an exponential function of the altitude h. His equation follows:

$$n = A e^{-(mgh/kT)} \tag{2.8}$$

[1] E. M. Conwell, *Proceedings of the IRE*, Vol. 46, June 1958, pp. 1281–1300.

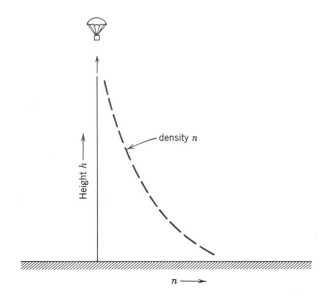

Fig. 2.4. Arrangement for measuring air density
as a function of altitude or height.

where n is the number of molecules per cubic meter, A is a constant, m is the mass of a molecule, g is the gravitational constant, and k is Boltzmann's constant which relates temperature T to its equivalent energy. Note that mgh is the potential energy $w(h)$ of a molecule at altitude h. Therefore, Eq. 2.8 can be written as

$$n = A\,e^{-(w(h)/kT)} \tag{2.9}$$

This exponential density distribution, which is known as the Boltzmann distribution, has been theoretically and experimentally verified. However, the atmospheric temperature and composition vary with altitude, so the Boltzmann distribution given by Eq. 2.9 is more easily applicable to semiconductors than it is to the atmosphere.

Note two things about the ideal atmospheric model.

1. The Boltzmann equation gives the density at any altitude h providing the density at any given reference $h = h_o$ is known. The reference may be at sea level, on a mountain top, or at the center of the earth, assuming a hole could be dug to the center of the earth, and the temperature is uniform. For example, if the density of air is 5×10^{25} molecules per cubic meter at sea level, the constant $A = 5 \times 10^{25}$ if the reference for $h(h = 0)$ is at sea level.

2. The Boltzmann distribution predicts the density only if the molecules are not restricted or forbidden for some reason. For exam-

ple, the density of air molecules in Moffat tunnel in Colorado may be 4×10^{24} molecules per cubic meter, but the air density ten meters above the tunnel inside the mountain is zero if the soil and rocks inside the mountain completely exclude the air.

When the gas molecule analogy is applied to the free electron density problem in a semiconductor, the availability of occupiable electron energy levels or states must be considered and must weigh, or multiply, the Boltzmann density distribution function. For example, electrons cannot exist in the forbidden band. Also, as previously mentioned, each electron energy level or shell in an atom will accept only a limited number of electrons. When these factors are included and an integration is performed over the range of conduction band energy levels, the total number of electrons in the conduction band can be obtained for a unit volume of the material as follows:

$$n_i = A \, T^{3/2} \, e^{-(W_g/2kT)} \tag{2.10}$$

where

T is temperature in °K

k is the Boltzmann constant $= 1.38 \times 10^{-23}$ joule / °K

W_g is the forbidden band or *gap* energy (the energy required to break a covalent bond) in joules

A is a constant

The gap energy W_g decreases as the temperature increases, because the crystal lattice spacing increases due to thermal expansion. This atomic spacing, and W_g, is a linear function of temperature over the useful temperature range. Thus W_g can be written as

$$W_g = W_{g0} \, (1 - CT) \tag{2.11}$$

where

C is the gap-energy temperature coefficient (2.8×10^{-4} for silicon and 3.9×10^{-4} for germanium)

W_{g0} is the gap energy at 0°K

Then, Eq. 2.10 can be written

$$n_i = A \, T^{3/2} e^{-W_{g0}(1-CT)/2kT}$$
$$= A \, e^{(W_{g0}C/2k)} \, T^{3/2} e^{-(W_{g0}/2kT)} \tag{2.12}$$

Table 2.2   Constant A' and gap energy at 0°K.

Material	W_{go} electron volts	Constant A'
Silicon	1.20	3.88×10^{22}
Germanium	0.782	1.76×10^{22}

But $e^{W_{go}C/2k}$ is a constant, so it can be combined with A to give a new constant A'. Therefore,

$$n_i = A' \, T^{3/2} \, e^{-(W_{go}/2kT)} \tag{2.13}$$

The constant A' can be either derived theoretically or determined experimentally. Commonly accepted values of W_{go} and A' for silicon and germanium are given in Table 2.2. Since each free electron leaves a missing, or broken, covalent bond in the crystal lattice, the hole density p_i is equal to the free electron density n_i in the intrinsic crystal, and thus Eq. 2.13 can also be used to determine free-hole density in the intrinsic crystal. Since the gap energies are given in elecron volts, these values must be multiplied by the electronic charge $q = 1.6 \times 10^{-19}$ to convert them to joules.

The free-electron density can now be calculated for temperature near 300°K (approximately room temperature), which is the normal operating range. For example, the density of free electrons in pure, or intrinsic, silicon at 300°K is 1.7×10^{16} electrons / m³. The conductivity of pure silicon at 300°K is 0.5×10^{-3} S / m, and the resistivity is 200 ohm-meter.

The free-carrier density, n_i or p_i in germanium and silicon, is plotted as a function of temperature in Fig. 2.5.

As one would expect, free carriers are continually being generated in a semiconductor, and their density would continually increase if recombinations did not occur. A *recombination* is the process of filling a hole with a free electron, thus eliminating both. Since the opportunities for recombination are proportional to the product of the number of free electrons and the number of holes, the recombination rate is proportional to this product *pn*. Thus, at a given temperature, the recombination rate must equal the generation rate so the free-carrier density remains essentially constant at the value predicted by Eq. 2.13. Therefore, if the temperature of an intrinsic semiconductor was increased enough to cause the rate of charge-carrier generation to double, the carrier density would increase until the rate of recombination would also double. However, each carrier density would increase by only a factor of $\sqrt{2}$.

The variation of resistivity of a semiconductor with temperature is utilized in a temperature-sensitive resistor known as a *thermistor*. In

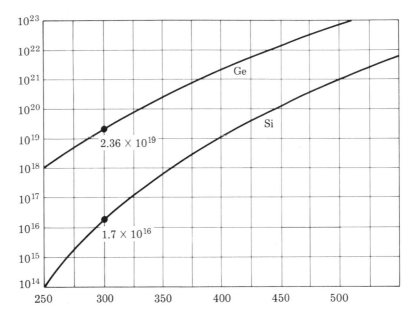

Fig. 2.5. Free-carrier density, n_i or p_i in germanium and silicon.

addition to its resistivity, the resistance of a thermistor depends upon its length and cross-sectional area.

2.4
DOPED SEMI-
CONDUCTORS

The conductivity of a semiconductor may be greatly increased if small amounts of specific impurities are introduced into the crystal. For example, if a few pentavalent atoms, such as arsenic or antimony which have five valence electrons, are added to each million semiconductor atoms, the impurity atoms form covalent bonds with their neighbors, as shown in Fig. 2.6. The pentavalent atom, after furnishing four electrons for the covalent bonds, has an extra valence electron which will not fit into the lattice arrangement. This extra electron is loosely bound to the atom at 0°K and requires very little thermal energy to become a free electron. In fact, as shown in Fig. 2.7, this extra electron is located at the level W_d and is about 0.01 eV below the conduction band. Therefore, nearly all the extra electrons gain sufficient additional energy to become free electrons in the conduction band when the crystal temperature is above about 50°K. Observe that essentially each impurity atom provides a free electron without creating a hole. Thus, the pentavalent atom is known as a donor because it donates or provides a free electron, and the semiconductor is said to be *doped* or *extrinsic*. Also note that the donor provides a *fixed* positive charge in the crystal lattice and the crystal is

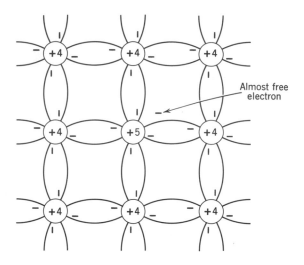

Fig. 2.6. Semiconductor crystal with a donor impurity.

electrically neutral. The donor-doped crystal is known as n -type crystal because most of the charge *carriers* are negative.

Example 2.2

The effectiveness of the donor impurity in increasing the conductivity of a semiconductor will be emphasized by an example. Let us consider a silicon crystal which has approximately 1.7×10^{16} free electrons which result from thermally broken covalent bonds, per cubic meter at 300°K, according to our previous calculation. Then let us assume that one part arsenic per million parts silicon was added

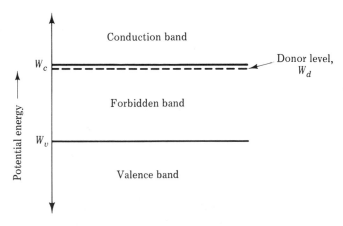

Fig. 2.7. Energy level diagram for an n-doped semiconductor.

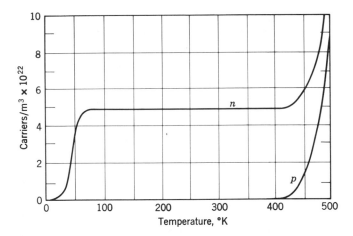

Fig. 2.8. Charge-carrier density in a donor-doped semiconductor as a function of temperature.

when the crystal was formed. Since there are approximately 5×10^{28} silicon atoms per cubic meter, the density of arsenic atoms is 5×10^{22} atoms/m^3. Then assuming all the donors to be activated (electrons are free), the conductivity of the doped silicon is 960 S/m, which is nearly 2×10^6 times as high as the 300°K conductivity of intrinsic silicon which we calculated to be 5×10^{-4} S/m.

As noted, the loosely bound electrons of the donor atoms become activated at about 50°K, and for the 5×10^{22} atoms/m^3 doping concentration of the preceding example, the thermally generated carriers are few compared with the donors until the temperature reaches about 450°K. Figure 2.8 is a sketch of the free-electron and hole densities in this doped silicon as functions of temperature.

In a donor-doped crystal, the free electrons are known as the *majority* carriers, and the holes are known as the *minority* carriers. The majority-carrier density n_n is the sum of the density due to the donors, N_d, and the intrinsic density n_i. Thus,

$$n_n = N_d + n_i \tag{2.14}$$

The conductivity is

$$\sigma = q(\mu_n n_n + \mu_p p_n) \tag{2.15}$$

where p_n is the minority-carrier density in the n-type crystal.

When the density of donor atoms N_d is large in comparison with the intrinsic-carrier density n_i, as in Example 2.2, the total carrier density, at normal temperatures, is essentially equal to the doping density N_d. Then Eq. 2.15 can be simplified to

$$\sigma \simeq q\mu_n N_d \tag{2.16}$$

The minority-carrier density is much lower than you may have suspected. As previously discussed, the rate of carrier recombination is proportional to the product of the free electrons and holes, but the rate of carrier generation is dependent only on the temperature. Therefore, when additional free electrons are introduced into the crystal by doping, the rate of recombination increases until the population of holes is reduced to the point where the recombination rate again equals the rate of thermal generation. For example, if the doping suddenly increased the free-electron density by a factor of 100, the recombination rate would initially increase by a factor of 100 but would then decrease as the hole population decreases until the normal rate of recombination, which is equal to the rate of thermal generation, is reached. This normal rate is reached when the hole density reaches one percent of its value for the intrinsic material. An analogy of the reduction of minority-carrier density may be the reduction of free-girl density in San Diego after the U.S. Navy comes ashore.

Since the rate of carrier generation is the same for both intrinsic and doped semiconductors, assuming the temperature to be the same, it should be apparent from the above discussion that the steady-state recombination rates must be the same. Therefore, since the recombination rate is proportional to the product of the free electrons and holes, the following relationship holds:

$$p_n n_n = p_i n_i = p_i^2 = n_i^2 \tag{2.17}$$

Therefore, in a donor-doped crystal, the minority-carrier density p_n may be written

$$p_n = \frac{p_i^2}{n_n} = \frac{n_i^2}{n_n} \tag{2.18}$$

Since n_n is very nearly equal to the doping concentration, or density, N_d,

$$p_n = \frac{n_i^2}{N_d} = \frac{p_i^2}{N_d} \tag{2.19}$$

P-doped semiconductors can be produced by adding a trace of trivalent material such as indium or gallium to the molten tetravalent germanium or silicon. Then, as the crystal forms, each impurity atom fits into the crystal structure but lacks one electron in forming covalent bonds with its neighbors (see Fig. 2.9). Therefore, each trivalent atom provides

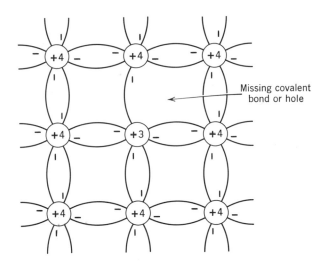

Missing covalent
bond or hole

Fig. 2.9. An acceptor-doped semiconductor crystal.

a hole in the semiconductor crystal. At temperatures above about 50°K, the valence electrons have enough energy to transfer from a neighboring atom to fill a hole. Therefore, the hole moves freely through the crystal while a fixed negative charge remains with the trivalent atom. This atom is therefore known as an acceptor, since it accepts an electron from its neighbors, and the symbol for acceptor-doping concentration or density is N_a. The trivalent-doped crystal is called p-doped because most of the charge carriers are holes which behave as positive charges.

Since very little energy is needed to move an electron from the valence shell of a neighboring tetravalent atom into the hole produced by an acceptor, the acceptor level is very near the valence band, as shown by the energy level diagram of Fig. 2.10. Thus, as noted, at temperatures above about 50°K, the holes created by the acceptors circulate freely and behave as free positive charges. Therefore, the probability that the acceptor holes are filled at normal temperatures is about the same as the probability that the valence band holes are filled. Thus at normal doping levels of a few parts per million, the holes associated with an acceptor atom are nearly all filled and holes are moving at random through the crystal.

From the preceding discussion, the following relationships can be written for a p-doped semiconductor:

$$p_p \simeq N_a \tag{2.20}$$

$$n_p = \frac{n_i^2}{N_a} = \frac{p_i^2}{N_a} \tag{2.21}$$

$$\sigma = qN_a\mu_p \tag{2.22}$$

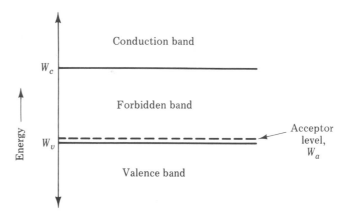

Fig. 2.10. Energy level diagram showing the acceptor level.

Since the conductivity of a semiconductor, especially silicon, can be controlled over such wide limits by doping, a complete circuit (including resistors, capacitors, and conductors) can be made in a single piece or chip of material by controlling the area and concentration of doping. In the following chapters, you will learn how diodes and transistors are formed from p-doped and n-doped semiconductors and how thus they may be integral parts of the circuit formed in the single or monolithic chip.

If a semiconductor is doped with equal concentrations of donor and acceptor atoms, the free electrons from the donors fill the holes created by the acceptors, and the material behaves as though it were intrinsic. Also, a semiconductor may be doped with unequal concentrations of donors and acceptors. Then the effective doping is the difference between the two concentrations and has the characteristic of the higher concentration. Therefore, the effective donor doping is $N_D = N_d - N_a$ and the effective acceptor doping $N_A = N_a - N_d$.

**2.5
GALLIUM
ARSENIDE
SEMI-
CONDUCTORS**

A semiconductor crystal can be formed from a compound known as a 3-5 compound composed of trivalent and pentavalent atoms. Gallium arsenide is the prime example of this type of compound and the only one in commercial use at the time of this writing. The gallium arsenide forms a crystal very similar to the silicon or germanium crystal, except the gallium and arsenic atoms alternate in the crystal lattice so that each gallium atom is surrounded by four arsenic atoms, and each arsenic atom is surrounded by four gallium atoms. Covalent bonds are formed as in a germanium or silicon crystal. The extra electrons from the pentavalent arsenic atoms fill the holes produced by the trivalent gallium atoms so that the crystal has the same general properties as an intrinsic semiconductor composed of tetravalent atoms.

The gallium arsenide (GaAs) crystal can be p-doped by adding small amounts of group II atoms, such as zinc, with two valence electrons. These atoms replace the trivalent gallium atoms and provide an extra hole in addition to accepting the extra electron from an arsenic atom neighbor. Also, the crystal can be n doped by adding a small quantity of group VI atoms, such as selenium, with six valence electrons. These atoms replace arsenic atoms and provide a free electron in addition to denoting an electron to fill the hole of a neighboring gallium atom.

The GaAs crystal is used to make semiconductor devices which are in some respects superior to either germanium or silicon. These advantages arise from the following characteristics of GaAs:

1. The forbidden band, or gap, energy is 1.40 eV at 25°C compared with about 1.1 eV for silicon and 0.67 eV for germanium. This higher gap energy provides satisfactory performance of GaAs devices up to about 300°C with present technology, and this limit could be increased to 400°C or above if some aging problems can be overcome. The upper temperature limit for silicon devices is about 200°C, and germanium is useful up to about 100°C. The intrinsic GaAs crystal is a much better insulator at normal temperatures than silicon or germanium because of the higher gap energy. This characteristic is desirable for making high-quality integrated circuits and low-loss varactor diodes which are discussed in Chapter 3.

2. GaAs has much higher electron mobility than either silicon or germanium. Therefore, the upper frequency limit of a transistor, which is proportional to charge-carrier mobility, can be much higher if the transistor is made of GaAs. The theoretical electron mobility in a pure, perfect GaAs crystal is 1.1 m^2/V-s, which is about three times as high as that of germanium and six times as high as electron mobility in silicon. Present refining processes can produce GaAs with about 0.8 m^2/V-s electron mobility, which is about twice that of germanium.

3. Electrons which return from the conduction band to the valence band transform their excess energy into electromagnetic radiation. In a GaAs crystal, this radiation is in the visible spectrum (light). In contrast, the radiation from silicon and germanium is in the infrared (or heat) region. Therefore, GaAs can be used in low-voltage, solid-state display devices. An example of such a device is the light-emitting diode which is discussed in Chapter 3. Some fundamental properties of germanium, silicon, and gallium arsenide are given in Table 2.3.

Table 2.3.   Some fundamental properties of germanium, silicon, and GaAs.

Specific Property		Silicon	Germanium	GaAs
Atomic number		14	32	
Atomic weight		28.06	72.6	
Density (25°C) kg$/$m^3		2.33×10^3	5.33×10^3	5.32×10^3
Melting point, °C		1420	936	1238
Relative dielectric constant E/E_v		12	16	11
Intrinsic carrier density n_i (300°K)$/$m^3		1.7×10^{16}	2.4×10^{19}	1.4×10^{12}
Gap energy, W_{go}, electron volts:	0°K	1.2	0.782	1.63
	300°K	1.12	0.67	1.40
Carrier mobility (300°K) m$^2/V$-s	μ_n	0.135	0.39	1.10
	μ_p	0.05	0.19	0.05
Diffusion constant (300°K) m$^2/s$	D_n	33.8×10^{-4}	98.8×10^{-4}	43×10^{-3}
	D_p	13×10^{-4}	46.6×10^{-4}	13×10^{-4}

Another promising 3-5 compound is indium phosphide (InP). This compound, like GaAs, has high carrier mobility and other desirable characteristics for the production of devices which will amplify and process signals in the GHz frequency, or millimeter wave, region. The InP devices have efficiencies almost twice as high as GaAs devices in this ultrahigh frequency range. The InP devices are just beginning to

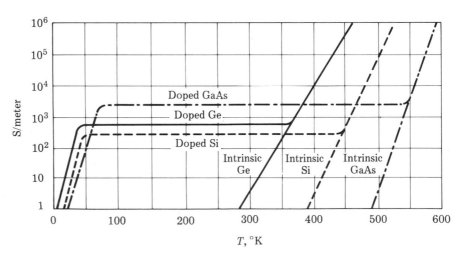

Fig. 2.11. *Conductivity of intrinsic and doped (10^{22} donors/m^3) Ge, Si, and GaAs as a function of temperature.*

emerge from the laboratories at the time of this writing. Other 3-5 compounds which are being developed in the leading research laboratories show great promise of fulfilling special needs in the vast semiconductor market.

As a summary of the characteristics of intrinsic and doped semiconductors, the conductivity of intrinsic and doped germanium, silicon, and gallium arsenide is sketched as a function of temperature in Fig. 2.11. Note that the conductivity scale is logarithmic. Also note that the conductivity of the doped semiconductor *decreases* slightly as temperature increases, over the moderate temperature range, because of the reduced mean-free path of the carriers between collisions. Thus the carrier mobility μ decreases somewhat as temperature increases. This effect occurs in nearly all conductors and results in the positive temperature coefficient of resistance. The n-type doping was assumed to be 10^{22} donors/m^3 for all materials.

Problems   *Section 2.1*

    2.1   Of the tetravalent crystals shown in Fig. 2.2, which would you expect to be:

          a.   The best conductor?
          b.   The best insulator?
          c.   The best semiconductors?

Section 2.2

    2.2   If the density of the free electrons in pure silicon at room temperature is 1.7×10^{16} electrons/m^3, find the resistance of the bar given in Example 2.1 if it is made of silicon. **Answer: 10 MΩ.**

    2.3   Using the value of resistivity determined for germanium (0.45 ohm-meter) at 300°K, what is the approximate ratio of the resistivity of silicon to that of germanium at this temperature? **Answer: 4444.**

Section 2.3

    2.4   If a thermistor is made of intrinsic (pure) silicon, calculate its resistivity at $T = 400$°K. **Answer: 7.5 ohm-m.**

    2.5   By what factor is the resistance of the silicon thermistor decreased if its temperature is raised from 300°K to 400°K? **Answer: 272.**

Section 2.4

2.6   Verify that the conductivity of the doped silicon in Example 2.2 is 960 mho/m.

2.7   In Example 2.2, where the doping concentration was assumed to be 5×10^{22} donors/m^3 and free-electron density in the intrinsic crystal is 1.7×10^{16}, determine the minority-carrier (or hole) density in the doped silicon.
       Answer: 5.8×10^9 holes/m^3.

2.8   If you were making an integrated monolithic circuit, how would you dope for:

       a.   A conductor?
       b.   An insulator?
       c.   A resistor?
       d.   A capacitor?

Section 2.5

2.9   Calculate the end-to-end resistance of a rectangular bar of intrinsic GaAs crystal at 300°K if $w = 2$ mm, $t = 1$ mm, and $\ell = 1$ cm.
       Answer: $R = 1.94 \times 10^{10}$ Ω.

2.10   There are 4.43×10^{22} atoms per ml in gallium arsenide at 300°K. If donor-type atoms are added until the total number of free electrons is equal to $2 n_i$, what will be the free-hole density and what will be the ratio of intrinsic atoms to donor atoms?
       Answer: $P_i = 0.5 n_i$; 3.16×10^{16}.

2.11   We have a piece of pure silicon.

       a.   Find the value of n_i at 200°K.
       b.   Find the resistivity of this silicon at this temperature.
       c.   The silicon is a square bar 1 cm long, 0.1 cm thick, and 0.1 cm wide. What is the resistance between the two ends of this bar at 200°K?

2.12   Repeat Problem 2.11 at 250°K.

2.13   Repeat Problem 2.11 at 350°K.

2.14   Make a plot of resistivity versus temperature for pure silicon. Use the temperature range from 200°K to 400°K.

2.15   Make a plot of resistivity versus temperature for doped silicon which has $N_d = 10^{19}$ atoms/m^3. Use the temperature range from 200°K to 400°K.

2.16 Assume that you are developing an electronic fever thermometer which uses a thermistor in a bridge circuit to sense the temperature. You want the resistance of one leg of the bridge to decrease by only 10 percent as the temperature rises from 95°F to 105°F. A fixed resistor can, of course, be connected either in series or in parallel with the thermistor. If the thermistor is made of intrinsic germanium, determine the resistances of both the thermistor and the fixed resistor if the total resistance of the combination is to be 10 kΩ at 95°F.

2.17 You intend to mass-produce the thermometer of Problem 2.16 and recognize that you could possibly save the cost of the fixed resistor by doping the germanium to produce the desired resistance change. Can this technique be used? If so, determine the approximate n-type doping density required.

3

Diodes

A diode is an electronic device that readily passes current in one direction but does not pass appreciable current in the opposite direction. A semiconductor diode is formed when a piece of n-type semiconductor is connected to a piece of p-type semiconductor. In actual production, a single crystal of semiconductor is formed with part of the crystal doped with acceptor impurities and the other part of the crystal doped with donor impurities.

Diodes are used in a number of electronic applications such as converting ac to dc, detecting radio signals, performing computer logic, and other useful functions. In this chapter, we will study the basic characteristics of diodes and examine some of their uses.

3.1
THE p-n
JUNCTION

The basic mechanism of a p-n junction is illustrated in Fig. 3.1. Assume two pieces of semiconductor (one p- and one n-type material) exist as shown in Fig. 3.1a. The p material contains a certain density (N_a) of acceptor (+3 valence) atoms. These atoms are represented in the diagram by the negative signs with the circles around them. These atoms are fixed at a given location in the crystal structure. Associated with each acceptor atom is a free hole which is represented by a positive sign in the diagram. Of course, some of the intrinsic atoms have lost electrons due to thermal agitation. These thermal electrons and the holes they produced are represented, respectively, by the negative sign in the square and the positive sign in the square. Of course, thermally generated carriers are indistinguishable from carriers produced by doping. Thus, the total density of majority carriers (p_p) is equal to the density

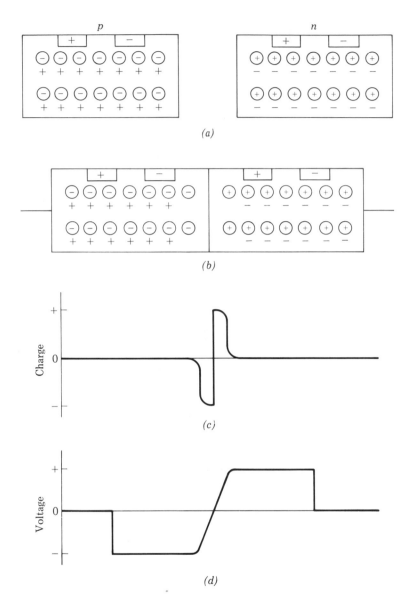

Fig. 3.1. The p-n junction: (a) representation of p- and n-type material;
(b) representation of p-n junction; (c) charge distribution of the p-n junction;
(d) potential distribution of the p-n junction.

of acceptor atoms (N_a) plus the density of thermal holes (p_i). As we have already noted, since $N_a \gg p_i$, $p_p \simeq N_a$. The density of minority carriers (n_p) are due to the thermally generated electrons, but as noted in Eq. 2.21, $n_p \ll n_i$.

The n material (Fig. 3.1a) contains a given density (N_d) of donor ($+5$ valence) atoms. The donor atoms (fixed in the crystal structure) are represented in the diagram by the positive signs with the circles around them. The associated free electrons are represented by negative signs. Again, the thermally generated electrons and holes are represented by the negative and positive signs in the squares. In the n material, the density of majority carriers (n_n) is equal to ($N_d + n_i$). However, since $N_d \gg n_i$, $n_n \simeq N_d$. The p (or minority) carriers are due entirely to the thermally generated carriers. However, from Eq. 2.19 we note that $p_n \ll p_i$ if $N_d \gg p_i$. Of course, both pieces of crystal are electrically neutral.

When the p-type and n-type crystals are joined together, assuming their crystal structure to be continuous, a charge redistribution occurs, a shown in Fig. 3.1b. Some of the free electrons from the n material migrate across the junction and combine with the free holes in the p material. Similarly, free holes from the p material migrate across the junction and combine with free electrons in the n material. Therefore, as a result of the redistribution, the p material acquires a net negative charge and the n material acquires a net positive charge, as shown in Fig. 3.1c. This electric charge produces an electric field and a potential difference between the two types of material, as shown in Fig. 3.1d. As previously mentioned, the diode manufacturing process does not actually employ the joining technique, but usually employs selective doping of a continuous crystal.

**3.2
DIFFUSION
AND
RECOMBINA-
TION**

The process by which the charges cross the junction is known as *diffusion*. This process may be visualized by considering the action of the electrons from the n piece of material. When the two pieces of material are joined, there is a concentration of free electrons in the n material, but very few in the p material. The random motion of the electrons will cause some electrons to pass from the n to the p region. Because there are fewer electrons in the p region, fewer electrons will tend to pass from the p to the n region. If no other forces existed, the diffusion process would continue until the concentration of electrons would be uniform throughout the material. This process would be the same process which occurs when two containers of dissimilar pure gases are joined. However, because of the charges on the electrons, these electrons are attracted back toward the n region, which becomes positively charged as the electrons leave it. Similarly, the holes are attracted toward the negatively charged p region. Hence, an electric field is established (by the diffusion process) which inhibits the diffusion of carriers.

**3.3
DIFFUSION
CURRENTS
AND DRIFT
CURRENTS**

Since diffusion is a common process, mathematical relationships for this process have been established. In simple terms, the rate at which electrons (gas atoms, etc.) will cross a given reference plane of unit cross-sectional area is proportional to the distribution gradient (the rate at which the density of electrons changes with distance) of the electrons. This relationship can be expressed mathematically as

$$\frac{dn}{dt} = D_n \frac{dn}{dx} \tag{3.1}$$

where dn/dt is the rate at which electrons will cross a given unit cross-sectional area, D_n is the constant of proportionality called the *diffusion constant*, and dn/dx is the distribution gradient.

Since electrical current is the rate at which charge passes a given reference, we can use Eq. 3.1 to find the diffusion current density. Thus,

$$J_n = q\frac{dn}{dt} = qD_n\frac{dn}{dx} \tag{3.2}$$

where

J_n = current density due to electrons in A/m^2

q = the charge of an electron in C

D_n = the diffusion constant for electrons in m^2/s

The diffusion constants for electrons in silicon, germanium, and gallium arsenide are given in Table 3.1.

Holes also diffuse in a semiconductor. Thus, an equation similar to Eq. 3.2 can be developed for current flow due to hole diffusion. Since dp/dx must be negative for current flow in the positive x direction,

$$J_p = -qD_p\frac{dp}{dx} \tag{3.3}$$

where

J_p = current density due to holes in A/m^2

D_p = the diffusion constant for holes in m^2/s

Table 3.1. Diffusion constants at 300°K in m^2/s.

	Ge	Si	GaAs
D_n	98.8×10^{-4}	33.8×10^{-4}	174×10^{-4}
D_p	46.8×10^{-4}	13.0×10^{-4}	17.4×10^{-4}

dp / dx = the hole distribution gradient

The total current due to diffusion is

$$I_{\text{diff}} = qA \left(D_n \frac{dn}{dx} - D_p \frac{dp}{dx} \right) \tag{3.4}$$

where A is the cross-sectional area of the material. Again, the diffusion constants for holes in silicon, germanium, and gallium arsenide are listed in Table 3.1. Of couse, the diffusion constant is a function of the material and temperature as well as the type of carrier.

In the actual n-p junction, as stated previously, both types of semiconductive materials are neutral before the junction is made. After the junction is made, the n material loses electrons due to diffusion and gains holes.

However, as the electrons enter the p region, recombination with the numerous holes occurs. In fact, the hole population next to the junction is said to be "depleted" and this region is known as the *depletion region*, since practically all of these holes are filled with electrons. The original electrical charge neutrality of this region is replaced with negative ions wherever a +3 valence atom exists. Similarly, any holes which "travel" (or are *injected*) into the n material are immediately removed by recombination with the free electrons near the junction. In the n material, the free electrons are depleted, and positive ions occur wherever a +5 valence atom exists. As a result of this action, the p material becomes negative and the n material becomes positive, as shown in Fig. 3.1d, and an electric field or potential hill is established at the junction. In the steady-state condition, this field is just strong enough to inhibit the diffusion of the electrons from the n material and the holes from the p material.

The potential hill at the junction is caused by the diffusion of the majority carriers, while the minority carriers which are swept across the junction by the electric field tend to reduce the height of the potential hill. As a result, an *equilibrium* condition is reached in which the flow of minority carriers is equal to the flow of majority carriers. The current resulting from the diffusion of the majority carriers is known as the *injection current* or *majority current*, and the current resulting from the minority carriers is known as the *saturation current* or *minority current*. The minority, or saturation, current is known as a *drift current* because the minority carriers drift through the depletion region because of the electric field $\mathscr{E}$. This is the same motivation for current flow that occurs in any conductor to which a voltage is applied. Therefore, the current density due to free electron drift is

$$J_n = \sigma \mathscr{E} \tag{3.5}$$

where $\sigma = q\,(\mu_n n + \mu_p p)$ as given in Eq. 2.15. The total current due to drift is therefore

$$I_{\text{drift}} = qA\,(\mu_p p + \mu_n n)\,\mathscr{E} \qquad (3.6)$$

The diffusion constant D and the mobility μ for a given material are closely related because they both are limited by collisions between the carriers and the vibrating atoms. Their relationship is given by

$$\frac{\mu_n}{D_n} = \frac{\mu_p}{D_p} = \frac{q}{kT} \qquad (3.7)$$

where k is the Boltzmann constant $= 1.38 \times 10^{-23} J\,/°K$. Eq. 3.7 is known as the *Einstein relation*. At room temperautre, $T = 290°K$, and

$$\frac{q}{kT} \simeq 40 \text{ V}^{-1} \qquad (3.8)$$

As previously mentioned, the charge distribution of Fig. 3.1c causes a potential difference across the p-n junction (Fig. 3.1d). This potential difference is a few tenths of a volt at room temperature. It seems possible that a current would flow in an external circuit if a conductor were connected to the open ends of the p-n combination, since they are at different potentials. This supposition is *not* true, because the contact difference of potential at the junctions of the crystal and the external conductor causes the total emf around the closed circuit to be zero, and no current will flow. For example, if a conductor is brought into contact with an n crystal, the net diffusion of electrons will be from the crystal to the conductor until the crystal becomes positive with respect to the conductor, and equilibrium is reached. Of course, the opposite is true if the conductor is connected to a p-type crystal.

The directions of the injection current I_I and the saturation current I_S are shown in Fig. 3.2b. In addition, the lengths of the current arrows indicate the relative magnitude of these currents. The net current flow in this case is zero, $|I_I| = |I_S|$. The majority carriers must have enough kinetic energy to "climb" the potential "hill," whereas all minority carriers just "slide down" this potential hill gaining kinetic energy in traversing the junction.

If an external battery is connected with the positive terminal to the n-type material and the negative terminal to the p-type material, the potential difference across the junction will increase, as shown in Fig. 3.2c. The height of the potential "hill" is increased by the amount of the external battery voltage. Then only those majority carriers which have a very large amount of kinetic energy can "climb" the potential "hill." As a consequence, the injection current I_I is essentially eliminated, as shown in Fig. 3.2c. Since the minority carriers still require no energy to

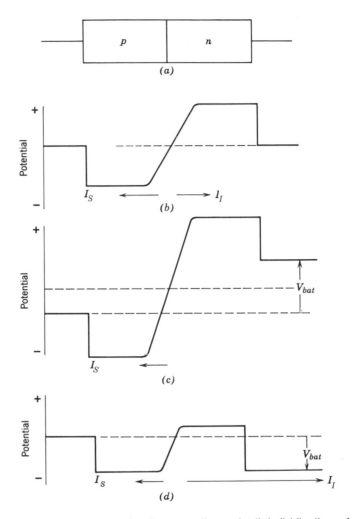

Fig. 3.2. Effect of an external voltage on the potential distribution of a p-n junction: (a) the p-n junction; (b) potential distribution with no external battery; (c) potential distribution with a reverse-bias external battery; (d) potential distribution with a forward-bias external battery.

cross the junction, the current I_S remains the same as in Fig. 3.2b. The external current flow is, therefore, mainly due to the thermally generated minority carriers of the semiconductor. This current I_S is known as the *reverse current* of the diode.

If the external battery is connected with the negative terminal to the n-type material and the positive terminal to the p material, the potential difference across the junction is decreased, as shown in Fig. 3.2d. The potential "hill" is reduced by the magnitude of the external battery voltage, neglecting the IR drop in the crystals. As a result, a large number of the majority carriers are able to cross the junction. Therefore, the injection current I_I is greatly increased, as shown in Fig. 3.2d. Since the minority carriers can still traverse the junction with no loss of energy (these carriers still gain kinetic energy, but not as much as in Fig. 3.2b and Fig. 3.2c), the current I_S does not change. The *forward current* of the diode $I_I - I_S$ is, therefore, quite large. In fact, because of the large number of majority carriers as compared to the minority carriers, the forward current is usually thousands or millions of times larger than the reverse current.

**3.4
MINORITY
CARRIER
DIFFUSION**

The holes that are injected from the p region across the depletion region to the n region diffuse into the n region as illustrated in Fig. 3.3. As they diffuse away from the junction they recombine with the free electrons, which are the majority carriers, so the free-hole density decreases with distance x from the junction. If the N region is long enough, the hole density approaches the hole density $p_{no} = n_i^2 / N_D$ that is due to thermal generation.

Since the recombination rate is proportional to the excess hole density,

$$\text{The net recombination rate} = \frac{p - p_{no}}{\tau} \qquad (3.9)$$

where τ is defined as the *excess carrier lifetime*. This excess carrier lifetime is a function of the temperature and the physical properties of

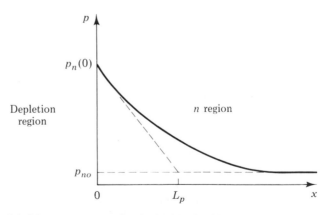

Fig. 3.3. Minority-carrier distribution in the N-region while forward bias is applied to the diode.

the semiconductor material. τ usually ranges between 10^{-9} seconds and 5×10^{-7} seconds. Equation 3.9 may be rewritten as

$$\text{Net recombination rate} = \frac{p}{\tau} - \frac{p_{no}}{\tau} \tag{3.10}$$

where the recombination rate $= p/\tau$ and the generation rate $= p_{no}/\tau$.

Equations 3.9 and 3.10 hold for free electrons as well as free holes. They are generally applicable to carrier generation and recombination situations.

An expression may now be derived for the hole density in the n region, or any excess carrier density in any given region, as a function of distance. Observe the rectangular sample of n-type material through which diffusion current I_p is flowing in Fig. 3.4. The hole current I_p that enters the imaginary box of length Δx is $I(x)$ and the current that exits at $x + \Delta x$ is $I(x + \Delta x)$. Then the net loss of current in the box is $I(x) - I(x + \Delta x)$ and the net loss of free holes per unit volume in the box is

$$\Delta p = \frac{1}{qA\,\Delta x} \left(I(x) - I(x + \Delta x) \right) \tag{3.11}$$

If Δx is sufficiently small,

$$I(x + \Delta x) = I(x) + \frac{\partial I_x}{\partial x} (\Delta x) \tag{3.12}$$

The partial derivative is used because I may vary in both time and x. Then, substituting Eq. 3.12 into Eq. 3.11,

$$\Delta p = -\frac{1}{qA} \frac{\partial I_p}{\partial x} \text{ holes per second per cubic meter} \tag{3.13}$$

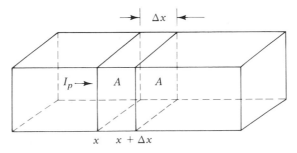

Fig. 3.4. A small sample of the N-type semiconductor.

Using the diffusion current Eq. 3.4 for holes in Eq. 3.13, the outward flow is

$$\Delta p_n = D_p \frac{\partial^2 p_n}{\partial x^2} \qquad (3.14)$$

When this hole flow is added to the net recombination rate, the result is $\partial p / \partial t$, which is the rate at which holes disappear from the volume $A \, \Delta x$. Then

$$\frac{\partial p_n}{\partial t} = D_p \frac{\partial^2 p_n}{\partial x^2} + \frac{p_n - p_{no}}{\tau} \qquad (3.15)$$

This equation is known as a *minority carrier diffusion equation*. The classical solution to this equation under the steady-state condition is

$$p_n = c_1 e^{-x/(D_p\tau)^{1/2}} + c_2 e^{x/(D_n\tau)^{1/2}} + p_{no} \qquad (3.16)$$

The $(D_p\tau)^{1/2}$ is known as the *hole diffusion length* L_p, shown in Fig. 3.3. Similarly, the diffusion equation may be written for electrons diffusing through a p region.

$$n_p = c_3 e^{-x/L_n} + c_4 e^{x/L_n} + n_{po} \qquad (3.17)$$

where $L_n = (D_n\tau)^{1/2}$. These equations are based on the assumption that the majority carrier density, n_n or p_p, is constant throughout the material so the recombination rate is n / τ or p / τ. This relationship holds only in low-level injection where the injected carrier density $p_n(0)$ is small in comparison with n_n, or $p_n(0) \leq 0.1n_n$.

Let us now apply Eq. 3.16 to the diffusion of holes through the n region shown in Fig. 3.3. Since p_n cannot increase exponentially with x, c_2 must be 0. c_2 is nonzero only when holes are diffusing in the negative x direction. Then

$$p_n = c_1 e^{-x/L_p} + p_{no} \qquad (3.18)$$

Using the boundary condition $p_n = p_n(0)$ at $x = 0$, $c_1 = p_n(0) - p_{no}$ and

$$p_n = (p_n(0) - p_{no})e^{-x/L_p} + p_{no} \qquad (3.19)$$

Similarly,

$$n_p = (n_p(0) - n_{po})e^{-(-x/L_n)} + n_{po} \qquad (3.20)$$

Since the free hole density decreases exponentially with distance in the n region, the diffusion current must also decrease exponentially with x. One may then ask, "How can the current then be continuous through the n region and the diode, as required by Kirchoff?" The

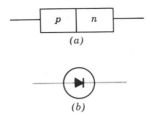

Fig. 3.5. The p-n diode: (a) physical representation of a diode;
(b) diode symbol.

answer is that when a hole recombines with a free electron, it leaves a
fixed positive charge, an unneutralized donor atom, in the vicinity. The
positive charge produces an electric field which causes a drift current
to flow in such a direction as to neutralize the field. Thus the drift
current increases by the same amount as the diffusion current decreases
with increasing x. At x = 0, the current is entirely diffusion current, and
as x becomes large in comparison with the diffusion length, the current
is essentially all drift current.

All that has been said about the n region is also true for the p region
except the carrier types are reversed. The symbol for a p-n diode is given
in Fig. 3.5. The arrow points in the direction of positive current flow
when forward bias is applied to the diode.

**3.5
THE DIODE
EQUATION**

When no external voltage, or *bias*, is applied to a diode and no current
flows in the external circuit, the injection current, which is a diffusion
current through the depletion region, must be equal to and in opposite
direction from the saturation current, which is a drift current through
the same region. We may use Eqs. 3.4 and 3.6 to equate these currents.
Let us first consider only the currents resulting from hole flow to and
from the n material.

$$-qAD_p \frac{dp}{dx} = qA\mu_p p\mathscr{E} \qquad (3.21)$$

Then

$$\frac{dp}{dx} = -\frac{\mu_p}{D_p} p\mathscr{E} \qquad (3.22)$$

Rearranging terms and using the Einstein relationship,

$$\frac{dp}{p} = -\frac{q}{kT} \mathscr{E} \, dx \qquad (3.23)$$

Integrating both sides across the depletion region,

$$\int_{p_{po}}^{p_{no}} \frac{dp}{p} = \frac{q}{kT} \int_{-\ell}^{o} -\mathscr{E} \, dx \qquad (3.24)$$

where $x = 0$ is at the right-hand side of the depletion region as given in Fig. 3.3, and the left-hand side of the depletion region is at $x = -\ell$. But the negative of the integral of the field intensity through the depletion region is the potential difference or potential hill V_{ho}. Also p_{po} is essentially equal to N_A and p_{no} is the minority carrier density in the n-region $= n_i^2 / N_D$.

Performing the indicated integration in Eq. 3.24,

$$\ln \frac{N_A}{p_{no}} = \frac{q}{kT} V_{ho} \qquad (3.25)$$

where V_{ho} is the height of the potential hill with no bias applied.
Then

$$p_{no} = N_A \, e^{-qV_{ho}/kT} \qquad (3.26)$$

When bias voltage V is applied to the junction because of an external voltage, the potential difference across the junction becomes $V_{ho} - V$ and the minority carrier density at the n side of the junction becomes $p_n(0)$, as shown in Fig. 3.3. Then Eq. 3.18 becomes

$$p_n(0) = N_A \, e^{-q(V_{ho}-V)/kT} \qquad (3.27)$$

With the aid of Eq. 3.26, Eq. 3.27 may be written

$$p_n(0) = p_{no} \, e^{qV/kT} \qquad (3.28)$$

Similarly, at the edge of the p region

$$n_p(0) = n_{po} \, e^{qV/kT} \qquad (3.29)$$

Substituting the expression for $p_n(0)$ in Eq. 3.28 into Eq. 3.19,

$$p_n = p_{no} (e^{qV/kT} - 1) e^{-x/L_p} + p_{no} \qquad (3.30)$$

Similarly, using Eq. 3.20,

$$n_p = n_{po} (e^{qV/kT} - 1) e^{x/L_n} + n_{po} \qquad (3.31)$$

Since the component of diode current due to holes injected into the n region is equal to the diffusion current at the edge of the depletion region, we may obtain this component of current by using the diffusion Eq. 3.4 evaluated at $x = 0$ and Eq. 3.30 to obtain

$$I_p = \frac{-qAD_p}{L_p} \, p_{no} \left(e^{qV/kT} - 1 \right) \qquad (3.32)$$

Similarly, the component of diode current due to electrons injected into the p region is

$$I_n = \frac{qAD_n}{L_n} \, n_{po} \left(e^{qV/kT} - 1 \right) \qquad (3.33)$$

The total diode current $I = I_p + I_n$. Therefore,

$$I = qA \left(\frac{D_p \, p_{no}}{L_p} + \frac{D_n \, n_{po}}{L_n} \right) \left(e^{qV/kT} - 1 \right) \qquad (3.34)$$

We previously defined the saturation current I_S as the diode current that flows when sufficient reverse bias is applied to essentially eliminate the injection current. Then $e^{qV/kT}$ is negligible and Eq. 3.34 shows that

$$|I_S| = qA \left(\frac{D_p \, p_{no}}{L_p} + \frac{D_n \, n_{po}}{L_n} \right) \qquad (3.35)$$

Then

$$I = I_S \left(e^{qV/kT} - 1 \right) \qquad (3.36)$$

Example 3.1

A given silicon diode has $N_A = 5 \times 10^{21}$ and $N_D = 10^{22}$ SI units. Let us assume that the lifetime of excess carriers is 1 μs in both regions. The cross-sectional area of the diode is 4×10^{-6} m². We will find the diode current as a function of the bias voltage V at 300°K.

Figure 2.5 shows that $n_i = 1.48 \times 10^{16}$ so $n_{po} = (1.48 \times 10^{16})^2 / 5 \times 10^{21} = 4.38 \times 10^{10}$ and $p_{no} = 2.19 \times 10^{10}$. From Table 2.3, $D_n = 33.8 \times 10^{-4}$, $D_p = 13 \times 10^{-4}$. Thus, $L_p = (D_p \tau)^{1/2} = (13 \times 10^{-10})^{1/2} = 3.6 \times 10^{-5}$ m and $L_n = (33.8 \times 10^{-10})^{1/2} = 5.8 \times 10^{-5}$ m. Then $I_S = 1.6 \times 10^{-19} \times 4 \times 10^{-6}$ $(13 \times 10^{-4} \times 2.19 \times 10^{10} / 3.6 \times 10^{-5} + 33.8 \times 10^{-4} \times 4.38 \times 10^{10} / 5.8 \times 10^{-5}) = 2.13 \times 10^{-12}$ A, and $I = 2.13 \times 10^{-12}$ $(e^{38.7V} - 1)$.

Figure 3.6 shows I as a function of V for two different current scales.

3.6
THE FORWARD-
BIASED DIODE

The diode equation is accurate only under the conditions of low-level injection because the majority carrier density was assumed to be constant in both the p region and the n region of the diode. In order to understand this more fully, let us consider the carrier densities of the diode with characteristics given in Example 3.1 when the forward cur-

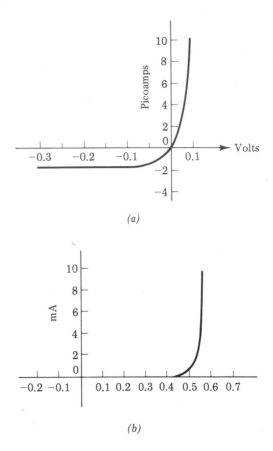

(a)

(b)

Fig. 3.6. Plots of theoretical diode current as a function of bias voltage V:
(a) picoamp range; (b) milliamp range.

rent is 10 mA and the bias voltage = .575 V. Using Eqs. 3.30 and 3.31, the minority-carrier densities are easily found and are given in Fig. 3.7.

As the holes p_n are injected into the n region and diffused away from the junction, as shown in Fig. 3.7, a net positive charge is created by these excess holes. However, the n region is conductive so the free electrons redistribute themselves and additional electrons are drawn from the external circuit so that the excess charge is neutralized. This redistribution of charge is a capacitive effect known as *diffusion capacitance* (which is discussed in section 3.11), but our immediate concern is that the majority-carrier density is actually $N_D + |p_n|$ as shown in Fig. 3.7. This increased density invalidates our assumption that $n_n = N_D$ unless the increase due to p_n is 0.1 N_D or less and the injection is *low level*. Similarly, electrons injected into the p region cause the majority carriers to rise above N_A by the magnitude of n_p. Note in Fig. 3.7 that the

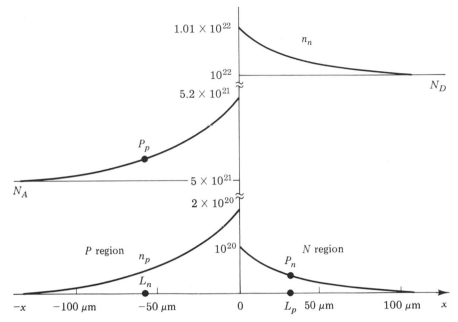

Fig. 3.7. Minority- and majority-carrier densities near the junction of a forward-biased diode.

more lightly-doped side has the greatest injected-carrier density. So we need to check the injection level on the more lightly-doped side. In this example, it is the p side and the injection level $L_i = n_p(0) / N_A = 2 \times 10^{20} / 5 \times 10^{21} = .04$ which is low level. For this diode, currents above 25 mA would cause high-level injection.

Since diode current is a function of the saturation current which is proportional to n_i^2, the diode current is a function of temperature. However, the diode current is often controlled primarily by the external circuit resistance and the forward-bias voltage V varies with temperature while the current remains relatively constant, as shown in Fig. 3.8a. If the current is held constant, the forward-bias voltage decreases about 2 mV with each degree K increase in temperature.

Typical forward-bias characteristics for Ge and GaAs may be compared with Si in Fig. 3.8b. Since the intrinsic carrier density in Ge is about 1,000 times as great as that of Si, the forward bias required for similar currents in similar diodes is about 0.35 volts less for Ge than for silicon. On the other hand, GaAs has about 10^{-4} times as many intrinsic carriers as Si, so its required forward bias is about 0.4 V greater than that of silicon.

The curves of Figs. 3.6 and 3.8 show that the diode is a nonlinear device. Therefore, either graphical analysis must be used to solve diode

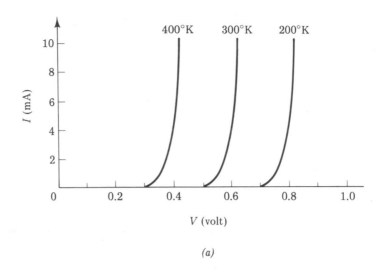

(a)

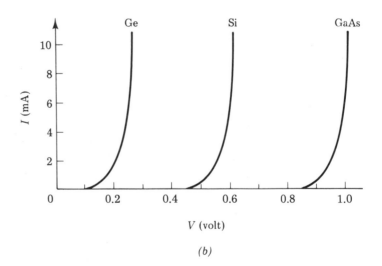

(b)

Fig. 3.8. Diode forward characteristic variation with (a) temperature and (b) materials at 300°K with similar doping densities.

circuits or an approximate linear model must be used to replace the actual diode in the circuit. The diode model may be very simple if the diode is assumed to be *ideal*. That is, the diode is assumed to have no current through it while it is reverse biased and no voltage across it while it is forward biased. The diode then acts as a perfect voltage-operated switch, and a linear circuit is drawn for each of the switch positions. This ideal diode model is adequately accurate when the voltage across the diode is negligible compared with the voltages across the other circuit elements during forward bias, and the diode reverse current does not cause significant voltages to appear across the other elements. When these conditions are not met, a piece-wise linear model is usually used for the diode.

A piece-wise linear model is shown for the forward-biased diode in Fig. 3.9. The diode is assumed to be nonconductive for forward bias voltages lower than V_o, as shown in Fig. 3.9a. Then the diode current is assumed to increase linearly along the dashed line which is tangent to the actual diode curve at the average value of forward current I_f. The linear circuit model of Fig. 3.9b may then be used to represent the diode when the currents vary about the average value I_f.

The value of V_o for the model of Fig. 3.9b is usually about .55 V to .6 V for silicon and about 0.2 V to .25 V for germanium diodes at 300°K. The value of forward conductance g_f for this model is the slope $\Delta I / \Delta V$ of the diode characteristic at the point $i = I_f$. Therefore, g_f may be obtained from the derivative of the diode Eq. 3.36 with respect to V evaluated at the value of V_f that will give $I = I_f$. Then

$$g_f = \frac{d\,(I_S\,e^{qV/kT}\quad I_s)}{dV}\bigg|_{V=V_f} = \frac{q}{kT}\,(I_S\,e^{qV/kT})\bigg|_{V=V_f} \qquad (3.37)$$

Since $I_f \gg I_S$ for the usual values, $I_f \simeq I_S\,e^{qV_f/kT}$ and Eq. 3.37 may be written

$$g_f = \frac{q}{kT}I_f \qquad (3.38)$$

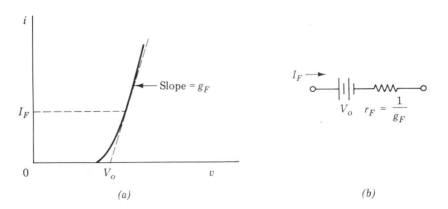

Fig. 3.9. Piecewise linear model for a forward-biased diode: (a) graphical model; (b) circuit model.

At normal room temperatures, $q/kT \simeq 40$. Therefore, at these temperatures

$$g_f \simeq 40\,I_f \qquad\qquad (3.39)$$

$$r_f \simeq \frac{25}{I_f\,(\text{mA})} \qquad\qquad (3.40)$$

Up to this point we have assumed that the voltage across the junction is the same as the voltage applied to the diode. However, the doped semiconductor has ohmic resistance because of its finite conductivity, and the externally applied voltage is the junction voltage plus the IR drops in the p region and the n region, as illustrated in Fig. 3.10.

Diodes are available in a wide range of current capacities. Both the forward-current capability and the saturation current are proportional to the cross-sectional area A as shown in Eqs. 3.34 and 3.35.

3.7 THE REVERSE-BIASED DIODE

The diode Eq. 3.36 shows that the theoretical diode current remains essentially constant at I_S for all values of reverse bias voltage magnitude greater than about 0.1 V. However, the effect of thermally-generated carriers in the depletion region was not considered in the derivation of Eq. 3.36. Only the minority carriers in the p and n electrically neutral regions that diffuse to the edges of the depletion region and are swept through it were considered.

The carriers that are thermally generated within the depletion region are swept out of the region by the intense electric field. These carriers add to the minority carriers diffusing into the junction and thus cause a *thermal generation component* of reverse current in addition to the saturation current as shown in Fig. 3.11a. As the reverse bias voltage increases, the width, and thus the volume of the depletion region,

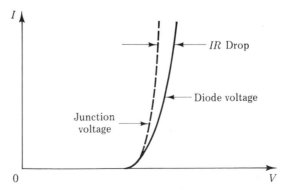

Fig. 3.10. Comparison of diode voltage and junction voltage.

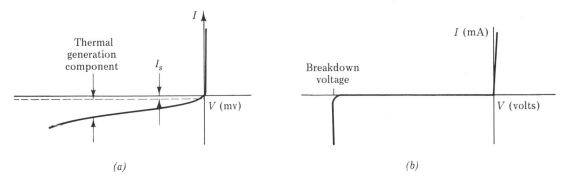

Fig. 3.11. Reverse characteristics of a silicon diode showing (a) the thermal
generation current component and (b) breakdown voltage. Note the
change in scales.

increases. Therefore, the thermal-generation component of saturation
current increases with increasing reverse bias, as indicated.

The thermal generation current is large in comparison with the saturation current in silicon and gallium arsenide diodes except at very
small values of reverse bias. In addition, there is some surface-leakage
current across the junction due to the broken covalent bonds at the
surface of the material. Therefore, the reverse current of a typical small
silicon diode may be of the order of a nanoampere rather than the
picoampere range obtained in Example 3.1. The diode equation (3.36)
still holds, however, where I_S is determined by Eq. 3.35. The saturation
currents of germanium diodes are typically of the order of micro-
amperes and therefore predominate over the thermal-generation and
surface-leakage currents. Therefore, the reverse current of germanium
diodes is almost independent of voltage, near the value of I_S, until
voltage breakdown is approached.

As the reverse bias is increased, a voltage is reached where the current first increases rapidly with voltage and then increases abruptly as
shown in Fig. 3.11b. This breakdown voltage is consistent for a given
diode but varies widely from diode to diode, depending upon the doping densities, as discussed later in Section 3.9. Voltage breakdown will
not damage the diode *unless* the current, which is controlled by the
external circuitry during breakdown, causes the diode to exceed its
maximum power dissipation capability.

**3.8
RECTIFIER-FILTER
CIRCUITS**

Diodes serve many useful functions in electronic circuits. One of their
most common uses is to convert ac power from the commercial power
lines to dc power, which is required for the operation of electronic
amplifiers and digital devices. These ac-to-dc converters are commonly
known as *dc power supplies* or simply *power supplies*.

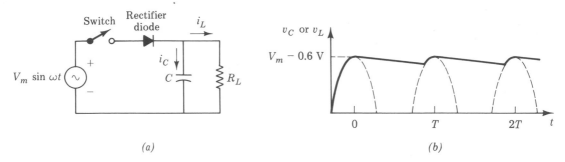

Fig. 3.12. (a) Half-wave rectifier-filter circuit. (b) Voltage waveform across the capacitor.

The basic element of a power supply is a diode *rectifier* and filter circuit shown in Fig. 3.12a. The rectifier diode is distinguished by its large current capability in comparison with the milliampere capability of the diodes previously considered. The current capability of a diode is proportional to its cross-sectional area, as you may recall.

Let us assume that the switch is closed at the time the source voltage is passing through zero in its positive direction. The diode is then forward biased and current passes through the diode, providing current i_L to the load resistance R_L and current i_C to charge the capacitor. The voltage across the load and the capacitor is about 0.6 V less than the source voltage if a silicon rectifier is used; therefore, the capacitor will charge to within about 0.6 V of the peak source voltage in the first half cycle, as shown in Fig. 3.12b. However, as the source voltage decreases and becomes less than the capacitor voltage, the rectifier is reverse biased so the capacitor cannot discharge through the diode but must discharge only through the load. If the time constant R_LC is long compared with the period T of the input power, the voltage across the load will not decrease significantly before the next positive half cycle of input voltage recharges the capacitor to $V_m - 0.6$ V. In Fig. 3.12b, $R_LC \simeq 6T$. While the diode is not conducting, i_C reverses direction and $i_C = i_L$. The capacitor stores energy from the source while the diode conducts and delivers power to the load while the diode is not conducting.

The voltage variations across the load over the input cycle are known as *ripple*. The ripple voltage may be reduced for a given capacitance and load resistance if the single *half-wave rectifier* of Fig. 3.12a is replaced by a *full-wave bridge* rectifier as shown in Fig. 3.13a. In this rectifier, during the positive half cycle, current i flows through diode D_1 to the load-capacitor combination and back through D_3 to the negative side of the source as shown. During the next half cycle, current flows through diode D_2 and in the same direction as before to the load and then back through diode D_4 to the source.

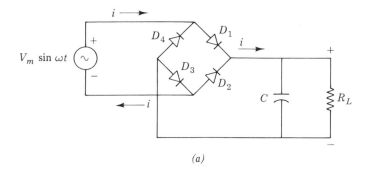

(a)

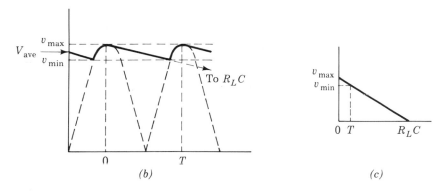

(b)   *(c)*

Fig. 3.13. (a) Full-wave bridge rectifier circuit. (b) Rectified output voltage. (c) Voltage and time relationships in the circuit.

The full-wave rectified output voltage is shown in Fig. 3.13b. The voltage that would appear across R_L if the capacitor were removed is indicated by the dashed sinusoidal segments. Note that the period T of this voltage is one-half the period of the half-wave rectifier output. The solid line shows the output voltage with the smoothing, or filter, capacitor included. The dc voltage component across the load is the average value which is approximately $(v_{max} + v_{min})/2$. The peak-to-peak ripple voltage is $(v_{max} - v_{min})$. The rms value may be approximated by assuming the ripple voltage to be somewhat sinuosidal. Then, the rms ripple $\simeq (v_{max} - v_{min})/2\sqrt{2}$.

We may easily develop approximate relationships among the ac input voltage, the dc output voltage, the ripple voltage, and the components R_L and C with the aid of Fig. 3.13c. The construction in this figure is extracted from Fig. 3.13b where v_{max}, v_{min}, 0, and T remain the same, but the time axis has been extended to show that if the capacitor were to continue to discharge at its initial rate, it would decrease to zero in one time constant $R_L C$. However, we will assume that the capacitor discharges linearly for the period T and is then abruptly recharged to v_{max}. This is a good approximation when the peak-to-peak ripple voltage is 0.1 v_{max} or less as it usually is. Then, noting the similar triangles in Fig. 3.13c,

$$\frac{v_{max} - v_{min}}{T} = \frac{v_{max}}{R_L C} \tag{3.41}$$

We usually would like to determine the required filter capacitance C to provide a desired, or tolerable, peak-to-peak ripple = $v_{max} - v_{min}$. Making this substitution and solving Eq. 3.41 for C,

$$C = \frac{T v_{max}}{R_L v_{ripple} \text{ (peak-to-peak)}} \tag{3.42}$$

Unless otherwise specified, the ripple voltage is assumed to be sinusoidal. Using v_{ripple} (peak-to-peak) = $2\sqrt{2}\, v_{ripple}$ and $T = 1/f$, we may write Eq. 3.42 as

$$C = \frac{v_{max}}{2\sqrt{2} f R_L v_{ripple}} \tag{3.43}$$

The load resistance R_L is not usually a physical resistor but is the ratio of the dc load voltage to the load current. $R_L = V_L / I_L$.

Example 3.2.

Let us assume that we need a dc power source that will provide about 160 volts to a load that draws 0.5 amps. The rms ripple voltage should not exceed 5 V. We need to determine the minimum capacitance required, the required rms voltage of the ac power source, and the maximum reverse bias that will be applied to the diodes. The load resistance $R_L = 160/.5 = 320\ \Omega$. The peak ripple voltage = $5\sqrt{2} = 7$ V, so the v_{max} applied to the filter capacitor and load is $160 + 7 = 167$ V. Let us choose a full-wave bridge rectifier. The load current flows through two diodes in series so the peak voltage applied by the source to the rectifier needs to be $167 + 2(0.6) = 168.2$ volts. Thus the rms voltage of the source needs to be $168.2/\sqrt{2} = 119$ V, which is a typical commercial power voltage, and no transformer is required. The required filter capacitance for the 5 V ripple is, using Eq. 3.43,

$$C = \frac{167}{2\sqrt{2}\,(120)\,(320)\,5} = 308\ \mu F$$

Observe from Fig. 3.12a that the source voltage is applied directly across the series combination of diodes D_1 and D_2, and also D_3 and D_4. When the input voltage is at its positive peak, D_1 and D_3 are conducting and have about 0.6 V across them, so essentially all of the 168.2 peak input volts appear as reverse bias across D_2 and D_4. On the next half cycle, the same 168.2 V *peak inverse voltage* appears across D_1 and D_3. Allowing a generous safety factor for reliable, or trouble-free operation, diodes having a *peak inverse voltage* rating of at least 200 V should be used.

The ratio of rms ripple voltage to the dc output voltage is defined as the *ripple factor* of the power supply. Thus

$$\text{ripple factor} = \frac{\text{rms ripple voltage}}{\text{dc output voltage}} \tag{3.44}$$

When the ripple factor is multiplied by 100, the *percent ripple* is obtained. For the power supply of Example 3.2, the ripple factor is $5/160 = .031$ or 3.1 percent.

You may observe from Fig. 3.12b that the ripple voltage decreases toward zero and the average, or dc, voltage across the load increases toward v_{max} as the load is decreaed toward zero (no load) or the load resistance R_L is increased toward infinity. Ideally the ripple factor would be zero for all values of load, and the dc output or load voltage would be constant, or independent of load. A term that is used to specify the variation of load voltage with changing load is *regulation*, which is defined as

$$\text{Regulation} = \frac{(\text{no-load voltage}) - (\text{full-load voltage})}{(\text{full-load voltage})} \tag{3.45}$$

The regulation of the power supply of Example 3.2 is $(168.2 - 160)/160 = .051$ or 5.1 percent. The leakage current of an electrolytic capacitor may cause 0.5 V or so to be dropped across each diode and the regulation would then be reduced to about $7.2/160 = .045$ or 4.5 percent.

It was just coincidental that the power line provided the desired dc voltage for the load on the power supply in Example 3.2. Usually a transformer is needed to either increase or decrease the available commercial voltage as shown in Fig. 3.14. When the transformer has a center tap, full-wave rectification may be obtained with only two diodes, as shown. The center tap is normally connected to the circuit ground or reference so that when terminal 1 of the transformer goes positive with respect to the center tap, the *half*-secondary voltage is applied through diode D_1 to the load resistance and filter capacitance, while diode D_2 is reverse biased. During the next half cycle diode D_2 conducts and diode D_1 is reverse biased. When terminal 1 is maximum positive, diode D_1 is conducting with only about 0.6 V across it so that almost all of the peak end-to-end secondary voltage is applied to diode D_2 as inverse voltage.

In addition to voltage transformation, the transformer provides elec-

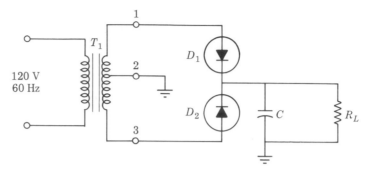

Fig. 3.14. Full-wave rectifier circuit using a center-tapped transformer.

tical isolation between the primary and secondary, so one side of both the primary and the load may be grounded. One side of the primary is always grounded in the commercial power system. The load must *float* with respect to ground in the bridge circuit of Fig. 3.13. The danger of electrical shock is increased greatly when the load must float.

The preceding discussion of power supplies neglects the power source resistance and some other important aspects of power supply design. See Chapter 16 for a more thorough and complete treatment of power-supply design.

**3.9
REFERENCE
DIODES**

In most diode applications, operation in the reverse-breakdown voltage region of the diodes is avoided. However, one class of diodes known as *Zener* or *reference* diodes are designed to operate in this breakdown region. In order to understand how these diodes operate, the physics of the junction must be considered.

The characteristic curve for a diode is shown by the solid line in Fig. 3.15. As the reverse voltage across the diode is increased, a point is reached where breakdown of the junction occurs and the reverse current through the diode increases abruptly. One theory attributes this rather abrupt current increase to the high potential gradient (or electric field) which exists at the junction. According to this theory, the high electric field is able to disrupt the covalent bonds and therefore greatly increase the minority carriers. This effect is known as *Zener* breakdown. Another accepted theory for the voltage breakdown is *avalanche* breakdown. This theory was founded by Townsend while he was studying the behavior of gases subject to electron bombardment. Accordingly, this theory was originally applied to gaseous conduction, but it applies just as well to the semiconductor.

According to the avalanche theory, a few carriers are generated in the intrinsic semiconductor material owing to thermal action, as previously

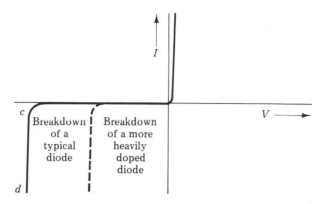

Fig. 3.15. Effect of doping density breakdown potential of p-n junction.

discussed. These carriers are accelerated by the high electric field near the junction until high velocities are acquired. A carrier with sufficient energy can produce an electron-hole pair when this carrier collides with a neutral atom. The new carriers so produced are free to be accelerated and, in turn, to produce additional carriers. The origin of the term *avalanche* can be seen by the foregoing explanation.

Since the electrons are much more mobile than the holes, most of the carriers are produced by electron collisions. The electrons have many random collisions as they travel through the semiconductor. In order for the avalanche effect to manifest itself, the electrons must obtain sufficient energy in traveling one mean free path to produce ionization of the atoms in the semiconductor. Hence, the electrons must have a kinetic energy equal to or greater than the gap energy of the semiconductor for an avalanche to be produced.

The junction breakdown potential is a function of the doping densities of the semiconductor and actually involves both breakdown mechanisms. As previously noted, the depletion width varies inversely with the doping density. Hence, as the doping concentration increases, the depletion region becomes thinner. As the depletion region becomes thinner, the electric field intensity becomes higher for a given junction voltage. Higher electric field potentials produce higher electron energies per mean free path. Therefore, a heavily doped p-n junction will have a lower breakdown potential than a relatively lightly doped p-n junction. This effect is illustrated in Fig. 3.15.

The *Zener diodes* or *reference diodes* operate in a region (from c to d on the curve of Fig. 3.15) where the voltage is essentially independent of current.

Reference diodes are available with breakdown voltages ranging from about 3 V to well over 100 V. The avalanche appears to be the primary breakdown mechanism in diodes, with reference voltages above about

7 V, because the breakdown voltages of these diodes increases with increasing temperature. As temperature increases, the atoms in the crystal structure vibrate about their 0°K position. The higher the temperature, the greater the amplitude of vibration. Thus, the atoms effectively occupy a larger volume. Consequently, the probability of a free electron-atom collision increases with temperature. Thus, increased temperature means a reduced mean free path for the free electrons. As a result, a higher voltage will be required to produce breakdown at a higher temperature. This type of behavior produces a *positive temperature coefficient*. The temperature coefficient is normally given in millivolts /°C and indicates the change in breakdown voltage which accompanies a one-degree increase in temperature.

In contrast to the foregoing behavior, diodes with breakdown voltages less than about 5 V have negative temperature coefficients, which indicates that the Zener breakdown mechanism is predominant in this range. The increased kinetic energy of the valence electrons would aid the high field in producing carriers and thus cause a *negative* temperature coefficient.

Unfortunately, the reference voltage increases slightly as the current increases, so a reference diode has some internal incremental resistance. The important parameters specified for a reference diode are:

1. Breakdown potential

2. Temperature coefficient

3. Power rating

4. Incremental resistance, r_z

If a series of similar diodes with different breakdown potentials are examined, additional insight can be gained. Thus, a plot of temperature coefficient versus breakdown potential is given in Fig. 3.16. The transition from negative to positive temperature coefficient occurs at 6 V (for $I_z = 41$ mA) for this series. A plot of incremental resistance versus breakdown potential is shown in Fig. 3.17. Note that a minimum of resistance also occurs near the 6 V value. If cost is not an important factor, two 6-V reference diodes connected in series would give better performance than one 12-V reference diode. The temperature coefficient of the combination would be approximately zero, and the combined resistance of two 6-V diodes would be less than the resistance of a single 12-V diode.

The symbol for a Zener, or reference, diode is given in Fig. 3.18a and a model for this diode is given in Fig. 3.18b. This model is useful over the normal operating range of the reference diode. An example will be given to illustrate its use.

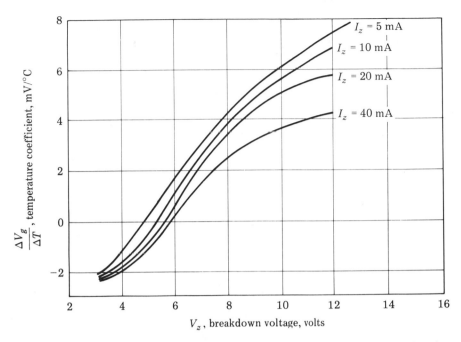

Fig. 3.16. Temperature coefficient versus breakdown voltage: 1N746 series. (Courtesy of Texas Instruments, Inc.).

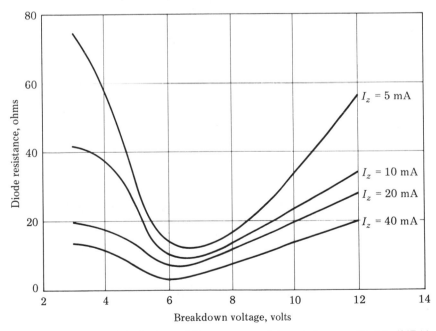

Fig. 3.17. Incremental diode resistance r_z versus breakdown voltage: 1N746 series. (Courtesy of Texas Instruments, Inc.).

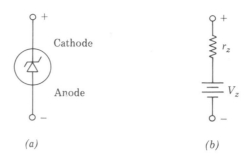

Fig. 3.18. (a) Reference-diode symbol and (b) an appropriate model for its
useful operating range.

Example 3.3

Let us use the reference diode with characteristics given in Fig. 3.19
to provide a relatively constant low-ripple voltage of 10 V to an
electronic circuit that draws a varying load current from 0 to 25 mA.
The power source is a rectifier-filter circuit that provides 18 V dc
output at 30 mA load current with 2 V peak-to-peak ripple. The
commerical power voltage fluctuations may cause the output voltage
to rise as high as 20 V at some times during the day.

The regulating circuit is given in Fig. 3.20a and the reference diode
is replaced by its model in Fig. 3.20b. The value of r_z is obtained by
observing from Fig. 3.19 that $\Delta V = 0.3$ V while $\Delta I = 40$ mA in the
breakdown region, so $r_z = 0.3/.04 = 7.5$ Ω. The minimum $R_L = 10$
V$/.025$ A $= 400$ Ω. The minimum reference-diode current must be
5 mA in order to maintain r_z at approximately 7.5 Ω. The series
resistor R_s controls the current to the reference diode and load. When
V_i is constant, the current I thru R_s must be essentially constant
because the load voltage is essentially constant at approximately 10 V.
But $I = I_L + I_z$ so I_z must increase as I_L decreases. When the load is
drawing $I_{Lmax} = 25$ mA, the reference diode must be drawing at least
$I_{zmin} = 5$ mA. Then $I_{min} = 25$ mA + 5 mA = 30 mA. This current must
flow through R_s when the minimum voltage occurs across R_s. The
minimum input voltage is the minimum dc voltage minus the peak
ripple voltage = $18 - 1 = 17$ V in this example. Thus $R_s = (17 - 10)$V$/$
30 mA = 233 Ω. Let us use the next lower ten percent tolerance size
of 220 Ω.

We could use conventional circuit-solving techniques to find the
dc and ripple voltages across the load for various conditions of input
voltage and load current. However, we can find approximate solutions more readily and perhaps gain greater insight into the operation
of the circuit. For example, if the input voltage remains constant, as
specified, the current through R_s is essentially constant. Therefore, as
the load current decreases from 25 mA to 0, the reference-diode

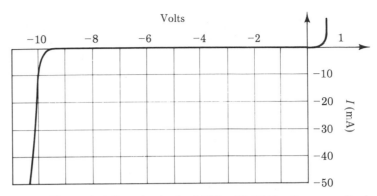

Fig. 3.19. *Characteristics of a typical 500-mW, 10-V reference diode.*

current increases by 25 mA. The output voltage increases by (25 mA) (7.5 Ω) = 0.1875 V and the regulation is $0.187 / 10 = .0187$ or 1.87 percent. Also, since 7.5 Ω is very small in comparison with the minimum load resistance, the voltage divider technique to find the peak-to-peak ripple across the load = $2 \times 7.5 / 227.5 = .066$ V and the ripple factor = $.066 / 2\sqrt{2} (10) = .023$ percent. When the input voltage rises to 20 V, the current through $R_s = (20 - 10) / 220 = 45.5$ mA. If the load current is 0, the Zener voltage, and thus the load voltage, is 10 V + (45.5 − 5) mA × 7.5 Ω = 10.3 V. The power dissipation of the Zener is 10 V × 45.5 mA = 455 mW.

A diode may be so heavily doped that the depletion region experiences Zener breakdown with only V_{ho} (no external bias) across the junction. The characteristics of this diode are shown in Fig. 3.21. This diode is known as a *backward* diode because it conducts more readily in the reverse-bias direction than in the forward-bias direction.

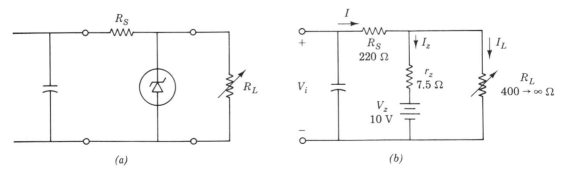

(a)   (b)

Fig. 3.20. *(a) Voltage-regulating circuit of Example 3.3. (b) Reference diode replaced by its model.*

**3.10
JUNCTION
CAPACITANCE**

A *p-n* junction and a charged capacitor are similar. As previously noted, the stored charge in the region of the junction results from the removal of free electrons from the n region, which leaves fixed positive donors. Similarly, the filling of the missing covalent bonds of the acceptor atoms in the p material produces fixed negative charges. The removal of the free or mobile carriers near the junction produces a *depletion* region which supports fixed excess charges and an electric field (Figs. 3.1 and 3.22). Some relationships between the barrier potential and the depletion width, the junction capacitance and the barrier potential, the maximum field intensity and the doping concentrations, and so on will be developed with the aid of Fig. 3.22, representing an abrupt junction *planar* diode of unit cross-sectional area. The excess positive charge density in the n-doped crystal is approximately equal to qN_d and penetrates a distance W_n (Fig. 3.22b). Similarly, the excess negative charge density in the p-doped crystal is $-qN_a$ and it penetrates a distance W_p. As indicated by the dashed lines in Fig. 3.22, the excess-charge regions do not terminate abruptly, but little error will be introduced by assuming abrupt termination at the effective distances W_n and W_p from the junction. Conservation of charge requires that the total charge be zero. Then

$$qN_dW_n = qN_aW_p \qquad (3.46)$$

The electric field intensity (Fig. 3.22c) may be determined from Gauss' law, which states that the total electric flux *DA* passing through a given

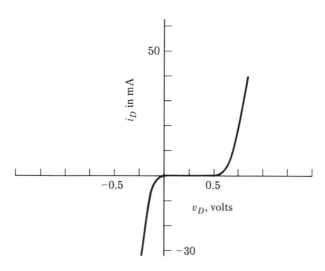

Fig. 3.21. Characteristics of a backward diode.

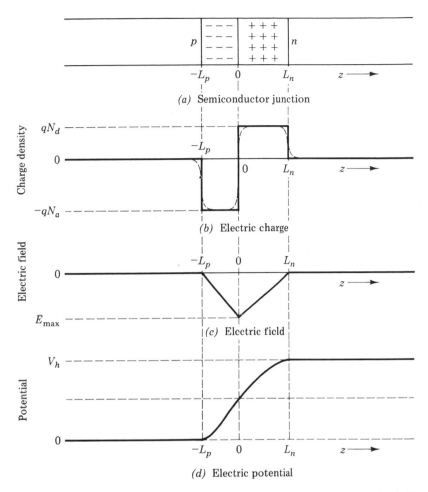

Fig. 3.22. Electric charge, field, and potential relationships in the depletion region associated with a p-n junction.

closed surface with area A is equal to the coulomb charge enclosed by the surface. Imagine that a ZY plane is located through the depletion region to the left of the junction. The total charge to the left (and to the right) of the plane is $-qN_a(W_p+x)$. All the electric flux lines originate on positive charges to the right of the plane and terminate on the negative charges to the left of the plane. Since the cross-section area is unity,

$$\mathscr{E} = \frac{D}{\epsilon} = -\frac{qN_a(W_p+x)}{\epsilon} \qquad (3.47)$$

where ϵ is the dielectric constant of the material.

Since the imaginary plane is to the left of the junction, x is negative and the electric field reduces to zero at $x = -W_p$. The maximum field exists at $x = 0$ and is given by

$$\mathcal{E}_{max} = -\frac{qN_aW_p}{\epsilon} = -\frac{qN_dW_n}{\epsilon} \tag{3.48}$$

The electric potential, using the p material to the left of W_p as the reference, is

$$V_h = -\int_{-W_p}^{W_n} \mathcal{E}\, dx \tag{3.49}$$

Since the potential difference V_h across the barrier is equal to the negative of the area under the $\mathcal{E}$ curve, inspection of Fig. 3.22c shows that

$$V_h = \frac{\mathcal{E}_{max}(W_p + W_n)}{2} = \frac{qN_aW_p(W_p + W_n)}{2\epsilon} \tag{3.50}$$

But from Eq. 3.48, $W_n = N_aW_p/N_d$ and

$$V_h = \frac{qN_aW_p^2(1 + N_a/N_d)}{2\epsilon} \tag{3.51}$$

Therefore,

$$W_p = \left[\frac{2\epsilon V_h}{qN_a(1 + N_a/N_d)}\right]^{1/2} \tag{3.52}$$

A similar solution for W_n yields

$$W_n = \left[\frac{2\epsilon V_h}{qN_d(1 + N_d/N_a)}\right]^{1/2} \tag{3.53}$$

Note that the depth of charge penetration, W_p or W_n, is proportional to the square root of the total barrier potential V_h and is roughly inversely proportional to the square root of the doping concentrations.

A concentration of charge exists at a junction and, also, a potential difference appears across the junction. Whenever these conditions exist, a capacitance also exists. The usual definition of capacitance C is charge Q, divided by voltage V. However, this junction capacitance is a function of the junction voltage and is therefore nonlinear. Thus, an incremental capacitance may be defined as

$$C = \frac{dQ}{dV} \tag{3.54}$$

For the junction, the charge and voltage are both functions of the distance W_p or W_n. Therefore, Eq. 3.54 may be written as

$$C = \frac{dQ/dW_p}{dV_h/dW_p} \qquad (3.55)$$

A voltage increase dV_h will cause an increased depth of penetration dW_p to the left of W_p and an increased depth of penetration dW_n to the right of W_n. But, dQ for the increase to the left of W_p is

$$dQ = qN_a dW_p \qquad (3.56)$$

or

$$\frac{dQ}{dW_p} = qN_a \qquad (3.57)$$

We can take the derivative of Eq. 3.51 to obtain

$$\frac{dV_h}{dW_p} = \frac{2qW_p N_a}{2\epsilon}\left[1 + \frac{N_a}{N_d}\right] \qquad (3.58)$$

The value of W_p from Eq. 3.52 is substituted into Eq. 3.58 to yield

$$\frac{dV_h}{dW_p} = \left[\frac{2qN_a\left(1 + \dfrac{N_a}{N_d}\right)V_h}{\epsilon}\right]^{1/2} \qquad (3.59)$$

Now, substitution of Eqs. 3.57 and 3.59 into Eq. 3.55 yields

$$C = \frac{qN_a}{\left[\dfrac{2qN_a\left(1 + N_a/N_d\right)V_h}{\epsilon}\right]^{1/2}}$$

$$= \left[\frac{qN_a}{2\left(1 + N_a/N_d\right)V_h}\right]^{1/2} \quad \text{farad}/\text{m}^2 \qquad (3.60)$$

Equation 3.60 can be written as

$$C \simeq KV_h^{-1/2} \qquad (3.61)$$

where K is a constant.

**3.11
DIFFUSION
CAPACITANCE**

In addition to the junction capacitance just considered, semiconductor diodes also exhibit an effect known as *diffusion capacitance*. When no external potential is applied, a profile of n carriers across the junction would appear as shown by the dashed curve in Fig. 3.23a. The relationship between the minority carrier density at the junction and the applied voltage is given by Eq. 3.29 (repeated here for convenience),

$$n_p(0) = n_{po}e^{qV/kT} \tag{3.29}$$

Note that when forward bias is applied, the n carriers which are injected from the n to the p material must increase. (Of course the p carriers in the n material also increase to neutralize the field in the p region.) This effect is illustrated by the solid curve in Fig. 3.23a. The additional minority carriers which result from forward bias (repesented by the area between the two curves in Fig. 3.23a) are often referred to as *stored charges*.

If the forward bias across the diode is suddenly reversed (Fig. 3.23b), there are a relatively large number of minority carriers near the junction. Consequently, these minority carriers diffuse to the junction and are swept back across the junction by the electric field due to the reverse-bias potential. As a result, the stored charges produce a reverse current through the diode, as shown in Fig. 3.23b. In addition to the minority carriers which are swept back across the junction, some of the stored charges are also lost by recombination. Therefore, the stored charges are rapidly depleted and the reverse current across the diode decreases to the normal saturation current I_S. Some diodes called *hot-carrier* diodes or *fast* diodes have comparatively few stored charges. Other diodes are purposely constructed so they have a large stored charge, but the reverse current drops to the I_S value very rapidly when

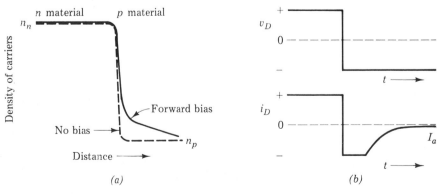

Fig. 3.23. Diffusion capactive effect: (a) n-carrier profile; (b) effect of diffusion capacitance.

the stored charge is depleted. These diodes are known as *snap diodes* and are used for frequency multiplication applications.

Problems   *Section 3.1*

3.1   Why must the electric field exist only in the depletion region at the p-n junction and not in the regions where free carriers exist, assuming no external connections are made to the diode?

Section 3.3

3.2   The distribution of free holes in a given piece of p-doped silicon is as given in Fig. 3.24. Determine the diffusion current that would flow through a $10^{-4}\,\mathrm{m}^2$ cross-sectional area, normal to x, at $T = 300°\mathrm{K}$.

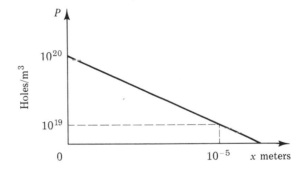

Fig. 3.24. p-distribution for Problem 3.2.

3.3   In the piece of p-doped silicon of Problem 3.2 and Fig. 3.24, what would be the magnitude of electric field intensity required at $x = 10^{-5}$ meter to reduce the total current to zero at that point? $(T = 300°\mathrm{K})$

3.4   Verify the Einstein relationship at $T = 300°\mathrm{K}$ using Tables 2.1 and 3.1.

Section 3.4

3.5   A given diode has $N_a = 10^{20}$ in the p region and $N_d = 5 \times 10^{-19}$ in the n region. If forward bias is applied so that $n_p(0) = 2 \times 10^{18}$ and the recombination time $T = 1\ \mu\mathrm{s}$, determine the current density due to the diffusion of electrons into the p material at $x = 0$. Assume the temperature to be 300°K. Is the injection low-level or high-level?

Section 3.5

3.6   A given silicon diode has $N_a = 2 \times 10^{21}$ and $N_d = 5 \times 10^{21}$ in SI units. The excess carrier lifetime is 0.3 μs in both regions and the cross-sectional area of the diode junction is 2×10^{-6} m². Determine I_S at 300°K and find the diode current when $V = 0.6$ V. Is the injection low-level?

Section 3.6

3.7   A given germanium diode has the same dimensions, doping densities, and carrier lifetimes as given in Example 3.1. Determine the bias voltage required to cause a 10 mA current to flow through this diode and compare it with the bias voltage required to cause 10 mA to flow through the diode of Example 3.1.

3.8   Determine the forward-bias voltage required to cause 10 mA to flow through the diode of Example 3.1 if its temperature is raised to 400°K.

3.9   A given forward-biased diode has 5 mA flowing through it.
   a.   What is its dynamic, or incremental, resistance and conductance if its temperature is 290°K?
   b.   What is its incremental conductance and resistance if $T = 400°K$?

3.10 If the diode of Example 3.1 has 2 Ω total axial resistance in the p-region and the n-region, what external forward voltage is required for 10 mA to flow through the diode at 300°K? What is the diode dynamic, or incremental, resistance under these conditions?

Section 3.8

3.11 A 200-Ω load resistance and a 1000-μF filter capacitor are connected through a silicon diode to the 120-V, 60-Hz commercial power line (Fig. 3.12).

    a. Determine the approximate dc load voltage and peak-to-peak ripple.

    b. Find the voltage regulation and the ripple factor for this power supply.

    c. What is the peak inverse voltage across the diode.

3.12 A power supply is needed to supply 15 V dc to an electronic load that draws 2 A current. The ripple factor may not exceed .03 or 3%.

    a. Draw the circuit diagram of a full wave rectifier-filter circuit using a center-tapped transformer. Determine the filter capacitance required and the rms voltage rating for the full secondary of the transformer using silicon diodes.

    b. Determine the peak inverse voltage across the diode rectifiers.

3.13 Use a full-wave bridge rectifier for the power supply of Problem 3.12.

Section 3.9

3.14 A 12-V reference diode of the IN746 series (Fig. 3.17) is used to provide a well-regulated, low-ripple dc voltage to an electronic circuit that draws 30 mA maximum current. The rectifier-filter circuit that precedes the regulator provides 20-V dc with 5% ripple at 20 V and 35 mA output.

    a. Determine the value of series resistance R_s that will provide a 5 mA minimum current through the reference diode.

    b. Find the maximum permissible rise in the input voltage if the power dissipation capability of the reference diode is 0.5 W and load current may be zero under certain conditions.

    c. Draw a model of a regulator circuit as given in Fig. 3.20b and determine the regulation of the regulated output. You may use reasonable approximations, but state them.

    d. Calculate the approximate percent ripple in the regulated output.

3.15 Repeat Problem 3.14 using two 6-V reference diodes from the IN746 series in series instead of the single 12-V diode.

Section 3.10

3.16 A silicon diode (at room temperature) has doping concentrations such that $N_a = 10^{21}/m^3$ and $N_d = 10^{22}/m^3$.
   a. What is the magnitude of the potential V_{ho}?
   b. What are the lengths of W_p and W_n?
   c. What is the maximum electric field intensity in the depletion region?

3.17 The cross-sectional area of the diode of Problem 3.16 is 10^{-6} m². Determine the incremental capacitance of this diode with external voltages of 0 V, -1 V, -5 V, and -25 V. Sketch this capacitance as a function of reverse bias.

4

Junction Transistors

Transistors are used in a wide variety of applications, including television, automatic control, satellite instrumentation, and medical electronics. The ability of the transistor to amplify electric signals accounts for its wide use.

The amplifier is actually an energy converter. The input signal merely controls the current that flows from the power supply or battery. Thus, the energy from the power supply is converted by the amplifier to signal energy.

**4.1
BIPOLAR
JUNCTION
TRANSISTORS**

When a semiconductor is arranged so that it has two p-n junctions as shown in Fig. 4.1a, it is known as a *bipolar junction transistor* or a *p-n-p transistor*. The electrical potential in the transistor as a function of distance along the axis is shown in Fig. 4.1b. Note that this potential is the same as the potential expected from two diode junctions. However, the electrical performance of a transistor is very different from the electrical performance of two diodes connected in a similar configuration. To illustrate this electrical behavior, consider a transistor connected as shown in Fig. 4.1c. The potential profile has now been altered as shown in Fig. 4.1d. The left-hand junction is forward biased, the potential hill is lowered, and carriers are injected into the n region which is known as the *base*. This base is made very thin, so nearly all the injected carriers diffuse across the base and are accelerated across the right-hand junction into the region known as the *collector*. The left-hand region is known as the *emitter* because it provides, or emits, the injected carriers. The current which flows out of the base lead

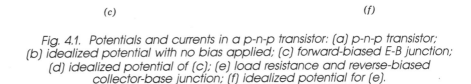

Fig. 4.1. *Potentials and currents in a p-n-p transistor: (a) p-n-p transistor; (b) idealized potential with no bias applied; (c) forward-biased E-B junction; (d) idealized potential of (c); (e) load resistance and reverse-biased collector-base junction; (f) idealized potential for (e).*

results from recombinations in the base. Therefore, the base is lightly doped, in addition to being thin, in order to minimize recombinations. The ratio of carriers flowing into the collector to those injected across the emitter-base junction is known as α (alpha), which ranges from about 0.90 to 0.998 in modern transistors.

The transistor is useless as an amplifier unless a load resistance, such as a headset or relay, is placed in the collector, or output, circuit. But a resistance in the collector circuit of Fig. 4.1c would cause the collector terminal to become positive with respect to the base, and would forward bias the collector-base junction. The injection current resulting from this bias would tend to cancel the initial collector current. However, if a battery is placed in the collector circuit as shown in Fig. 4.1e, the collector-base junction will remain reverse biased, providing the voltage drop across the load is less than the collector supply voltage V_{CC}. The reverse bias across the collector-base junction increases the height of the potential hill, as shown in Fig. 4.1f. Therefore, the electrons, or charge carriers, gain more kinetic energy as they are accelerated from the base into the collector region. This energy is dissipated as heat and causes the temperature of the transistor to rise.

From Fig. 4.1e, we note that the supply voltage V_{CC}, and also the voltage across the load resistor, may be very large in comparison with the input voltage across the forward-biased, emitter-base junction. The circuit is therefore an *amplifier* because it has voltage amplification or *gain*.

From the preceding discussion, it would appear that either end of the transistor could be used as the emitter. This would be true for the transistor configuration of Fig. 4.1. However, most transistors are made with collector junctions larger than emitter junctions for improved collector power dissipation and increased α (which will be discussed later), and thus α is higher in the normal, or forward, direction than in the reverse direction.

At this point, a few words should be said about conventional current and voltage directions. Actual current directions and potential polarities were shown in Fig. 4.1. However, IEEE standards require that currents which flow *into* a device be defined as *positive* and currents which flow *out* of the device be defined as *negative*. Therefore, the collector and base currents which usually flow out of the p-n-p transistor are negative currents because they flow in the opposite direction of the conventional positive currents. Also, a potential is considered positive if it is positive with respect to the common terminal or ground, which is the base terminal in Fig. 4.1. Voltage polarities will be given by + and − signs. A + sign on an ac voltage source indicates this terminal is positive when (sin ωt) is positive.

Most amplifiers must amplify voltage or current signals which vary with time, such as voice, music, or TV video signals. Through the use of the Fourier analysis, these time-varying signals can be reduced to the summation of a series of sinusoidal waveforms. Consequently, we will use sinusoidal signals in many of our examples, but we do not intend to infer that more complex waveforms cannot be handled. With this understanding, let us examine the behavior of a transistor with a sinusoidal input voltage when it is connected as shown in Fig. 4.2a. The conventional symbol for the p-n-p transistor is introduced in Fig. 4.2a.

The input signal will vary with time according to the expression $v_i = V_m \sin \omega t$. When t is zero, the voltage across the emitter to base junction will be zero. With zero voltage on this junction, the current flow across this junction will be zero. The only current flow across the base to collector junction will be the negligibly small saturation current of this reverse-biased junction. With no appreciable current flowing through R_L, there will be no voltage drop across R_L and $v_o = V_{CC}$. When the input voltage becomes more positive, the emitter base junction becomes forward biased. Then, holes continue through the base to the

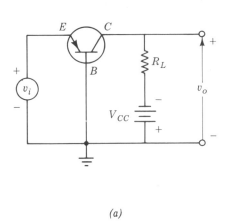

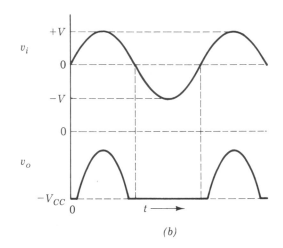

(a)

(b)

Fig. 4.2. Transistor amplifier with no bias voltage; (a) circuit; (b) input and output voltage waveforms.

negative collector. The collector current flowing through R_L produces a voltage drop across R_L, and v_o has the form shown in Fig. 4.2b. When v_i is negative there is essentially no current across either junction and $v_o = V_{CC}$.

The output voltage waveform in a linear amplifier should have the same shape as the input voltage waveform. Only the amplitude should be changed. Thus, the configuration shown in Fig. 4.2 must be modified if we desire a linear amplifier. The problem in this configuration occurred when the input junction became reverse biased. The circuit shown in Fig. 4.3a can be used to overcome this problem. In this circuit, a dc *bias* is applied to the input junction. The ac signal now rides on top of this dc signal (the law of superposition), and the emitter-to-base junction is always forward baised. Of course, if the input signal becomes too large, clipping will still occur in the output waveform. Then, either the input junction will become reverse biased on negative peaks or the voltage across R_L will become equal to V_{CC}, and there will be no voltage across the collector-to-base junction on positive input voltage peaks.

The capacitor C in Fig. 4.3 is used to block the dc bias current from flowing through the signal generator. While the dc is blocked, the ac signal from the signal generator v_s can pass readily through C_1. Similarly, C_2 is used to remove the dc component from the output signal. The capacitors must be large enough to offer negligible reactance to the ac signals. Note that the emitter circuit resistor R_E should have high resistance compared with the input resistance of the transistor in order to avoid shunting an appreciable part of the signal current i_s to ground.

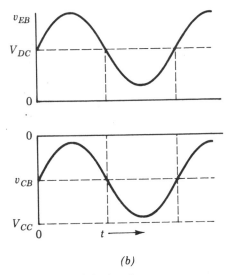

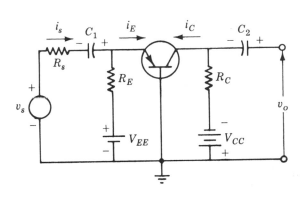

(a)　　　　　　　　　　　　　　　　　(b)

Fig. 4.3. Common-base ac amplifier using a p-n-p transistor: (a) circuit;
(b) transistor input and output voltage waveforms.

Then, V_{EE} must be larger than V_{EB} because of the $I_E R_E$ voltage drop across R_E. In order to emphasize these concepts, let us consider an example.

Example 4.1 | A transistor is connected as shown in Fig. 4.3. Let us determine proper circuit components if the average or bias current I_E is 2.0 mA.

Since the emitter-to-base junction is a forward-biased p-n junction, the dynamic input resistance r_e from emitter to base is the same as the dynamic forward-biased resistance r_f for a diode. Thus, from Eq. 3.40, we have $r_e \approx 25/I_E$ (in mA) $= 25/2 = 12.5\ \Omega$. If we choose the resistor R_E to be 100 times as large as r_e, then only one percent of the signal current i_s will be shunted through R_E. This value of R_E will be $12.5 \times 100 = 1{,}250\ \Omega$. The average (or dc) voltage drop across R_E is $I_E R_E = 1{,}250 \times .002 = 2.5$ V. If the transistor is silicon, the average voltage from base to emitter may be about 0.6 V. The battery voltage V_{EE} should be 0.6 V + 2.5 V = 3.1 V.

A suitable value for R_C can be determined if we first choose a value for V_{CC}, which must be below the voltage breakdown rating of the collector junction. Let us choose $V_{CC} = -20$ V. Now, if α for this transistor is 0.95, the average collector current I_C will have a value of $I_C = -\alpha I_E$ (the actual collector current is flowing *out* of the collector) $= -.95 \times 2 = -1.9$ mA. Normally, the voltage drop across R_C is chosen to be about one-half of V_{CC}. (This value is chosen so the signal variations of collector voltage can be symmetrical.) As a result, the average voltage drop across R_C should be about 10 V. Then, the value of R_C is 10 V $/I_C$ = 10 V $/$ 1.9 mA $\approx$ 5.25 kΩ. The available 10% tolerance value of 5.1 kΩ would probably be used.

Now, if the signal current, i_s has a peak value of 0.5 mA, the peak signal voltage from emitter to base v_{eb} is (neglecting the one percent loss of signal current through R_E) $i_s r_e = 5 \times 10^{-4} \times 12.5 = 6.25$ mV. The signal current component i_c in the collector circuit is $-\alpha i_s = -.95 \times 0.5$ mA $= -0.475$ mA. When this current flows through R_C, the voltage drop across R_C and therefore the output voltage v_o will have a peak magnitude of 2.37 V. The signal voltage gain of this amplifier is 2.37 V $/ 6.25 \times 10^{-3}$ V $= 380$.

In Example 4.1, several different current symbols were used. A few words should be said here about voltage and current symbols. The IEEE standards are based on the system given in Table 4.1.

The double subscripts associated with the voltage symbols (except the bias battery symbols) indicate the terminals between which the voltage is measured. For example, V_{BC} is the dc or average value of potential difference between the base and the collector. Therefore, $V_{BC} = -V_{CB}$. Sometimes a single subscript is used when the potential is with respect to ground or the chassis. For example, v_C is the total voltage between the collector and ground (or chassis or other ground reference) at any instant. If the base terminal is grounded, $v_C = v_{CB}$. As previously mentioned, the current flow into the transistor is positive if its sign is positive, whereas a negative sign indicates that current is flowing out of the transistor. Current arrows always indicate the direction of positive current flow; therefore, the current arrowheads always point toward the transistor, or device. Of course, ac or time-varying signal components that alternately change direction can also be represented by arrows or polarity signs. The arrows or polarity signs indicate the direction of the current or voltage, respectively, when it is positive.

So far, we have only considered p-n-p type transistors. Transistors are also constructed with the n-p-n configuration. In these transistors, electrons are injected into the p-type base region and proceed to the collector. Because of the doping reversal, all voltages and currents are also reversed. The symbol and a typical circuit for an n-p-n transistor

Table 4.1. Standard voltage and current notation.

Component	Symbol	Subscripts	Example
dc or average	Capital	Capital	V_{CB}, I_C
s domain or ω domain	Capital	Lower case	V_{cb}, I_c
Total (dc + time varying)	Lower case	Capital	v_{CB}, i_C
Instantaneous time varying	Lower case	Lower case	v_{cb}, i_c
Bias supply voltage	Capital	Double capital	V_{CC}, V_{EE}

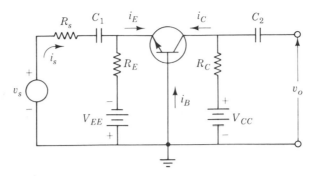

Fig. 4.4. Common-base n-p-n transistor amplifier.

are given in Fig. 4.4. In this circuit, i_E is negative and both i_C and i_B are positive.

4.2
THE
EBERS-MOLL
MODEL

Transistors are used as switches as well as for amplifiers. When used as switches, both junctions may be either forward biased or reverse biased. Also, a transistor may be used in the reverse direction with the collector junction forward biased and the emitter junction reverse biased.

A basic transistor model known as the *Ebers-Moll* model that is applicable for any configuration or use has been developed and is given in Fig. 4.5. In this model the diode symbols may be considered either actual or ideal diodes depending upon the accuracy desired. The dependent current generator $\alpha_F i_F$ accounts for the current which flows in the collector circuit as a result of the biased emitter-base junction. Similarly, the dependent current generator $\alpha_R i_R$ is the current which flows in the emitter circuit as a result of the biased collector junction. Note that in this model, either junction may be either forward or reverse biased. As observed in Fig. 4.5 the collector and emitter currents can be expressed as

$$i_C = i_R - \alpha_F i_F \tag{4.1}$$

$$i_E = -\alpha_R i_R + i_F \tag{4.2}$$

The currents i_F and i_R are the diode currents which result from the bias voltages across the emitter and collector junctions, respectively. These currents can be expressed by the familiar diode equation form as

$$i_C = I_{CS}\left(e^{qv_{CB}/kT} - 1\right) - \alpha_F I_{ES}\left(e^{qv_{EB}/kT} - 1\right) \tag{4.3}$$

$$i_E = -\alpha_R I_{CS}\left(e^{qv_{CB}/kT} - 1\right) + I_{ES}\left(e^{qv_{EB}/kT} - 1\right) \tag{4.4}$$

where I_{CS} and I_{ES} are known as the short-circuit saturation currents of the junctions.

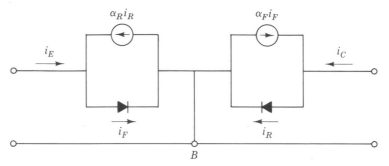

Fig. 4.5. Ebers-Moll model of a transistor.

The Ebers-Moll model is very general and can be quite accurate if diode resistances are included in the model. Computer simulation programs such as SCEPTR, TRAC and SPICE use the Ebers-Moll model with excellent results.

The Ebers-Moll Equations 4.3 and 4.4 may be expressed in more usable form if we multiply Eq. 4.4 by α_F and add the result to Eq. 4.3 to obtain

$$i_C + \alpha_F i_E = (1 - \alpha_F \alpha_R)\, I_{CS}\, (e^{qV_{CB}/kT} - 1) \tag{4.5}$$

Similarly, by multiplying Eq. 4.3 by α_R and adding the result to Eq. 4.4 we obtain

$$i_E + \alpha_R i_C = (1 - \alpha_F \alpha_R)\, I_{ES}\, (e^{qV_{EB}/kT} - 1) \tag{4.6}$$

Equations 4.5 and 4.6 may be rearranged into the following form:

$$i_C = -\alpha_F i_E + I_{CO}\, (e^{qV_{CB}/kT} - 1) \tag{4.7}$$

$$i_E = -\alpha_R i_C + I_{EO}\, (e^{qV_{EB}/kT} - 1) \tag{4.8}$$

where the saturation currents I_{CO} and I_{EO} are defined as

$$I_{CO} = (1 - \alpha_F \alpha_R)\, I_{CS} \tag{4.9}$$

$$I_{EO} = (1 - \alpha_F \alpha_R)\, I_{ES} \tag{4.10}$$

An alternate form of the Ebers-Moll model is given in Fig. 4.6 by using Eqs. 4.7 and 4.8. This model is more convenient than Fig. 4.5 because the dependent current source is expressed in terms of the terminal currents i_E and i_C rather than the diode currents i_F and i_R. The saturation currents I_{EO} and I_{CO} are known as the *open-circuit saturation currents* while I_{ES} and I_{CS} are the *short-circuit saturation currents*. These definitions are easily verified by applying the proper terminal condi-

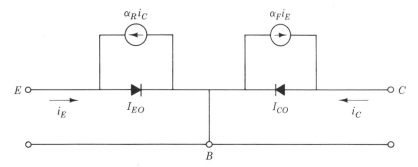

Fig. 4.6. *Alternate Ebers-Moll model for a PNP transistor.*

tions to Eqs. 4.7 and 4.8 and Eqs. 4.3 and 4.4. This verification is given as an exercise to the student (see Problem 4.3). This exercise demonstrates that I_{EO} and I_{CO} are the true diode saturation currents while I_{ES} and I_{CS} are larger because of the transistor action.

4.3 CHARACTER-ISTIC CURVES

An increased understanding of the transistor is obtained by observing the volt-ampere characteristics of the input and output ports. The input characteristics of a typical silicon transistor in the common-base configuration are given in Fig. 4.7. The input current i_E is plotted as a function of the input voltage v_{EB} with the collector voltage v_{CB} held constant. These curves are typical diode curves as might have been expected. However, you may observe that i_E is increased slightly, for any given value of v_{EB}, when the collector voltage v_{CB} is increased in magnitude from 0 to -20 V. The reason for this influence of collector voltage on the input characteristics is discussed later in Section 4.5.

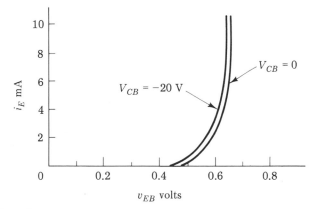

Fig. 4.7. *Input characteristics of a typical silicon transistor in the common-base configuration.*

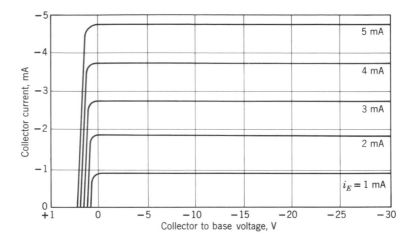

Fig. 4.8. Family of collector characteristics curves. (Note the change of scale for positive collector-to-base voltage.)

The output characteristics are obtained by holding the emitter current constant while the collector current i_C is plotted as a function of collector voltage v_{CB}. The resulting curve is known as a *collector-characteristic curve*. Each different value of emitter current will yield a different collector-characteristic curve. A set of several curves obtained from several representative values of emitter current is known as a *family* of collector-characteristic curves. A typical family or set of collector characteristics for a p-n-p transistor is shown in Fig. 4.8. The negative current indicates that the current is flowing out of the transistor where the reference direction is into the transistor. A set of curves for an n-p-n transistor might be identical to the set shown, except that the polarities of currents and voltages would be reversed.

From Fig. 4.8, it should be observed that:

1. The collector current is almost equal to the emitter current when reverse bias is applied to the collector junction.

2. The collector current is almost independent of the collector voltage when reverse bias is applied to the collector junction.

3. The collector current is rapidly reduced to zero and then reversed when forward bias is applied to the collector junction. This behavior occurs because the injection, or majority, current across the collection junction opposes the majority current across the emitter junction. The collector current is essentially the algebraic sum of these two majority currents.

The characteristic curves[1] given above are known as *static* curves because any point on these curves represents the dc or static current and voltage relationships of the transistor. However, when a signal is applied to a transistor amplifier, most voltages and currents vary with time. Therefore, it may seem that the static curves are of little use in determining the operating characteristics of an amplifier. Fortunately, this assumption is not true because the dynamic characteristics can be obtained simply by drawing a load line on the static collector characteristics. The voltage-current relationship in the collector circuit is given by the equation $v_{CB} = V_{CC} - i_C R_C$. Therefore, the load line intercepts the v_{CB} axis ($i_C = 0$) at $v_{CB} = V_{CC}$ and intercepts the i_C axis ($v_{CB} = 0$) at $i_C = V_{CC}/R_C$. This load line is known as a dc load line because the direct current must flow through the collector load resistor R_C. For example, the dc load line for a value of $R_C = 5$ kΩ has been drawn on a set of collector characteristics in Fig. 4.9a for a V_{CC} value of -20 V. The dc load will be considered as the only load at this time.

A dc bias point or *quiescent point* is selected someplace along the load line. The transistor operates at this quiescent point when no signal currents are present. If the expected input signals are symmetrical about the quiescent or q-point, such as sinusoidal signals, a good selection for the q-point would be near the center of the load line. The q-point is selected at $i_E = 2.0$ mA in this example.

The operating characteristics of the amplifier can readily be determined graphically. In order to illustrate the procedure, we will consider an example.

Example 4.2

The characteristic curves for a transistor are given in Fig. 4.9. If $R_C = 5$ kΩ and $V_{CC} = -20$ V, let us determine the characteristics of this amplifier.

First, the load line is drawn on the collector characteristic curves as shown. Then, a q-point is chosen on the load line. In this example, let $i_{Eq} = 2$ mA be the desired value. Now, the signal current waveform can be plotted at right angles to the load line. In this example, we have chosen a signal current of $i_s = 10^{-3} \sin \omega t$. At $\omega t = 0$ the transistor is operating at its quiescent value which, from Fig. 4.9b, we note is at $i_C = -1.9$ mA, $v_{CB} = -10.5$ V. When $\omega t = \pi/2$ radians, the value of i_s is 1 mA. This is represented as point A on the input signal and is projected to the load line as point B. We note that $i_E = 3$ mA, $i_C = -2.85$ mA, $v_{CB} = -6$ V. In this manner, plots may be obtained of i_C, v_{CB}, and i_E as functions of time. In this example, the peak-to-peak

---

[1] The common-base characteristics of Figs. 4.7 and 4.8 are seldom given by the manufacturers but can be readily obtained from a transistor curve tracer.

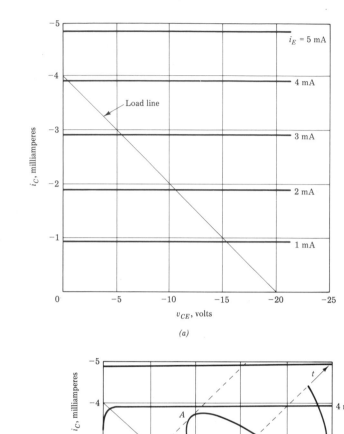

(a)

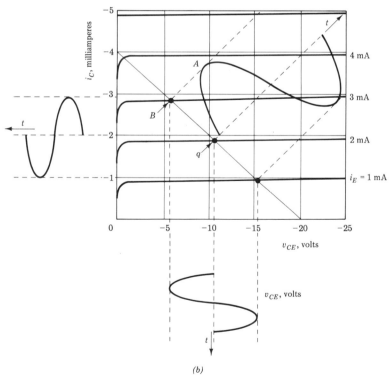

(b)

value of collector current (Δi_C) is $2.85 - 0.95 = 1.9$ mA. The current gain of this amplifier is $\Delta i_C / \Delta i_E = 1.9$ mA $/ 2$ mA $= 0.95$, which is essentially α for this transistor. The peak-to-peak magnitude of collector voltage (Δv_{CB}) is $15 - 6 = 9$ V.

4.4
THE
COMMON-
EMITTER
CONFIGU-
RATION

The input current requirement, and therefore the input power requirement, for a transistor amplifier may be reduced dramatically by using the emitter as the common or grounded terminal, as shown in Fig. 4.10. When the transistor is used as an amplifier, the emitter-base junction is forward biased and the collector-base junction is reverse biased as before. Using Eq. 4.7 under these conditions,

$$i_C = -\alpha_F i_E + I_{CO} \tag{4.11}$$

From Kirchoff's current law

$$-i_E = i_C + i_B \tag{4.12}$$

Using Eq. 4.12 to eliminate i_E in Eq. 4.11,

$$i_C = \alpha_F (i_C + i_B) + I_{CO} \tag{4.13}$$

Solving for i_C explicitly,

$$i_C = \frac{\alpha_F}{1 - \alpha_F} i_B + \frac{I_{CO}}{1 - \alpha_F} \tag{4.14}$$

Equation 4.14 may be written

$$i_C = \beta_F i_b + (\beta_F + 1) I_{CO} \tag{4.15}$$

where β_F is the forward *current amplification factor* or *short circuit current gain* of the common emitter amplifier, defined as

$$\beta_F = \frac{\alpha_F}{1 - \alpha_F} \tag{4.16}$$

When the transistor is operating in the reverse mode the subscript C is replaced by E and the subscript F is replaced by R in Eqs. 4.15 and 4.16. Since the transistor is normally used in the forward direction, β and α are usually used without subscripts when the forward mode is understood.

Fig. 4.9. (a) A 5-kΩ load line drawn on the collector characteristics of Fig. 4.8. (b) Dynamic operation along load line (Ex. 4.2).

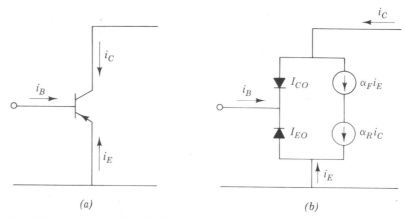

Fig. 4.10. p-n-p transistor in the common-emitter configuration: (a) transistor symbol; (b) Ebers-Moll model.

The input and collector characteristic curves of a typical n-p-n silicon transistor in the comon-emitter configuration are given in Fig. 4.11. You may observe that the typical diode-input curve is slightly dependent upon the magnitude of v_{CE}. At values of v_{CE} near zero this dependence becomes quite large. The reason for this is that both junctions are forward biased when v_{CE} approaches zero and β_R is small compared with β_F so the base current increases greatly as v_{CE} approaches zero. Note that the two junctions are equally biased at $v_{CE} = 0$.

The collector characeristics given in Fig. 4.11b are similar to the common-base set of Fig. 4.8 with the exception of the following differences:

1. Instead of i_E, the base current i_B is a parameter in the common-emitter set.

2. The collector current is essentially zero at $v_{CE} = 0$ for all values of base current because of the equal forward bias on both junctions which causes essentially equal currents to flow in opposite directions across the collector junction. As v_{CE} is increased in the reverse-bias direction, the collector current increases rapidly as the collector-junction injection current decreases toward zero.

3. The spacing between the collector characteristics is not as uniform and the slopes of the curves are greater in the common-emitter set. These effects are due to small variations in α_F which are caused by base-width modulation, and to other effects discussed in Section 4.6 which result in much larger changes in β_F.

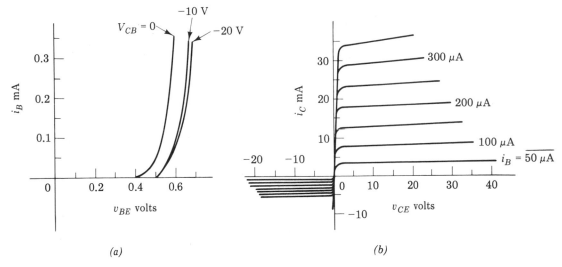

Fig. 4.11. (a) Input characteristics and (b) collector characteristics of a typical n-p-n silicon transistor in the common-emitter configuration.

**4.5
REGIONS OF
OPERATION**

The transistor may be used as either an amplifier or a switch. The common-emitter configuration is commonly used for either application. A simple circuit that may be used for either switching or amplifying is shown in Fig. 4.12a and the collector characteristics, including the load line, are given in Fig. 4.12b. The use of the transistor as an amplifier was discussed in Section 4.3 where the common-base configuration was employed. The common-emitter arrangement of Fig. 4.12a may similarly be used as a linear amplifier if the source voltage v_S is not too large and is unidirectional. For example, if $(v_{Smax} - 0.6 \text{ V})/R_B \leq$ 150 μA, assuming the V_{BE} diode drop to be about 0.6 V, and $(v_{Smin} - 0.6 \text{ V})/R_B \geq 10 \mu$A, the transistor operates as a *linear* amplifier, or in the linear region along the load line.

The transistor behaves as a switch rather than as an amplifier if v_S is a square wave having $(v_{Smax} - 0.6 \text{ V})/R_B \geq 175 \mu$A and $v_{Smin} < 0.5$ V, assuming the transmitter to be silicon, so $i_{Bmin} \approx 0$. During the positive half cycle of v_S the transistor is said to be *saturated* at the upper end of the load line in the vicinity of A, Fig. 4.12b, and nearly all of the voltage V_{CC} appears across R_L and the transistor switch is ON. On the other hand, during the negative half cycle of v_S the transistor is said to be CUT OFF at the lower end of the load line in the vicinity of B. Essentially no current flows through R_L so nearly all the voltage appears across the OPEN transistor switch that is said to be OFF.

The region A, Fig. 4.12a, is known as the *saturation region* because i_C is almost independent of i_B in this region since both junctions are

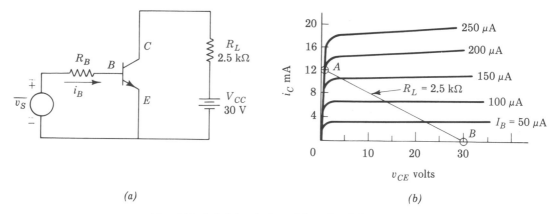

(a)

(b)

Fig. 4.12. (a) A basic transistor circuit that may be used for either amplification or switching. (b) The collector characteristics and load line for (a).

forward biased. The ON resistance of the transistor switch is a function of base current, however, because the collector characteristic curves for i_B greater than the value required for saturation do not pass through the same point as one may assume from Fig. 4.12b. This may be observed in Fig. 4.13a where the saturation region A is expanded by changing the v_{CE} scale. The ON resistance is obtained from the relationship

$$r_{ON} = \frac{v_{CE}}{i_C}\bigg| \text{ at load line and } I_B \text{ saturation} \tag{4.17}$$

In our example, $r_{ON} \simeq 0.3/.012 = 25$ Ω when $I_B = 200$ μA and $r_{ON} \simeq 0.1/.012 = 8.3$ Ω when $I_B = 500$ μA, as obtained from Fig. 4.13a. These characteristic curves are readily obtained from a transistor curve tracer and may also be provided by the transistor manufacturer.

The collector current does not reduce to zero when v_{BE} decreases to zero. The minimum value of i_C is approximately I_{CO} when the emitter junction is reverse biased, and it increases through I_{CS} as v_{BE} increases through zero to positive values, as illustrated in Fig. 4.13b. Therefore, the switch current in the cutoff region depends upon the base bias as well as the transistor type. The OFF resistance is determined from the relationship

$$r_{OFF} = \frac{v_{CE}}{i_C}\bigg| \text{ at load line in cut-off region} \tag{4.18}$$

In our example, if $i_C = I_{CS} = 4.5 \times 10^{-9}$ A and $v_{CE} \simeq 30$ V, from Fig. 4.13b, $r_{OFF} = 30/4.5 \times 10^{-9} = 6.7 \times 10^9$ Ω. Germanium transistors have much lower values of r_{OFF} than silicon or GaAs because of the relatively large values of I_{CO} for Ge.

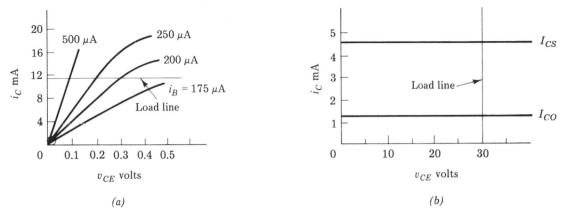

Fig. 4.13. (a) Expanded view of the saturation region.
(b) Expanded view of the cut-off region.

4.6
BASE WIDTH
MODULATION

The collector voltage has an influence on the collector and emitter currents because of the dependence of the depletion region width on the collector voltage, as illustrated in Fig. 4.14. In this figure, the vertical line at $x = 0$ represents the emitter junction. The depletion region at this junction is very narrow because of the small potential difference across it. The solid vertical line beyond $x = w$ represents the collector junction. Since the base is lightly doped compared with the collector, and since depletion region width is inversely proportional to the doping density, the collector-junction depletion region extends primarily into the base region. But the injected carriers which diffuse across the base region are swept into the collector as soon as they reach this depletion region; therefore, the effective base width (Fig. 4.14) is w. Now the current flow through the base is primarily by diffusion. The diffusion equation was developed in Section 3.3 (Eq. 3.3) and is repeated here for hole diffusion

$$J_p = -qD_p \frac{dp}{dx} \tag{4.19}$$

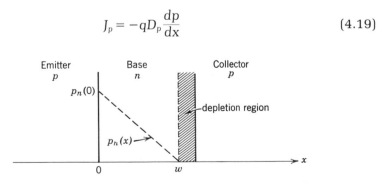

Fig. 4.14. Diagrammatic sketch showing the effect of depletion region width, and hence collector voltage, on base width.

But the diffusion current through the base region must be continuous, or constant, except for the small loss due to recombinations. Therefore, in the rectangular configuration assumed here, the slope of the $p_n(x)$ curve, dp/dx, must be essentially constant. Then the density of injected carriers $p_n(x)$ must decrease almost linearly from $p_n(0)$ at the emitter junction to essentially zero at the edge of the collector-junction depletion region. Figure 4.14 shows that this slope may be expressed as $p_n(0)/w$. This slope is negative since it is downward. Since the diffusion current is proportional to the slope $p_n(0)/w$ and the base width w decreases as the collector voltage increases, the emitter current increases as the collector voltage increases, assuming the emitter-junction voltage remains constant. Alpha (α) also increases with collector voltage because the narrower base region provides less opportunity for carrier recombination. This variation of effective base width with collector-base voltage is known as *base width modulation*.

The effect of collector voltage on α becomes more pronounced as the collector voltage becomes high and approaches avalanche breakdown of the collector junction. Then, carrier multiplication occurs across the collector junction because high-velocity carriers collide with atoms in the crystal and produce additional carriers. A sketch of α as a function of v_{CB} is given in Fig. 4.15 for a typical transistor with a collector-junction breakdown voltage of 80 V. Observe that α is unity at a collector voltage considerably below junction breakdown. In the transistor of Fig. 4.15, α is unity at approximately $v_{CB} = 60$ V.

We have not yet considered the component of base current that results from the injection of majority carriers in the base region across the emitter-base junction into the emitter region. This injection causes a component of current to flow in the base and emitter regions but does not affect the collector current. Therefore, α and β are reduced because

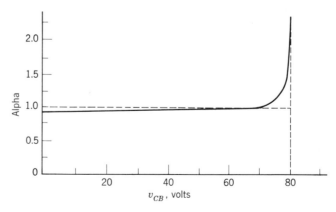

Fig. 4.15. Sketch of α as a function of v_{CB} for a typical transistor.

of this injection. Since the density of injected carriers is proportional to the doping density, as may be seen in Fig. 3.6, the number of carriers injected from base to emitter may be very small compared with the number injected from emitter to base, providing the base is very lightly doped in comparison with the emitter. The ratio of injection current from emitter to base to the total injection current across the emitter-base junction is known as *emitter efficiency*.

4.7
TRANSISTOR
RATINGS

Essentially the only cause of deterioration and destruction of a semiconductor device is heat, which may melt the solder connections, deteriorate the insulating materials, and change the crystal structure of the semiconductor. Therefore, manufacturers rate their semiconductor products in accordance with their power dissipation capability and maximum permissible temperature. Also, maximum voltage ratings are given and sometimes maximum current is specified. These ratings and their application to circuit design will be discussed in this section.

The ratings of a typical low-power n-p-n silicon transistor follow:

Absolute Maximum Ratings, 25°C

V_{CEO}	30 V
I_C	100 mA
Power dissipation, P_d^*	200 mW
Junction temperature, T_j	85°C

* Derate 3.33 mW / °C for ambient
temperatures above 25°C.

The safe operating area of this transistor may be marked on the collector characteristics by dashed lines as shown in Fig. 4.16. The transistor operation will be confined to this area if the dc load line remains *below* the maximum dissipation curve, which is drawn through all points where $v_{CE}i_C = P_d$ max. For example, if $v_{CC} = 30$ V, R_C min = 1.11 kΩ.

If the ambient, or surrounding, temperature is higher than 25°C, the maximum dissipation rating must be reduced, as specified by the manufacturer. For example, the transistor with the ratings given above must have its maximum dissipation rating reduced 3.3 mW for each degree C of ambient temperature above 25°C. Therefore, the maximum dissipation curve drawn on the collector characteristics should represent the maximum permissible dissipation at the highest expected ambient temperature. For example, if the transistor in Fig. 4.16 is to be enclosed in a metal cabinet and used on the desert in the summer, the ambient temperature may rise to 55°C (131°F). The maximum dissipation rating is then P_d (max) = 200 − 3.33 × 30 = 100 mW.

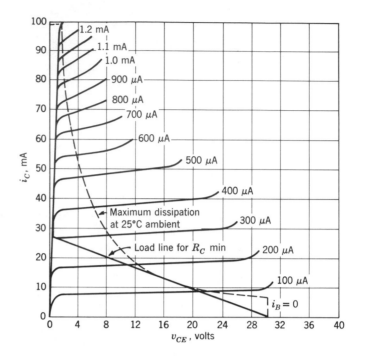

Fig. 4.16. Safe operating area for a typical 200-mW transistor.

Sometimes a derating curve is given instead of a derating factor. A derating curve for the transistor of Fig. 4.16 is given in Fig. 4.17. Observe that the slope of the derating curve is equal to the derating factor. The slope of the curve in Fig. 4.17 (above 25°C) is -200 mW$/60$°C $= -3.33$ mW$/$°C. The negative sign results from the negtive slope. The derating factor is, however, usually given as a positive number.

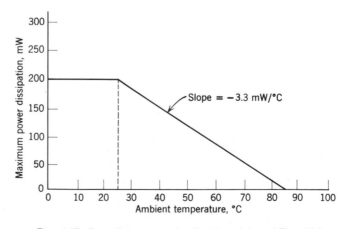

Fig. 4.17. Derating curve for the transistor of Fig. 4.16.

Most of the power dissipation in a transistor occurs at the collector junction because of the relatively large voltage across that junction. The junction temperature rises above the ambient temperature because of this dissipation. The rise in junction temperature is equal to the power dissipation divided by the derating factor. For example, if the transistor of Fig. 4.16 is operating with $V_{CE} = 10$ volts and $i_C = 10$ mA, the rise in junction temperature above the ambient is $v_{CE}i_C$/derating factor = 100 mW / 3.3 mW per °C $\simeq$ 30°C.

Heat flow, which results from a temperature difference, is analogous to current flow, which results from a potential difference. Therefore, a *thermal resistance* Θ_T has been defined as the ratio of the temperature rise to the power dissipation, or

$$\Theta_T = \frac{\Delta T}{P_d} \qquad (4.20)$$

and

$$\Delta T = \Theta_T P_D \qquad (4.21)$$

Note that if ΔT is replaced by V, Θ_T replaced by R, and P_d replaced by I, the foregoing two equations express Ohm's law. These equations show that the thermal resistance is the reciprocal of the derating factor, and is therefore the negative reciprocal of the slope of the derating curve. The thermal resistance between the junction and the ambient surroundings of the transistors in Fig. 4.17 is $\Theta_T = 1 / 3.33 = 0.3$°C / mW.

The junction temperature is the ambient temperature T_a plus the temperature rise due to power dissipation. Therefore, using Eq. 4.21

$$T_j = T_a + \Delta T = T_a + \Theta_T P_D \qquad (4.22)$$

For example, if the ambient temperature is 40°C and the transistor of Fig. 4.17 is dissipating 100 mW, the collector junction temperature $T_j = 40$°C + 0.3 × 100°C = 70°C. Observe and verify that at any point on the derating curve, the ambient temperature plus the corresponding power dissipation times the thermal resistance Θ_T gives the maximum permissible junction temperature.

The maximum current rating of a transistor, if given, usually indicates either the current at which the maximum dissipation curve crosses the saturation voltage, as shown in Fig. 4.16, or the current at which β falls below the minimum specified by the manufacturer.

The maximum voltage rating is not as simply specified for a transistor as for a diode. The transistor may *appear* to break down at the voltage for which $\alpha = 1$, although this voltage may be considerably below the avalanche breakdown voltage, as shown in Fig. 4.15. The reason the transistor appears to break down is because the collector current approaches infinity as α approaches unity, as seen by Eq. 4.14, repeated below.

$$i_C = \frac{\alpha}{1-\alpha} i_B + \frac{I_{CO}}{1-\alpha} \tag{4.23}$$

Observe that i_C is equal to infinity for $i_B = 0$ or for any positive value of i_B when $\alpha = 1$. The *apparent* breakdown voltage when $i_B = 0$ is known as the sustaining voltage or *maximum* V_{CEO}. The first two subscripts indicate the electrodes to which the voltage is applied and the third subscript indicates the conditions at the third terminal (base) which, in this case, is open. However, also observe from Eq. 4.23 that finite positive values of collector current may be obtained for values of α greater than unity if the base current is negative. In fact, if $i_B = -I_{CO}$, Eq. 4.23 shows that the collector current

$$i_C = I_{CO}\left(\frac{-\alpha}{1-\alpha} + \frac{1}{1-\alpha}\right) = I_{CO} \tag{4.24}$$

for any value of α. Of course, I_{CO} increases rapidly because of carrier multiplication as the avalanche breakdown voltage is approached. This avalanche breakdown voltage is known as the maximum V_{CBO} because there is no current across the emitter junction when the emitter is open and the base current is automatically held to $-I_{CO}$.

When neither the base nor the emitter circuits are open and the base circuit has a finite resistance R, as shown in Fig. 4.18a, part of I_{CO} flows as a negative base current i_B, and the remaining part of I_{CO} flows across the emitter junction. This negative base current reduces the collector current, as compared with the open base configuration, and causes the apparent breakdown voltage, *maximum* V_{CER}, to be higher than maximum V_{CEO}. The improvement in apparent breakdown voltage depends upon the value of R. Maximum improvement occurs when $R = 0$ because this value gives the maximum negative i_B. The apparent breakdown voltage with $R = 0$ is known as maximum V_{CES} where S means the base is shorted to the emitter. The value of base resistance must be specified when the maximum V_{CER} is given.

Because of the internal resistance r_b not all the I_{CO} flows in the base circuit when $R = 0$. Therefore, the negative base current can be increased, and hence, the apparent breakdown voltage increased, if a reverse-biasing voltage is included in the base circuit, as shown in Fig. 4.18b. The apparent breakdown voltage with this reverse voltage applied is known as *maximum* V_{CEX} and is essentially equal to V_{CBO}.

The collector characteristics for a typical n-p-n transistor, including the avalanche region, are given in Fig. 4.19a, and the breakdown characteristics of this transistor are given in Fig. 4.19b. The various *maximum*

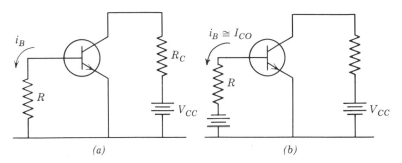

Fig. 4.18. *Negative i_B flows through resistance R in the base circuit.*

voltage ratings for this transistor are indicated along the voltage axis for a typical transistor.

4.8
BASIC
AMPLIFIERS

The transistor circuits previously considerd have used a bias battery and a current-limiting resistor to provide forward bias to the emitter junction. However, in the common-emitter configuration, the battery V_{BB}, which supplies forward bias to the base, has the same polarity with respect to the emitter as the battery V_{CC}, which supplies reverse bias to the collector. Therefore, a single battery, or power supply, can supply the proper bias to both junctions, as shown in Fig. 4.20. Since the dc voltage across the bias resistor R_B is $V_{CC} - V_{BE}$, the bias circuit resistance can be determined, using Ohm's law, from the relationship

$$R_B = \frac{V_{CC} - V_{BE}}{I_D} \qquad (4.25)$$

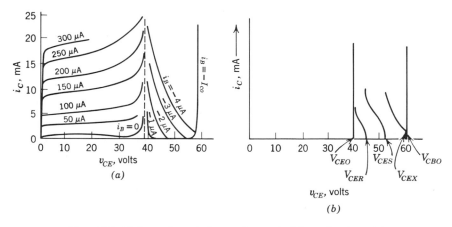

Fig. 4.19. *Voltage breakdown characteristics of a transistor.*

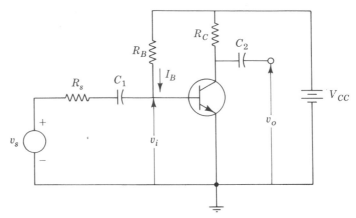

Fig. 4.20. A common-emitter amplifier that uses fixed bias obtained from the collector supply voltage V_{CC}.

This type of bias is known as *fixed bias* because the base bias current is *fixed* by the supply voltage V_{CC} and the bias resistor R_B, since V_{BE} is usually so small compared with V_{CC} that it has negligible influence on the base current. A more general type of bias circuit is treated in Chapter 7.

In order to gain additional experience and understanding of the transistor as a common-emitter amplifier with fixed bias, we will consider the characteristics of the circuit of Fig. 4.20 more carefully.

Example 4.3

Let us assume that the transistor in the circuit of Fig. 4.20 has the collector characteristics and load resistance $R_C = 2.5$ kΩ given in Fig. 4.12b. We need to determine the other component values to provide satisfactory operation of the circuit as an audio-frequency amplifier. In order to obtain maximum amplitude output signals, assuming sinusoidal or symmetrical input signals, we will choose the q-point at the center of the load line where $I_C = 6$ mA, $V_{CE} = 15$ V and $I_B \approx 90$ μA. The value of R_B required to establish this q-point is, using Eq. 4.25 and assuming the transistor to be silicon with $V_{BE} = 0.6$ V, $R_B = (30 - 0.6)/9 \times 10^{-5} = 327$ kΩ. We would use a 330-kΩ resistor, which is the nearest stock size ten percent tolerance resistor, and that is precisely the value we would have calculated had we neglected V_{BE}. The amplifier load R_C may be a headset or we may just want to observe the output voltage v_o with an oscilloscope when R_C is a resistor. The small-signal or *incremental* input resistance of the transistor may be determined from our knowledge of the dynamic resistance of a forward-biased diode given in Eq. 3.40. The q-point current across the forward-biased emitter junction is $I_E \simeq 6$ mA, so the resistance between emitter and base terminals, as viewed from the emitter, is $r_e = 25/6$ mA $= 4.16\ \Omega$. However, when viewed from the base terminal, the q-point current $I_B = 90\ \mu$A so $r_b = 25/.09$ mA $= 277\ \Omega$, assuming the transistor is at normal room temperature. Therefore, the signal current lost through the 330-kΩ bias resistor R_B is negligible and the current amplification, or gain, is approximately 6 mA$/90\ \mu$A $= 67 = \beta$. The output voltage $v_o = i_c R_C$ and the input voltage $v_i = i_b r_b$. Therefore, the voltage gain $= v_o/v_i = i_c R_C/i_b r_b = \beta R_C/r_b = 67\ (2.5$ k$\Omega)/277\ \Omega = 605$. However, the voltage gain from the driving source to the load is $v_o/v_s = i_c R_C/i_b (R_s + r_b) = \beta R_C/(R_s + r_b)$. Let us assume that $R_s = 500\ \Omega$. Then the voltage gain $v_o/v_s = 67\ (2.5$ k$\Omega)/(277 + 500) = 216$.

Each coupling, or dc blocking, capacitor in the circuit should be large enough to have negligible reactance in comparison with the total resistance R_T in series with it if the capacitor is to have negligible effect on the amplifier gain. This requirement may be met by letting $x_c = 0.1\ R_T$ at the lowest frequency of interest. Let us assume in this example that the lowest frequency of interest is 50 Hz. Then $x_{C_1} = 1/\omega C_1 = 0.1 (R_s + r_b)$ and $C_1 = 1/\omega (.1)(R_s + r_b) = 10/314(777) = 41\ \mu$F. The useful frequency range of the amplifier is considered to extend down to the frequency f_ℓ at which $x_C = R_T$ or $\omega_\ell = 1/R_T C$. The total impedance as seen by v_s is $\sqrt{2} (R_s + r_b)$ in magnitude, so both i_b and i_c signal currents are reduced by the factor $1/\sqrt{2}$ at frequency f_ℓ compared with midfrequencies where x_C is negligible. Since the output power is $i_{c^2} R_C$, the power is reduced to ½ at f_ℓ, which is known as either the *lower half-power frequency* or the *lower cutoff frequency*. For this example, $f_\ell = \frac{1}{2} \pi\ (777)(4.1 \times 10^{-5}) = 5$ Hz.

4.9
TRANSISTOR
CAPACITANCES

We learned in Chapter 3 that a diode has two different types of capacitance associated with it. One capacitance, known as *junction capacitance*, results from the stored charge in the depletion region and is dependent upon the height of the potential hill or voltage across the junction. The relationship between capacitance and junction voltage was derived for an abrupt junction, Eq. 3.61, and is repeated here for convenience.

$$C_j = KV_h^{-1/2} \tag{4.26}$$

where K is a constant for a given diode but is proportional to the cross-sectional area of the junction and increases with doping density. The emitter junction of a transistor behaves as an abrupt-junction diode and follows the relationship of Eq. 4.26.

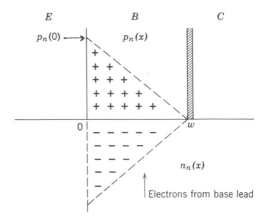

Fig. 4.21. Representation of stored charge in the base that results in diffusion capacitance.

The collector junction of a transistor is generally a *graded junction* where the doping density changes gradually through the depletion region. This type of junction is discussed in Section 5.7. At this point we are concerned only with the variation of junction capacitance with junction voltage which is given as

$$C_j' = K'V_h^{-1/3} \tag{4.27}$$

Since the potential hill V_h is much higher when reverse bias is applied as compared with forward bias and the collector region is not usually as heavily doped as the emitter region, the collector junction capacitance is usually small in comparison with the emitter junction capacitance when the transistor is used as an amplifier.

The other type capacitance discussed in Chapter 3 is *diffusion capacitance* which results from the carriers injected across the junction. This type of capacitance is proportional to the current through the junction. In a junction transistor almost all of the injected carriers are from the emitter into the base region, as shown in Fig. 4.21. The injected carriers tend to produce an excess charge (positive for a p-n-p) in the base, but the electric field, which begins to build up, causes electrons to flow through the base connecting lead to neutralize the charge. Note that these neutralizing charges do not necessarily combine with the injected carriers. In fact, they do not appreciably increase the recombination rate as long as their density is small in comparison with the base-doping concentration which is a requirement for low-level injection. The capacitive current which flows in the emitter-base circuit does not contribute to the collector current. Therefore, as the frequency increases and this capacitive current becomes significant, α decreases.

The dynamic or ac value of the diffusion capacitance can be deter-mined as a function of the effective base width w of the transistor with the aid of Fig. 4.22. If the injected carrier density is increased by Δp from p_o, the stored charge is increased by ΔQ, where ΔQ is the increase in average charge density times the effective volume of the base region. Then, since the average increase in charge density is $q\Delta p/2$,

$$\Delta Q = \frac{q\Delta pwA}{2} \tag{4.28}$$

where A is the effective cross-sectional area of the base region. The diffusion capacitance is

$$C_d = \frac{\Delta Q}{\Delta v_{EB}} \tag{4.29}$$

where Δv_{EB} is the change in voltage across the capacitor or the ac voltage v_{eb}.

$$\Delta v_{EB} = v_{eb} = i_e r_e \tag{4.30}$$

where r_o is the dynamic, or incremental, equivalent resistance of the emitter junction. Then, using Eqs. 4.28, 4.29, and 4.30,

$$C_d = \frac{q\Delta pwA}{2i_e r_e} \tag{4.31}$$

Now, i_e is the ac component of diffusion current. Therefore, we may obtain i_e from Eq. 3.3 as given below.

$$i_e = AqD_p\frac{\Delta p}{w} \tag{4.32}$$

Substituting this value of i_e into Eq. 4.31,

$$C_d = \frac{w^2}{2D_p r_e} \tag{4.33}$$

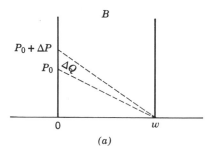

(a)

Fig. 4.22. Sketch to assist in the determination of diffusion capacitance.

The value of diffusion capacitance may be calculated for a uniform base transistor at a given q-point if the effective base width is known. For example, a p-n-p silicon transistor with $w = 10^{-5}$ meters and $I_E = 1$ mA has $C_d = (10^{-10}) / [2 \times 13 \times 10^{-4} (25)] = 1.54 \times 10^{-9}$ F.

The total capacitance associated with a junction is the junction capacitance plus the diffusion capacitance. These are shunt capacitances and therefore limit the high-frequency response of an amplifier and the switching speed of a transistor switch. The high-frequency response characteristics of an amplifier are examined in Section 5.8.

Problems   Section 4.1

4.1   In the amplifier circuit of Fig. 4.3 the emitter bias current $I_E = 4$ mA. With $V_{CC} = -20$ V, determine suitable values for R_E, R_C, and V_{BB}. Determine the voltage amplification of the circuit.

4.2   Compare the power dissipated in the resistor R_C in Problem 4.1 to that dissipated in R_C in Example 4.1.

Section 4.2

4.3   Verify the relationships between the short-circuit saturation currents and the open-circuit saturation currents by solving Eqs. 4.3 and 4.4 simultaneously using the terminal conditions $i_C = 0$ and v_{EB} reverse biased to find I_{EO}, etc.

4.4   Use the model of Fig. 4.6 to solve for I_{ES} and I_{CS} using the proper terminal conditions.

Section 4.3

4.5   Using the transistor curves of Fig. 4.9 with $V_{CC} = -20$ V and $R_L = 10$ kΩ, draw a load line on the collector characteristics, select a suitable q-point for symmetrical input signals, and graphically determine the current gain or α. Using the dynamic or incremental input resistance of the emitter diode, determine the voltage gain.

Section 4.4

4.6   Prove that $1 / (1 - \alpha_F) = \beta_F + 1$.

4.7   Determine the values of β_F for $\alpha_F = 0.99$ and $\alpha_F = 0.995$. What percent increase in β_F results from this 0.5% (approximate) increase in α_F?

Section 4.5

4.8 For the transistor and load resistance of Fig. 4.13, determine r_{ON} when $I_B = 250$ μA and r_{OFF} when $V_{BE} = 0$.

Section 4.6

4.9 A given n-p-n silicon transistor has emitter doping density $N_d = 10^{22}$ donors/m³ and emitter diffusion length $L_h = 3 \times 10^{-5}$ m. The uniformly doped base has $N_a = 2 \times 10^{19}$ acceptors/m³ and $w = 4 \times 10^{-6}$ m. Determine the emitter efficiency of this transistor. What is the approximate value of α_F if the diffusion length in the base is $L_e = 10^{-4}$ m and carrier multiplication in the collector depletion region is negligible?

Section 4.7

4.10 A given silicon transistor has a maximum power dissipation capability of 400 mW at an ambient temperature of 25°C or less. Its maximum junction temperature is given as 175°.

     a. Sketch the derating curve for this transistor.
     b. Determine its derating factor and thermal resistance.

4.11 A given transistor having a maximum dissipation capability of 500 mW at 25°C ambient temperature and a derating factor of 2 mW/°C is operating at the q-point $V_{CE} = 10$ V and $I_C = 8$ mA. What is the maximum permissible ambient temperature for this circuit?

4.12 Draw a 5-kΩ load line on the characteristics of Fig. 4.19a using $V_{CC} = 50$ V and determine the suitability of using this transistor as an amplifier under these conditions by sketching i_C vs. i_B along the load line.

Section 4.8

4.13 Use the amplifier with characteristics given in Fig. 4.12 as an amplifier with $V_{CC} = 30$ V and $R_C = 4.7$ kΩ in the circuit of Fig. 4.20. Choose a suitable q-point for symmetrical input signals and determine values for the fixed-bias resistor R_B, the amplifier input resistance for small, or incremental, signals, and the incremental voltage gains v_o/v_i and v_o/v_s using $R_s = 500$ Ω.

4.14 What capacitance would C_1 have in the amplifier of Problem 4.13 if the lower cut-off frequency is to remain at 5 Hz? See Example 4.3.

Section 4.9

4.15 A given uniform-base n-p-n silicon transistor has an effective base width $w = 10^{-5}$ meters. Determine the diffusion capacitance for this transistor if $I_E = 1\text{mA}$. Would you expect this n-p-n transistor to have better or worse high-frequency characteristics than a similar p-n-p transistor working at the same q-point?

5

Transistor Characteristics and Models

The transistor was introduced in Chapter 4 and its use as either an amplifying device or a switch was developed with the aid of the Ebers-Moll model and the static characteristic curves. The major focus was on the basic principles of operation of the transistor. The effects of nonlinearity, base-width modulation, transistor capacitance, and ohmic resistance were not seriously considered because they were not included in the basic Ebers-Moll model and were not immediately evident from the characteristic curves. This chapter is devoted to the development of models that will include all of these effects for the junction transistor when used as a linear amplifier. These models will lead to the satisfactory design of the various types of transistor amplifiers.

5.1 A SMALL-SIGNAL MODEL

The Ebers-Moll model is not a *linear model* because it includes diodes which are nonlinear devices. Therefore, the powerful tools of linear circuit analysis cannot be applied to the Ebers-Moll model. However, when the transistor is used as an amplifier and its operation is restricted to its linear range, linear models may be developed that permit the use of linear circuit theory in both the analysis and design of electronic circuits.

We took the first step toward the development of a linear model in Chapter 4 when we used the slope of the transistor input characteristic at the q-point to determine the small-signal input conductance or resistance of an amplifier. This resistance may replace the forward-biased emitter diode in the common-base configuration of the Ebers-Moll model as shown in Fig. 5.1. The common-base Ebers-Moll model of Fig.

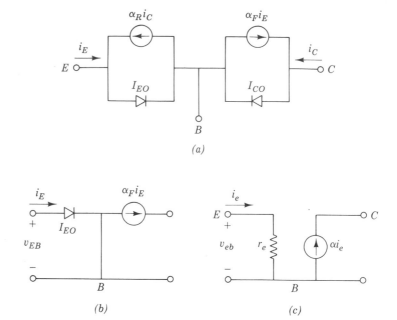

Fig. 5.1. (a) The Ebers-Moll model given in Fig. 4.6. (b) Ebers-Moll model
for the special case of forward-biased emitter junction, reverse-biased
collector junction, and saturation current I_{co} neglected.
(c) The small-signal, or incremental, model with the
forward-biased emitter junction replaced by r_e.

4.6 is given in Fig. 5.1a. This model may be simplified to that shown in Fig. 5.1b when the transistor is used as an amplifier, since the emitter junction is always forward biased and the collector junction is always reverse biased. The *linear* incremental, or small-signal, model is given in Fig. 5.1c where the hypothetical, or equivalent, resistance $r_e = \Delta v_{BE}/\Delta i_E$ has replaced the forward-biased emitter diode. Also the lower-case subscripts are used for the currents and voltages because this model is only valid for incremental changes in currents and voltages, not the total values. In addition, the arrangement of the circuit elements has been changed to that of an ideal dependent current source. The sub F has been dropped on α because forward mode is assumed in amplifier operation.

The incremental model for the common-emitter transistor configuration in the amplifier mode may be obtained from a simple modification of the common-base model of Fig. 5.1c. The common-emitter version is given in Fig. 5.2a. Here the base and emitter terminals have been interchanged so the input current i_b is smaller by the factor $\beta + 1$, as discussed in Chapter 4.

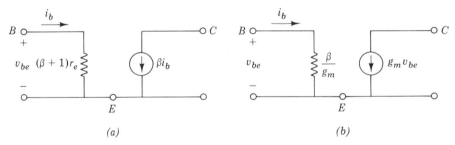

Fig. 5.2. (a) Simple current-controlled current-source model of a transistor in the common-emitter amplifier configuration. (b) Voltage-controlled current-source model of the common-emitter configuration.

The voltage-controlled current-source version of Fig. 5.2a is given in Fig. 5.2b where the collector current is given in terms of the input voltage v_{be} and the mutual conductance or *transconductance* g_m is, by definition, the ratio of the output signal current i_c to the input signal voltage v_{be}. Of course $g_m v_{be}$ must be equal to βi_b. The proof that $\beta / g_m = (\beta + 1) r_e$ is left as an exercise for the student.

The amplifier may be considered linear only for the range of input signal amplitudes over which the segment of diode curve in the vicinity of the q-point is approximately a straight line, as illustrated in Fig. 5.3. We will investigate this linearity mathematically in order to gain a quantitative feel for the magnitude of v_{eb} over which i_b may be consid-

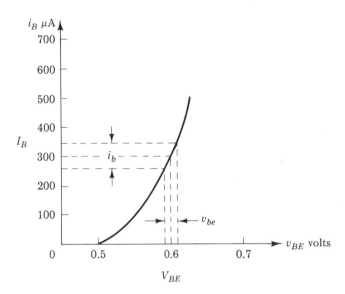

Fig. 5.3. Illustration of the range over which i_b is essentially a linear function of v_{be}.

ered a linear function of v_{eb} and also improve our understanding of the modeling process.

The variation of the transistor voltages and currents about their q-point values may be readily analyzed with the aid of a Taylors series expansion. The infinite Taylors series is

$$f(x) = f(a) + (x-a)f'(a) + \frac{(x-a)^2}{2!}f''(a) + \frac{(x-a)^3}{3!}f'''(a) + \ldots \quad (5.1)$$

Applying this series to the input variables of the transistor,

$$i_B(v_{BE}) = I_B + (v_{BE} - V_{BE})\frac{di_B}{dv_{BE}}\Big|_{I_B} + \frac{(v_{BE}-V_{BE})^2}{2!}\frac{d^2 i_B}{dV_{BE}}\Big|_{I_B} + \ldots \quad (5.2)$$

But $i_B - I_B = i_b$ and $v_{BE} - V_{BE} = v_{be}$ so Eq. 5.2 may be written

$$i_b = v_{be}\frac{di_B}{dv_{BE}}\Big|_{I_B} + \frac{v_{be}^2}{2!}\frac{di_B}{dv_{BE}}\Big|_{I_B} + \frac{v_{be}^3}{3!}\frac{d^3 i_B}{d v_{BE}^3}\Big|_{I_B} + \ldots \quad (5.3)$$

Using the diode equation

$$i_B = (1-\alpha)I_{EO}(e^{qV_{BE}/kT} - 1), \quad (5.4)$$

$$\frac{di_B}{dv_{BE}} = \frac{q}{kT}i_B$$

and

$$\frac{d^2 i_B}{dv_{BE}^2} = \left(\frac{q}{kT}\right)^2 i_B$$

and so on. Therefore Eq. 5.3 may be written

$$i_b = v_{be}\left(\frac{q}{kT}I_B\right) + \frac{v_{be}^2}{2!}\left(\frac{q}{kT}\right)^2 I_B + \frac{v_{be}^3}{3!}\left(\frac{q}{kT}\right)^3 I_B + \ldots \quad (5.5)$$

Observe that the first term on the right of Eq. 5.5 is the linear term and $(q/kT)I_B = (q/kT)I_E(1-\alpha) = g_e(1-\alpha) = 1/r_e(\beta+1)$. Therefore, our model of Fig. 5.2a is accurate providing the sum of the higher order terms on the right-hand side of Eq. 5.5 are small in comparison with the linear term.

Let us determine the magnitude of v_{be} that will cause the second-order term on the right side of Eq. 5.5 to be one-tenth as large as the linear term. Then

$$\frac{v_{be}^2}{2!}\left(\frac{q}{kT}\right)^2 I_B = 0.1\,v_{be}\left(\frac{q}{kT}\right)I_B \quad (5.6)$$

Simplifying,

$$v_{be} = \frac{0.2}{q/kT} \tag{5.7}$$

At normal room temperatures $q/kT \simeq 40$, so $v_{be} = 0.2/40 = 5$ mV maximum in order to limit the second-order term to one-tenth the magnitude of the linear term.

When the input signal is sinusoidal, the second-order term produces second harmonics, the third-order term produces third harmonics, etc., in the amplifier output. The variation of β with collector current also produces distortion that may either add to or subtract from that caused by the diode nonlinearity. The total amplifier distortion is usually considerably less than the diode distortion because of the driving source resistance and the ohmic resistance of the diode. These resistances tend to linearize the relationship between the driving-source voltage and the base current. Then, if β is essentially constant over the signal range, the output current and voltage may have low distortion even though v_{be} is highly distorted. However, the linear model is accurate only for small, or incremental, signals.

**5.2
A LOW-
FREQUENCY
MODEL**

The transistor models of Section 5.1 are low-frequency models because the transistor capacitances that are effective in reducing the high-frequency amplification or gain of the amplifier are not included. A more accurate low-frequency model is developed in this section by including the effects of base-width modulation in the model.

Base-width modulation causes both the base current and the collector current to be functions of collector voltage as well as base voltage with respect to the emitter. Assuming the signals to be small and using the concept of partial derivatives, we may expand the first term on the right side of Eq. 5.3 to include the effects of v_{ce}.

$$i_b = v_{be} \left. \frac{\partial i_B}{\partial v_{BE}} \right|_{I_B} + v_{ce} \left. \frac{\partial i_B}{\partial v_{CE}} \right|_{I_B} \tag{5.8}$$

Similarly, we may write an expression for i_c as

$$i_c = v_{be} \left. \frac{\partial i_C}{\partial v_{BE}} \right|_{I_C} + v_{ce} \left. \frac{\partial i_C}{\partial v_{CE}} \right|_{I_C} \tag{5.9}$$

It is convenient to introduce a set of conductances or g-parameters:

$$g_\pi = \left. \frac{\partial i_B}{\partial v_{BE}} \right|_{I_B} \qquad\qquad g_\mu = -\left. \frac{\partial i_B}{\partial v_{CE}} \right|_{I_B}$$

$$g_m = \left. \frac{\partial i_C}{\partial v_{BE}} \right|_{I_C} \qquad\qquad g_o = \left. \frac{\partial i_C}{\partial v_{CE}} \right|_{I_C} \tag{5.10}$$

Then Eqs. 5.8 and 5.9 may be written

$$i_b = g_\pi v_{be} - g_\mu v_{ce}$$

$$i_c = g_m v_{be} + g_o v_{ce}$$

(5.11)

We may observe from the definitions of these parameters in Eqs. 5.10 that

$$g_\pi = \left(\frac{q}{kT}\right) I_B$$

(5.12)

and

$$g_m = \left(\frac{q}{kT}\right) I_C$$

(5.13)

Therefore, since $I_C = \beta_F I_B$,

$$g_m = \beta g_\pi$$

(5.14)

You may recall that these are the parameters used in the simple current-source model of Fig. 5.2b.

The parameters g_o and g_μ result from base-width modulation, as illustrated in Fig. 5.4. Since the collector current is essentially equal to the minority carrier diffusion current in the base, which is proportional to $dp / dx = P_n(0) / w$, the collector current may be increased by the ratio $(P_n(0) / w') / (P_n(0) / w) = w / w'$ by increasing v_{CE} sufficiently to reduce the base width from w to w'. The same collector current increase would be obtained by increasing v_{BE} so that $P_b(0)$ increases to $P_b(0)'$, while holding the collector voltage constant to keep the base width at w so the slopes $P_b'(0) / w$ and $P_b(0) / w'$ are equal.

The ratio of the effectiveness of v_{CB} to the effectiveness of v_{EB} in controlling collector current is known as the *base-width modulation factor* η. In order to gain some feel for the magnitude for η, let us assume that the base width is changed by ten percent as the collector voltage is changed by five volts in the neighborhood of its q-point value. Then the collector current i_C would increase ten percent with v_{BE} held constant and $di_C / dv_{CE} = 0.1\, I_C$. With v_{CE} constant the change in v_{BE} required for $0.1\, I_C = (di_C / dv_{BE}) I_C \Delta v_{BE} = (q / kT) I_C \Delta v_{BE}$. Then $\Delta v_{BE} = 0.1 / (q / kT)$. At normal room temperature $q / kT \simeq 40$ and $\Delta v_{BE} = 0.1 / 40 = 2.5 \times 10^{-3}$. In this example, $\eta = \Delta v_{BE} / \Delta v_{CE} = 2.5 \times 10^{-3} / 5 = 0.5 \times 10^{-3}$. Note from the relationships of Eq. 5.10 that

$$g_\mu = \eta g_\pi$$

(5.15)

and

$$g_o = \eta g_m$$

(5.16)

since $g_m = \beta g_\pi$, $g_o = \beta g_\mu$.

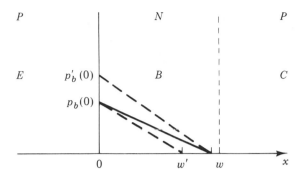

Fig. 5.4. Illustration of the effects of v_{BE} and v_{CB} on the slope p_b/w.

The circuit model of Fig. 5.5a will yield Eqs. 5.11 when nodal analysis is applied to the circuit. However, $g_\mu \ll g_o \ll g_\pi \ll g_m$, as shown in Eqs. 5.14, 5.15, and 5.16 so the model of Fig. 5.5a may be simplified as shown in Fig. 5.5b with negligible loss of accuracy. This model is known as a π model because its form is similar to the Greek letter π.

Example 5.1 | Let us determine the g parameters for a transistor having $\beta = 100$ at the q-point $I_C = 2$ mA at normal room temperature so $q/kT = 40$. Then $g_m = 40\, I_C = .08$ S, $g_\pi = g_m/\beta = .08/100 = 0.8$ mS, and using the value of η determined in the preceding example of this section where $\eta = 0.5 \times 10^{-3}$, $g_o = \eta g_m = 0.5 \times 10^{-3}\,(.08) = 4 \times 10^{-5}$ S and $g_\mu = \eta g_\pi = (g_o/\beta) = 4 \times 10^{-7}$ S.

5.3
THE HYBRID-π
MODEL

Although the π model includes the effects of base-width modulation, it does not include the transistor capacitances so it is strictly a low-frequency model. Also, the ohmic resistances of the active regions have not been included. The ohmic resistances of the emitter and collector regions may usually be neglected. But the ohmic resistance of the base

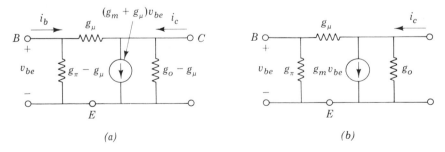

Fig. 5.5. (a) Circuit model based on Equations 5.11.
(b) The approximate π model.

region is not generally negligible because the base region is lightly doped and the cross-sectional area of the base in the transverse direction of base-current flow is very small.

The effective ohmic base resistance r_x and the transistor capacitances C_π and C_μ have been added to the model of Fig. 5.5b to produce the *hybrid-π* model of Fig. 5.6. This model may be used to predict the performance of the transistor with adequate accuracy over the entire useful frequency range of the transistor when connected to a given source and load. In this model the capacitor C_μ is the collector junction capacitance, but C_π includes both the emitter junction capacitance and the diffusion capacitance. Observe that the voltage v is the actual junction voltage that controls the collector current and v_{be} is this junction voltage plus the $i_b r_x$ drop in the base region.

The resistance r_x is usually known as the *base-spreading resistance*. It is not fixed in value but it is a function of both collector current and frequency for a given transistor. The reason for this variation may be understood with the aid of Fig. 5.7. In this figure the cross section of a typical cylindrical-shaped transistor is shown to illustrate the flow direction of the majority carrier base current. Observe that the longest flow path from the base terminal is to the center or axis of the transistor. Thus the greatest $i_B r$ drop occurs along this path and the minimum forward bias across the emitter junction occurs at the center of the transistor. The $i_B r$ drops are small when the q-point currents are small, so the current density is fairly uniform across the emitter junction. However, as the q-point current is increased the $i_B r$ drop increases and the injection current density near the periphery of the base region increases more rapidly than it does at the center. Thus the value of r_x decreases as the q-point current increases because the axial current, which is the current between the emitter and collector, crowds toward the perimeter of the transistor where the base-current flow paths are shorter.

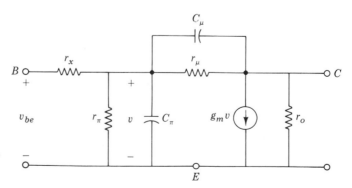

Fig. 5.6. *Hybrid-π model of a bipolar junction transistor (BJT).*

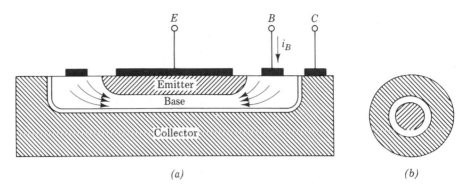

Fig. 5.7. (a) Cross section of a typical transistor. (b) Plan view showing the cylindrical shape.

At a given q-point current, the transistor currents increase as the frequency becomes high enough so the capacitive currents through C_π and C_μ become significant. Then perimeter crowding increases. The value of r_x decreases as the frequency increases in the same manner as it increases when the q-point currents are increased. However, r_x is usually considered to be a fixed resistance having a value that you may readily determine after studying the next section, 5.4.

**5.4
TWO-PORT
NETWORK
THEORY**

The hybrid-π model may be used to find the voltage gain, current gain, frequency response, or bandwidth and so on, of any small-signal electronic circuit and is indeed the most generally used model. However, there are other models that are easier to use for some specific applications. In this section we will introduce the concept of two-port network theory that will broaden our understanding of the modeling process and will yield the additional models that will be most useful in our continuing study of electronics.

Let us suppose that the box in Fig. 5.8a contains a linear network consisting of passive linear elements and perhaps one or more dependent sources such as $g_m v$ or βi_b. However, the network in the box

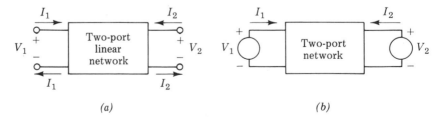

Fig. 5.8. (a) Two-port linear network. (b) Two-port network driven by two ideal voltage sources.

must not contain independent active (power) sources of the same frequency as the externally applied sources. This rule allows dc power sources ($\omega = 0$) such as V_{CC} and other bias sources. The forward terminal network is called a *two-port* network because it has an input port, always on the left, and an output port, always on the right, each consisting of two terminals. Kirchoff's current law requires that the current entering one terminal of a port is equal to the current leaving the other terminal.

An ideal voltage source is connected to each port of the network in Fig. 5.8b. This does not eliminate the use of nonideal sources because the source impedances may be placed within the box. The external sources are assumed to be sinusoidal generators having the same frequency ω, so we may meaningfully use impedance concepts for the elements within the box. Then we may use the superposition theorem to solve for the terminal currents in terms of the applied voltages. This superposition yields

$$I_1 = y_i V_1 + y_r V_2$$
$$I_2 = y_f V_1 + y_o V_2 \tag{5.17}$$

The circuit model that represents Eqs. 5.17 is given in Fig. 5.9. Since all of the terms in Eqs. 5.17 are currents, the ys must all be admittances having the following definitions:

y_i = the input admittance when the output is shorted

y_r = the reverse transfer admittance I_1 / V_2 with input shorted

y_f = the forward transfer admittance I_2 / V_1 with output shorted

y_o = the output admittance when the input is shorted

These definitions may be verified by observation of either Eqs. 5.17 or the model of Fig. 5.9. This model is known as the y-parameter model. It is especially useful for the analysis and design of radio-frequency

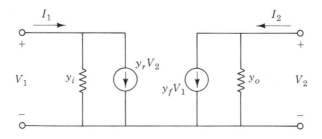

Fig. 5.9. A y-parameter model.

amplifiers where tuned circuits are employed. The transistor manufacturer usually gives the y-parameters as functions of frequency and q-point current for the transistors intended for RF amplifier use.

In writing the Eqs. 5.17 we used the voltages V_1 and V_2 as the independent variables and solved for the currents I_1 and I_2 as the dependent variables. We may just as well have used the currents as the independent variables and the voltages as the dependent variables to obtain, using the superposition concept,

$$V_1 = z_i I_1 + z_r I_2$$

$$V_2 = z_f I_1 + z_o I_2 \qquad (5.18)$$

The circuit model that results from this pair of equations is given in Fig. 5.10. This model is known as a *z-parameter* circuit model because the parameters are all impedances with the following definitions that may be visually verified from either the Eqs. 5.18 or Fig. 5.10.

z_i = the input impedance when the output is open

z_r = the reverse transfer impedance V_1/I_2 with the input open

z_f = the forward transfer impedance V_0/I_1 with the output open

z_o = the output impedance when the input is open

One may hastily imagine that the voltage sources of the z-parameter model could be obtained by merely transforming the current sources of the y-parameter model to voltage sources. This is incorrect because the y-parameters must all be either measured or calculated under short-circuit terminal conditions while the z-parameters must be either measured or calculated under open-circuit terminal conditions.

There are four more ways of choosing the independent and dependent terminal variables but only two ways have significance as tran-

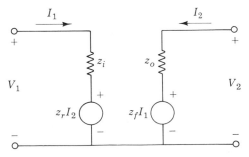

Fig. 5.10. A z-parameter circuit model.

sistor models. These two, in addition to the y- and z-parameters, are presented in matrix form below for comparison purposes:

$$\begin{bmatrix} I_1 \\ I_2 \end{bmatrix} = \begin{bmatrix} y_1 & y_r \\ y_f & y_r \end{bmatrix} \begin{bmatrix} V_1 \\ V_2 \end{bmatrix} \qquad \begin{bmatrix} I_1 \\ V_2 \end{bmatrix} = \begin{bmatrix} g_i & g_r \\ g_f & g_o \end{bmatrix} \begin{bmatrix} V_1 \\ I_2 \end{bmatrix}$$

$$\begin{bmatrix} V_1 \\ V_2 \end{bmatrix} = \begin{bmatrix} z_1 & z_r \\ z_f & z_o \end{bmatrix} \begin{bmatrix} I_1 \\ I_2 \end{bmatrix} \qquad \begin{bmatrix} V_1 \\ I_2 \end{bmatrix} = \begin{bmatrix} h_i & h_r \\ h_f & h_o \end{bmatrix} \begin{bmatrix} I_1 \\ V_2 \end{bmatrix} \qquad (5.19)$$

Of these, the h-parameter model is important and is developed in detail in the next section. Examples of how the parameters are obtained and used are included in that section.

**5.5
THE
H-PARAMETERS**

The two-port network parameters may be used to model any passive or active linear network that meets the requirements given in Section 5.4 including all the transistor configurations, providing the transistor signals are small. A second subscript is used in the transistor models to indicate the common terminal. For example, the h-parameter equations for a common-emitter transistor may be written

$$V_{be} = h_{ie}I_b + h_{re}V_{ce}$$

$$I_c = h_{fe}I_b + h_{oe}V_{ce} \qquad (5.20)$$

The h-parameter circuit model is given in Fig. 5.11. You may observe that this model contains one voltage source and one current source and is therefore a *hybrid* of the z-parameter and y-parameter models, so the h indicates *hybrid*.

The h-parameters are important because they may easily be obtained by the use of a transistor curve tracer and are usually given for a typical q-point by the manufacturer on the transistor data sheet. The reason for their easy obtainability from the input and collector characteristic

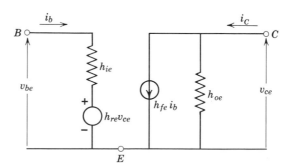

Fig. 5.11. Common-emitter h-parameter model.

curves may be understood from the definitions of the h-parameters obtained by observation of either Eqs. 5.20 or Fig. 5.11 and given below for the common-emitter configuration:

$$h_{ie} = \frac{V_{be}}{I_b}\bigg|_{V_{ce}=0} = \text{input impedance with output shorted}$$

$$h_{re} = \frac{V_{be}}{V_{ce}}\bigg|_{I_b-0} = \text{open circuit reverse voltage gain}$$

$$h_{fe} = \frac{I_c}{I_b}\bigg|_{V_{ce}=0} = \text{short circuit forward current gain}$$

$$h_{oe} = \frac{I_c}{V_{ce}}\bigg|_{I_b=0} = \text{output admittance with input open}$$

The method of obtaining these h-parameters from the characteristic curves is illustrated by the following example.

Example 5.2

Let us determine the common-emitter h-parameters for the transistor with the typical characteristics given in Fig. 5.12. Since the sinusoidal voltages and currents are small variations about the q-point values, $h_{ie} = V_{be}/I_b$ may be interpreted as $\Delta v_{BE}/\Delta i_B$ with v_{CE} held constant at the q-point value. Therefore, h_{ie} is the reciprocal of the slope of the input characteristic at the q-point. Let us choose the q-point at $I_B = 30$ μA, $I_C = 3$ mA, and $V_{CE} = 20$ V for our example. However, we have no input characteristic for $V_{CE} = 20$ V in Fig. 5.12a, but we note that the slope of the curve for $V_{CE} = 1$ V is essentially the same as that of the $V_{CE} = 40$ V curve. Therefore, we may use the slope of either of the given curves, assuming the slope of an intermediate curve would be the same. Then $h_{ie} \simeq \Delta v_{BE}/\Delta i_B$ at $i_B = 30$ μA = $(.62 \text{ V} - .58 \text{ V})/(40 \text{ μA} - 20 \text{ μA}) = .04 \text{ V}/20 \text{ μA} = 2,000 \text{ }\Omega$.

Since $h_{re} = V_{be}/V_{ce}$ with $I_b = 0$ is the same as $\Delta v_{BE}/\Delta v_{CE}$ with i_B held constant at the q-point value, we may also obtain h_{re} from a family of input characteristics. In Fig. 5.12a we only have two curves, so must use them. Then, $h_{re} \simeq (.62 - .60) \text{ V}/(40 \text{ V} - 1 \text{ V}) = .02/39 = 5.1 \times 10^{-4}$. Observe that h_{re} is the same as η defined in Section 5.3.

The short-circuit current gain $h_{fe} = I_c/I_b$ with $V_{ce} = 0$ is the same as $\Delta i_C/\Delta i_B$ with v_{CE} held at the q-point value. In this example, using Fig. 5.12b, we find that $h_{fe} \simeq (4 \text{ mA} - 2 \text{ mA})/(40 \text{ μA} - 20 \text{ μA}) = 100$ at the given q-point. Observe that $h_{fe} = \beta$. These are small-signal values taken about the q-point. A dc β or h_{FE} (cap subscripts) has also been defined as I_C/I_B. In this example, $h_{FE} = 3 \text{ mA}/30 \text{ μA} = 100$, the same as h_{fe}. Usually h_{FE} is not equal to h_{fe}.

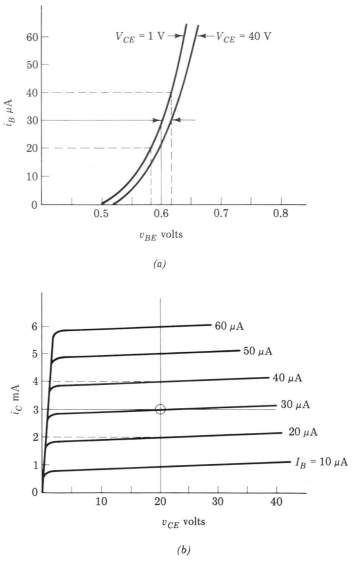

Fig. 5.12. Typical common-emitter (a) input characteristics and
(b) collector characteristics.

The output admittance $h_{oe} = I_c/V_{ce}$ with $V_{be} = 0$ is the same as $\Delta i_C/\Delta v_{CE}$ with i_B held constant at the q-point value. Therefore h_{oe} is the slope of the collector characteristic that passes through the q-point. In this example we see from Fig. 5.12b that the slope of the $I_B = 30$ μA collector characteristic is $h_{oe} \simeq (3.2 - 2.8)$ mA/$(40 - 2)$ V = 0.4 mA/38 V = 1.05×10^{-5} mho.

The h-parameters may be obtained with greater accuracy by applying small ac signals to the transistor and measuring the required current and voltage responses indicated by the definitions of the parameters. The h-parameters are complex and vary with frequency when the frequency is so high that junction and diffusion capacitances cannot be neglected. At these frequencies direct measurement must be used to obtain the h-parameters as well as the y- and z-parameters. However, the h-parameters are usually obtained or given for low frequencies only and the hybrid-π parameters are obtained from these h-parameters to permit the use of the superior hybrid-π circuit when the high-frequency amplifier characteristics are needed.

5.6 DETERMINATION OF HYBRID-π PARAMETERS   The parameters for any circuit model may be obtained in terms of any other model. As an example of how this may be done, let us obtain the common-emitter h-parameters in terms of the hybrid-π parameters by applying the definitions of the h-parameters to the low-frequency hybrid-π circuit of Fig. 5.13.

To find h_{ie} we visualize a short on the output terminals and see from Fig. 5.13 that the input impedance with the output shorted is r_x plus the resistance of r_π and r_μ in parallel. But $r_\mu \gg r_\pi$ so we may neglect r_μ and write

$$h_{ie} = r_x + r_\pi \tag{5.21}$$

A review of Fig. 5.5 will remind us that the approximation we made in Eq. 5.21 cancels the approximation we made in simplifying the model of 5.5a to that of 5.5b. Therefore Eq. 5.21 is precisely correct.

We may find h_{re} by observing that with the input open the ratio $V_{be}/V_{ce} = r_\pi/(r_\pi + r_\mu)$ by the voltage divider formed by r_π and r_μ. But again considering the approximations made in r_μ from Fig. 5.5,

$$h_{re} = \frac{r_\pi}{r_\mu} = \eta \tag{5.22}$$

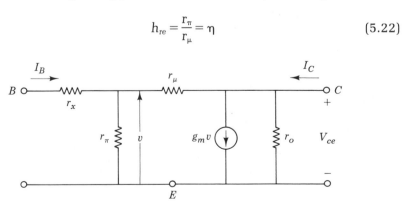

Fig. 5.13. Low-frequency hybrid-π model.

The current sources of the hybrid-π and h-parameter models must be the same. Therefore,

$$h_{fe}I_b = g_m V \tag{5.23}$$

but $V = I_b r_\pi$ so

$$h_{fe} = g_m r_\pi = \beta \tag{5.24}$$

Finally, with the input open we apply the voltage V_{ce} to the output terminals and find that

$$I_c = g_o V_{ce} + g_m V \tag{5.25}$$

But $v = h_{re} V_{ce} = \eta V_{ce}$, so

$$I_c = g_o V_{ce} + \eta g_m V_{ce} \tag{5.26}$$

However, from Eq. 5.6, $\eta\, g_m = g_o$. Therefore,

$$h_{oe} = \frac{I_c}{V_{ce}} = 2\,g_o \tag{5.27}$$

We may now obtain the hybrid-π parameters in terms of the h-parameters by rearranging the equations above. From Eq. 5.31

$$r_x = h_{ie} - r_\pi \tag{5.28}$$

From Eq. 5.24

$$r_\pi = \frac{h_{fe}}{g_m} = \frac{\beta}{g_m} \tag{5.29}$$

From Eq. 5.22

$$r_\mu = \frac{r_\pi}{h_{re}} \tag{5.30}$$

From Eq. 5.27

$$g_o = \frac{h_{oe}}{2} \qquad \text{or} \qquad r_o = \frac{2}{h_{oe}} \tag{5.31}$$

Therefore, with our knowledge that $g_m = (q\,/\,kT)\,I_C \approx 40\,I_C$ at normal room temperatures, we may obtain all of the low-frequency hybrid-π parameters with either the use of a transistor curve tracer or from the h-parameters normally provided by the transistor manufacturer.

5.7
THE
GRADED-BASE
TRANSISTOR

Up to this point we have assumed that the flow of carriers across the base region from emitter to collector is entirely by diffusion. However, in our discussion of diffusion capacitance we observed that the minority carriers in the base that have been injected from the emitter must be essentially neutralized by majority carriers, as illustrated in Fig. 5.14 for a p-n-p transistor. As previously mentioned, the additional majority carriers (electrons) that rush into the base to neutralize the injected charge do not appreciably increase the recombination rate in the base so long as their density is small in comparison with the doping density or concentration in the base. The electrons, like the holes, diffuse toward the collector, but the collector junction barrier potential prevents their entering the collector. Thus, a small negative charge exists at the collector junction and a net positive charge exists near the emitter junction as a result of the electron diffusion, as shown in Fig. 5.14. This charge distribution creates an electric field which is just sufficient to offset the diffusion of electrons while aiding the diffusion of holes, and thus it improves the performance of the transistor. The electric field causes a drift component of current in addition to the diffusion current and reduces the stored charge required for a given collector current, thus reducing the diffusion capacitance. In addition, the current gains α and β are increased as a result of the electric field because the minority carriers are accelerated through the base region and therefore have less

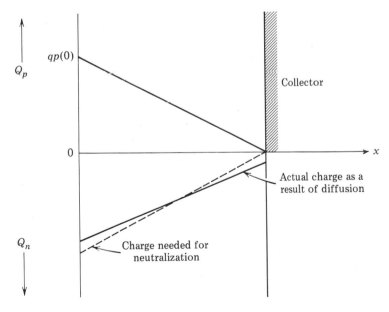

Fig. 5.14. Stored charge in the base of p-n-p transistor.

time to recombine with the majority carriers. However, the electric field intensity is proportional to the collector current over the range of low-level injection, so β increases strongly with i_c in the normal operating range and the transistor nonlinearity is increased as a result of the electric field.

The development of the *graded-base* transistor caused a major advance in the high-frequency capabilities of the transistor. The graded base provides a built-in electric field in the base region and also provides a small collector junction capacitance C_{jc}. The electric field accelerates the carriers through the base and thus reduces the stored charge, or diffusion capacitance, for a given collector current.

The term *graded base* means that the doping density in the base region is not uniform but decreases as one traverses the region from the emitter junction to the collector junction. This doping distribution, or profile, which is shown in Fig. 5.15a, is obtained by diffusing the doping atoms into a heated semiconductor crystal. For example, let us assume that a graded base n-p-n transistor is to be constructed. Then a bar of lightly doped n-type crystal is heated in an oven and the trivalent, or p, dopant is applied, probably in gaseous form, to one surface of the bar. The acceptor atoms diffuse into the semiconductor in a manner very similar to the diffusion of carriers from the emitter into the base when forward bias is applied to the emitter junction. However, the temperature must be high (near the melting point of the crystal) to permit the doping atoms to diffuse, and then the diffusion rate of these atoms is very slow in comparison with the diffusion of charge carriers. Since the effective doping density is the difference between the donor impurity concentration N_d and the acceptor impurity concentration N_a, the end of the bar exposed to the trivalent impurity soon becomes

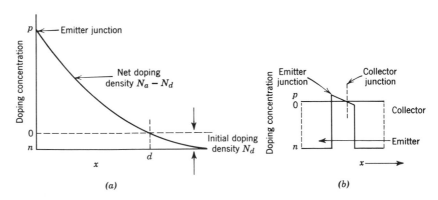

Fig. 5.15. *Doping characteristics of a graded-base transistor: (a) doping profile in base region; (b) doping profile in transistor.*

p-doped and the doping density decreases exponentially as the distance increases into the bar, as shown in Fig. 5.15a. At some distance d, the acceptor atom density is equal to the initial donor atom density and the crystal appears intrinsic. This point, which moves more deeply into the material as the diffusion process continues, becomes the collector-base junction when the diffusion is stopped by cooling the material, because the net doping changes from p-type to n-type at that point. The transistor is completed by replacing the trivalent doping material on the surface of the semiconductor with a heavily n-doped semiconductor which serves as the emitter. A process known as *alloying* is usually used to attach the emitter. Alloying is similar to welding except the continuous crystal structure is carefully maintained and the junction between the n-type and p-type materials is a plane surface. Also, the lightly n-doped collector material is usually replaced by heavily n-doped material, as shown in Fig. 5.15b, from a short distance to the right of the collector junction. This heavily doped material improves the conductivity of the collector region and decreases the saturation voltage for a given collector current. Either alloying or diffusion techniques can be used to increase the doping concentration in the collector. Of course, a p-n-p transistor can be produced by using the opposite type dopants. Finally, the leads are attached and the transistor is heated in an atmosphere of either oxygen or nitrogen to form a thin layer of silicon dioxide or silicon nitride (for a silicon semiconductor) on the surface of the semiconductor. This layer is a good insulator which protects the surface from contamination and reduces surface leakage. This process is known as *passivation*.

As previously mentioned, the graded base produces a built-in electric field in the base because the majority carriers in the base diffuse toward the collector but are not able to enter the collector because of the potential barrier. Therefore, they accumulate and an electric field builds up until the drift current, due to the electric field, is equal to the diffusion current. This electric field aids the flow of minority carriers in the base which are injected from the emitter into the base. The field therefore has the same effect as increasing the diffusion constant which reduces the diffusion capacitance. The built-in electric field also increases β, but since this field is not dependent upon the magnitude of collector current, β is essentially independent of i_C over the normal operating range because the built-in field is strong compared to the field due to injected carriers. Therefore, the graded-base transistor has better linearity.

The graded collector junction in the graded-base transistor has very light doping on both sides of the junction in the vicinity of the junction. Therefore, the depletion region is wide on both sides of the junction, for a given collector voltage, as compared with the uniformly-doped tran-

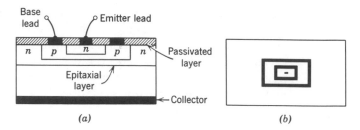

*Fig. 5.16. Construction details of a typical transistor:
(a) cross-sectional view; (b) top view.*

sistor. Therefore, the collector voltage breakdown is high and the junction capacitance comparatively low, which is an important contribution to the high-frequency performance. The lightly doped layer on the collector side of the collector junction is known as an *epitaxial* layer.

The only disadvantage of the graded-base transistor is the low emitter-breakdown voltage which results from the comparatively heavy base doping at the emitter junction and the very heavy emitter doping required to give high emitter efficiency. This low breakdown voltage is a handicap only when the transistor is used for switching and the emitter junction needs to be reverse biased. However, in these applications a diode can be connected in series with the input lead, and the diode will withstand the inverse voltage and protect the emitter-base junction in the transistor.

The physical layout of a typical graded-base, epitaxial, passivated transistor is shown in Fig. 5.16. The geometry of the transistor illustrates that the transistor is designed for high β and good collector power dissipation when the transistor is operated in the normal mode, in contrast with the *inverse mode* in which the collector is used as the emitter and vice versa.

**5.8
TRANSISTOR
HIGH-
FREQUENCY
CHARACTER-
ISTICS**

The values of the capacitances C_π and C_μ were not found in Section 5.7 because these capacitances do not appear explicitly in the *h*-parameter model or in any other two dependent-source models. The main advantage of the hybrid-π model is the close correlation between the model elements and the physical regions of the transistor, thus permitting the inclusion of the junction and diffusion capacitances explicitly in the model. This section develops a method for determining the values of these capacitances in addition to formulating the high-frequency characteristics of the bipolar junction transistor.

The complete hybrid-π model is given in Fig. 5.17. It is driven by an ideal sinusoidal current source I_b and has a shorted output. As the frequency of I_b is increased, the emitter junction voltage V is decreased because of the shunting effect of C_π and C_μ. Therefore, the collector

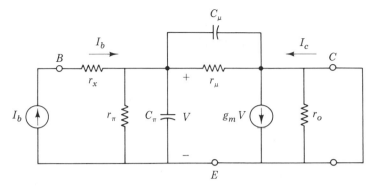

Fig. 5.17. Hybrid-π model, including shunt capacitances, with current-source input and shorted output.

current $I_c = g_m V$ is decreased. With the output shorted, r_μ is in parallel with r_π and C_μ is in parallel with C_π so, since $V = I/y$,

$$V = \frac{I_b}{g_\pi + j\omega(C_\pi + C_\mu)} \tag{5.32}$$

where g_μ has been neglected because $g_\mu \ll g_\pi$. Beta cutoff frequency f_β is defined as the frequency at which the short-circuit current amplification factor drops to $.707\,\beta_o$, where β_o is the low-frequency value of β. This frequency occurs when the j term in the denominator of Eq. 5.32 is equal to the real term. Then

$$\omega_\beta (C_\pi + C_\mu) = g_\pi = \frac{g_m}{\beta_o} \tag{5.33}$$

and

$$\omega_\beta = \frac{g_m}{\beta_o (C_\pi + C_\mu)} \tag{5.34}$$

A sketch of the magnitude of β as a function of frequency is given in Fig. 5.18. The frequency at which $|\beta| = 1$ is defined as the transition frequency f_T. Since the reactance of $(C_\pi + C_\mu)$ is inversely proportional to frequency, $|\beta|$ is also inversely proportional to frequency for frequencies well above f_β where this reactance is small in comparison with r_π. Therefore, the high-frequency part of the curve is a straight line if logarithmic scales are used. If this straight-line portion of the curve is extended upward until it intersects the β_o line, intersection occurs at f_β. Therefore, while β decreases by the factor β_o, the frequency must increase by the factor β_o and

$$f_T = \beta_o f_\beta \tag{5.35}$$

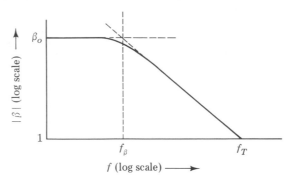

Fig. 5.18. Sketch of β as a function of frequency.

And from Eq. 5.34 we can see that

$$\omega_T = \beta_o \omega_\beta = \frac{g_m}{C_\pi + C_\mu} \tag{5.36}$$

The transition frequency f_T is often known as the current-gain bandwidth product because, as may be seen in Fig. 5.18,

$$f_T = \beta f|_{f_\beta < f < f_T} \tag{5.37}$$

From Eq. 5.36 we may write

$$C_\pi + C_\mu = \frac{g_m}{\omega_T} \tag{5.38}$$

Transistor manufacturers almost always give f_T as a function of I_C and C_{ob} as a function of V_{CE}. The capacitance C_{ob} is defined as the output capacitance in the common-base configuration. This C_{ob} is the collector junction capacitance C_μ plus the transistor header and lead capacitance. Therefore, C_{ob} is the *practical* substitute for C_μ, and C_π may be found from a rearrangement of Eq. 5.38.

$$C_\pi = \frac{g_m}{\omega_T} - C_{ob} \tag{5.39}$$

Sometimes it is helpful to separate the diffusion capacitance from the emitter-junction capacitance, the sum of the two being equal to C_π. This may be accomplished by obtaining f_T for two different values of I_C and calculating C_π for each of these values. With the knowledge that the junction capacitance C_{je} is essentially independent of I_C while the diffusion capacitance C_d is essentially proportional to I_C we may write two simultaneous equations involving $C_{je} + C_d = C_\pi$ and solve for both C_{je} and C_d for any desired value of I_C. As you may expect, C_π approaches C_{je} as I_C approaches zero, and C_π approaches C_d as I_C becomes large.

Occasionally the *alpha-cutoff* frequency f_α is given instead of f_T. Alpha cutoff occurs when $\alpha = 0.707 \; \alpha_o$ due to the shunting capacitances. At that frequency $\beta \simeq .7/(1/\alpha_o - .7) = 2.3$. Therefore $f_T \simeq 2.3 \; f_\alpha$.

Problems   *Section 5.1*

5.1   Prove that $(\beta + 1) \, r_e = \beta / g_m$.

5.2   Show that the distortion of a common-emitter amplifier is determined primarily by the linearity of β as a function of i_C in the special case where the driving source resistance is large in comparison with the input resistance of the transistor.

Section 5.2

5.3   A given transistor has $\beta = 150$ and $r_o = 2 \times 10^{-5}$ S at $I_C = 5 \times 10^{-3}$ A. Determine the values for g_m, g_π, r_π, g_μ, and r_μ at this value of I_C.

5.4   Determine the errors in the element values in Fig. 5.5b compared with the more accurate element values of Fig. 5.5a for the transistor of Problem 5.3.

Section 5.4

5.5   Show that the definitions of the y-parameters are correct by applying the specified terminal conditions to the Equations 5.17.

5.6   A given transistor has $y_{ie} = 6.7 \times 10^{-4}$ S, $y_{re} \simeq 0$, $y_{fe} = 6.4 \times 10^{-2}$ S, and $y_o = 10^{-5}$ S at the q-point $I_C = 2$ mA, $v_{CE} = 15$ V. Use the y-parameter model of Fig. 5.9 to determine the voltage gain V_2/V_1, the current gain I_2/I_1, the input resistance V_1/I_1, and the output resistance V_2/I_2 of this transistor when used as an amplifier with $R_L = 5$ kΩ and driving source resistance $R_S = 1$ kΩ. Assume that all the y-parameters are purely resistive.

Section 5.5

5.7   Determine the common-emitter h-parameters with reasonable accuracy for the transistor having the characteristics given in Fig. 4.11 at the q-point $I_B = 200$ μA, $V_{CE} = 15$ V.

5.8   A given transistor has $h_{ie} = 1500$ Ω, $h_{fe} = 100$, $h_{oe} = 3.2 \times 10^{-5}$ mho, and $h_{re} = 2 \times 10^{-4}$ at $I_C = 2$ mA, $V_{CE} = 10$ V. Use the h-parameter model of Fig. 5.11 to determine the voltage gain v_o/v_i, the current gain i_o/i_i, the input resistance v_i/i_i, and the output resistance v_o/i_o of this transistor when used as an amplifier with driving source resistance $R_S = 1\ k\Omega$ and $R_L = 5\ k\Omega$.

Section 5.6

5.9   Determine the hybrid-π parameters and draw the hybrid-π model for the transistor of Problem 5.8 at the specified q-point.

5.10 Calculate the voltage gain v_o/v_i, the current gain i_o/i_i, the input resistance v_i/i_i, and the output resistance v_o/i_o of the amplifier specified in Problem 5.8 using the hybrid-π model and the parameters determined in Problem 5.9. Compare these results with those obtained in Problem 5.8.

Section 5.7

5.11 A given silicon n-p-n transistor having a uniformly doped base has $N_a = 10^{18}$ holes/m³ in the base and $w = 10^{-5}$ m. Assuming the transistor to be of prismatic shape and the injected electron density $n_p(0)$ at the edge of the base region is 5×10^{16} carriers/m³, determine the electric field intensity in the base region and the total axial current density through the base.

5.12 A given graded-base n-p-n silicon transistor has a net doping density $p_p(0) = 10^{20}$ at the emitter edge of the base region and effective base width $w = 10^{-5}$ m. Assuming the net base doping density to decrease linearly with axial distance x, determine the electric field intensity in the base when $I_C = 0$.

Section 5.8

5.13 A given transistor has $f_T = 300$ MHz at $I_C = 5$ mA and $C_{ob} = 3$ pF at $V_{CE} = 10$ V. Determine the value of C_π and f_β if $\beta_o = 120$.

5.14 The transistor of Problem 5.13 has $f_T = 200$ MHz at $I_C = 2.0$ mA and $V_{CE} = 10$ V. Determine the emitter junction capacitance and the diffusion capacitance at both $I_C = 2$ mA and $I_C = 5$ mA for this transistor.

6

Transistor Amplifier Configurations

This chapter is devoted to the analysis of the three basic transistor configurations when used as amplifiers. The hybrid-π model is used to determine the voltage gain, current gain, power gain, and frequency response characteristics of each configuration. These parameters are then compared for the different configurations.

The purpose of an amplifier is usually to accept the weak signal from a low-level source such as a microphone, phonograph pickup, receiving antennas, or instrumentation transducer and increase the power level of the signal until it is capable of driving useful loads such as loud-speakers, TV picture tubes, or strip-chart recorders.

Several *stages* of amplification are usually required for high-gain amplification. In this chapter a *stage* consists of a single transistor and its associated components. In later chapters a pair of transistors may be used in a given stage. The *output stage* furnishes the required power to the actual load and receives its input, or *driving*, power from a pre-ceding transistor stage. The *input* stage receives its power from a low-level *transducer* such as a microphone and its load is the following amplifier stage. An *intermediate* stage has another amplifier stage as both its driving source and load. In any event, the load is usually represented by a resistor and the driving source is represented by either a voltage source or a current source.

Amplifiers may be classified as either *ac* amplifiers or *dc* amplifiers. The ac amplifiers use coupling, or dc blocking, capacitors to separate the signal voltage and current components from the bias components. On the other hand, *dc* or *direct-coupled* amplifiers must be used when

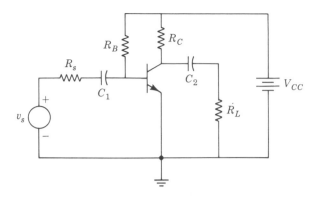

Fig. 6.1. Circuit diagram of a CE amplifier.

the signals vary so slowly that capacitors cannot be used in the circuit. Only ac amplifiers are considered in this chapter; dc amplifiers are treated in Chapter 10.

6.1 COMMON-EMITTER AMPLIFIER ANALYSIS

The common-emitter (CE) amplifier is the most commonly used configuration because of its superior gain. The circuit diagram of a CE amplifier is given in Fig. 6.1. Fixed bias is used for simplicity. An improved bias circuit is developed in Chapter 7.

With the load R_L capacitively coupled to the transistor, the load on the transistor is R_L in parallel with R_C for all frequencies at which the reactance of C_2 is negligible in comparison with R_L. This parallel resistance is known as the ac load resistance R_{ac} where

$$R_{ac} = \frac{R_C R_L}{R_C + R_L} \tag{6.1}$$

However, at $\omega = 0$, no current flows through C_2, so the dc load resistance is simply R_C. The frequency range over which X_{c2} is finite but not negligible is known as the *low-frequency range* and will be treated later.

An ac load line may be drawn on the collector characteristics as shown in Fig. 6.2. The dc load line is drawn as before from the x-axis intercept at V_{CC} to the y-axis intercept at V_{CC}/R_L. The desired q-point is then selected and the ac load line $\Delta v_{CE}/\Delta i_C = R_{ac}$ is drawn through the q-point. In Fig. 6.2 $V_{CC} = 20$ V, $R_{dc} = 2$ kΩ, so 20 V / 2 kΩ = 10 mA, which is the y-axis intercept. The q-point has been selected at $v_{CE} = 10$ V and $i_C = 5$ mA. The capacitively coupled load R_L is also 2 kΩ; therefore, the ac load on the transistor is $R_{ac} = 1$ kΩ and the ac load line having a slope of $-1/R_L = -10^{-3}$ is drawn through the q-point. The ac load line is easily constructed by finding its x-axis intercept $= V_{CE} + I_C R_{ac}$, where V_{CE} and I_C are the q-point values, and then drawing a straight line from

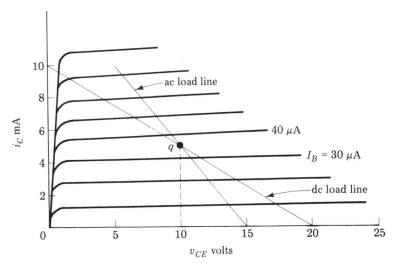

Fig. 6.2. Collector characteristics with both dc and ac load lines.

this intercept through the q-point. In Fig. 6.2 the x-axis intercept is 10 V + 5 mA (1 kΩ) = 15 V as shown.

A mid-frequency hybrid-π model of the CE amplifier shown in Fig 6.1 is given in Fig. 6.3. Mid-frequency is defined as the frequency range over which all reactances may be neglected. The reactances of the coupling capacitors are so low that the capacitors appear as short circuits. The reactances of the shunt capacitors in the hybrid-π models are so large that these capacitors appear as open circuits. A complete description of the amplifier requires that we determine the voltage amplification $K_v = V_o/V_i$, the current amplification $K_i = I_L/I_i$, the input impedance $Z_i = V_i/I_i$, and the output impedance $Z_o = V_o/I_o$ of the amplifier. The solution of the circuit of Fig. 6.3 requires an analysis technique such as nodal equations. However, r_μ is very large in comparison with

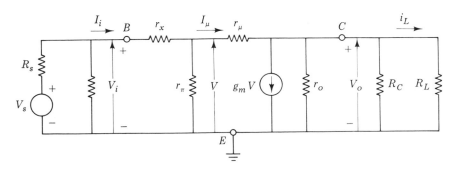

Fig. 6.3. Mid-frequency hybrid-π model of the amplifier given in Fig. 6.1.

the other resistances in the circuit model so we will simplify the circuit by assuming $r_\mu = \infty$ and will determine the validity of our assumption later. Also, the bias resistance R_B is large for fixed bias and may be neglected, but in general we may include R_B as part of the driving source resistance R_s' by applying Thevenin's theorem to the part of the circuit to the left of the transistor input terminals. With these simplifications, the circuit of Fig. 6.3 reduces to the circuit of Fig. 6.4, which does not require the solution of simultaneous equations.

Using the circuit model of Fig. 6.4,

$$R_i = r_x + r_\pi \tag{6.2}$$

$$I_i = \frac{V_i}{R_i} = \frac{V_i}{r_x + r_\pi} \tag{6.3}$$

$$V = \frac{V_i r_\pi}{r_x + r_\pi} \tag{6.4}$$

$$V_o = -\frac{g_m V}{g_o + G_C + G_L} = -\frac{g_m V_i r_\pi}{(r_x + r_\pi)(g_o + G_C + G_L)} \tag{6.5}$$

$$K_v = \frac{V_o}{V_i} = -\frac{g_m r_\pi}{(r_x + r_\pi)(g_o + G_C + G_L)} = -\frac{\beta_o}{h_{ie}(g_o + G_{ac})} \tag{6.6}$$

$$I_L = V_o G_L = -\frac{g_m V_i r_\pi G_L}{(r_x + r_\pi)(g_o + G_C + G_L)} \tag{6.7}$$

$$K_i = \frac{I_L}{I_i} = -\frac{g_m r_\pi G_L}{g_o + G_C + G_L} = -\frac{\beta_o G_L}{g_o + G_C + G_L} \tag{6.8}$$

The power gain or amplification is

$$K_p = \frac{V_o I_L}{V_i I_i} = K_v K_i \tag{6.9}$$

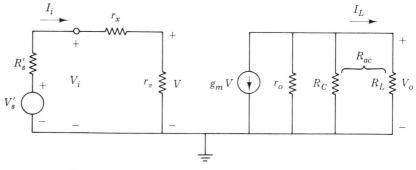

Fig. 6.4. Simplified hybrid-π model for the CE amplifier.

The voltage gain V_o/V_i as given above is most appropriate because the V_o of one amplifier is the V_i of the following amplifier when cascaded amplifier stages are used. However, the input amplifier is usually driven by a transducer with a source voltage V_s that is different than V_i as seen in Fig. 6.4. The voltage gain V_o/V_s may be more useful than the gain V_o/V_i. From Fig. 6.4 we may observe that

$$K_v = \frac{I_L R_L}{I_i (R_s + r_x + r_\pi)} \tag{6.10}$$

However, at mid frequencies $I_L = \beta_o I_i R_{ac}/R_L$ and $\beta_o = g_m r_\pi$. Therefore,

$$K_v = \frac{g_m r_\pi}{(R_s + r_x + r_\pi)(g_o + G_C + G_L)} \tag{6.11}$$

Often $g_o \ll (G_C + G_L)$. In this case, g_o may be neglected.

The output admittance y_o may be determined by applying a voltage V_o to the output terminals with the driving source voltage V_s turned off, and then calculating the current flow I_o into the output terminals, as shown in Fig. 6.5. In this model, $g_\mu = 1/r_\mu$ is included to show its effect on R_o. Also R_C has been included so the amplifier output resistance will also be obtained. $G_s'' = 1/(r_x + R_s')$ has been used to simplify the mathematics. Using nodal equations

$$I_o = (G_C + g_o + g_\mu) V_o + (g_m - g_\mu) V \tag{6.12}$$

$$0 = -g_\mu V_o + (G_s'' + g_\pi + g_\mu) V \tag{6.13}$$

Using the inequalities $g_\mu \ll g_o$, $g_\mu \ll g_\pi$ and $g_\mu \ll g_m$, we will obtain V in terms of V_o in Eq. 6.13 and solve for I_o/V_o in Eq. 6.12 to obtain

$$y_o = \frac{I_o}{V_o} = G_C + g_o + \frac{g_m g_\mu}{G_s'' + g_\pi} \tag{6.14}$$

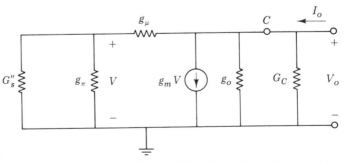

Fig. 6.5. Circuit model for determining R_o.

We may see from Eq. 6.14 that the output resistance of the amplifier is the parallel combination of R_C, r_o, and $(G_s'' + g_\pi)/g_m g_\mu$. We can also see from Eq. 6.14 that the output admittance of the transistor itself varies from g_o when the driving source resistance is zero, and G_s'' approaches infinity to $2g_o = h_{oe}$ when the input is open and $G_s'' = 0$, since $g_m/g_\pi = \beta_o$ and $\beta_o g_\mu = g_o$.

Let us now consider a CE amplifier example and determine the conditions under which we may use the simplified circuit of Fig. 6.4.

Example 6.1	The transistor having the characteristics given in Fig. 6.2 is used in the circuit of Fig. 6.1 with $R_C = 2$ kΩ and $R_L = 2$ kΩ. The q-point is at $I_C = 5$ mA, $I_B = 36$ μA, and $V_{CE} = 10$ V with $V_{CC} = 20$ V. $R_B = (20 - .6)\,\text{V}/36\,\mu\text{A} = 544$ kΩ and is very large compared with h_{ie}. Then, assuming normal room temperature, $g_m = 40\,I_C = 200$ mS, $\beta_o \simeq h_{FE} \simeq 5$ mA$/36$ μA $= 139$ and $h_{oe} = \Delta i_C/\Delta v_{CE} \simeq 0.2$ mA$/20$ V $= 10^{-5}$ S. Then $g_o = 10^{-5}/2 = 5 \times 10^{-6}$ S, $r_\pi = \beta_o/g_m = 139/0.2 = 695$ Ω, $\eta = h_{re} = g_o/g_m = 5 \times 10^{-6}/.2 = 2.5 \times 10^{-5}$ and $g_\mu = g_o/\beta_o = g_\pi h_{re} = 5 \times 10^{-6}/139 = 3.6 \times 10^{-8}$ S. In order to determine r_x, h_{ie} must either be given or determined from the curve tracer or other measurement. Let us assume that $h_{ie} = 850$ Ω. Then $r_x = 850 - 695 = 155$ Ω. Let us also assume that the driving source resistance $R_s = 1$ kΩ. We may now calculate the mid-frequency voltage gain, current gain, and power gain. From Eqs. 6.6, 6.8, and 6.9

$$K_V = -139/(850)(5 \times 10^{-6} + 10^{-3}) = -164$$

$$K_i = -(139)(5 \times 10^{-4})/(5 \times 10^{-6} + 10^{-3}) = -69$$

$$K_p = 164 \times 69 = 11,316$$

The negative signs indicate the output voltage or current is inverted with respect to the input signal. You may note that in calculating the circuit gains the effect of r_o, or g_o, was negligible because $g_o \ll (G_L + G_C)$. Let us now consider the effect on the circuit gains caused by neglecting r_μ. Observe from Fig. 6.3 that the emitter junction voltage $V = (I_i - I_\mu)r_\pi$. Therefore, the effect of r_μ on the circuit gains is negligible if $I_\mu \leq I_\pi/10$. Since $I_\pi = g_\pi V = V/695 = 1.44 \times 10^{-3}$ V, $I_\mu = (V - V_o)g_\mu$, but $V_o = -g_m V R_{ac}$ from Fig. 6.3, so $I_\mu = (1 + g_m R_{ac})g_\mu V = (1 + 200)\,3.6 \times 10^{-8}$ V $= 7.24 \times 10^{-6}$ V. Then $I_\mu/I_\pi = 7.24 \times 10^{-6}$ V$/1.44 \times 10^{-3}$ V $= 5 \times 10^{-3} = 0.5$ percent, which is definitely negligible.

Note carefully that although $g_\mu = h_{re}g_\pi$ where $h_{re} = 2.5 \times 10^{-5}$ in the preceding example, the ratio of current flow through these conductances was $(1 + g_m R_{ac})g_\mu/g_\pi = 5 \times 10^{-3}$ because in addition to the emitter-junction voltage the output voltage is forcing current through

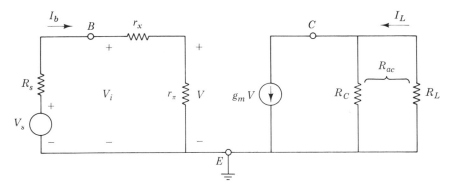

Fig. 6.6. Simplified hybrid-π model that is adequately accurate when $R_{ac} \ll r_o$.

g_μ, while only the emitter-junction voltage forces current through g_μ. Therefore, the criterion for neglecting r_μ is

$$(1 + g_m R_{ac}) g_\mu / g_\pi < .1 \tag{6.15}$$

This criterion is always met if $R_{ac} < .1\ r_o$ which is almost always the case. Therefore the model of Fig. 6.4 is generally used with r_o deleted, as shown in Fig. 6.6. Using this model the gain equations are

$$K_v = \frac{V_o}{V_i} = -\frac{g_m r_\pi R_{ac}}{r_x + r_\pi} = -\frac{\beta_o R_{ac}}{r_x + r_\pi} \tag{6.16}$$

$$K_i = \frac{I_L}{I_b} = -\frac{g_m r_\pi R_C}{R_C + R_L} = -\frac{\beta_o R_C}{R_C + R_L} \tag{6.17}$$

The upper cutoff, or half-power, frequency of the amplifier may be determined if we add the transistor capacitances to the circuit model of Fig. 6.6 as shown in Fig. 6.7. This circuit may be simplified if we observe that the total *capacitive* current flowing out of node V is

$$I_C = j\omega C_\pi V + j\omega C_\mu (V - V_o) \tag{6.18}$$

However, $V_o = -g_m V R_{ac}$ so

$$I_C = j\omega(C_\pi + (1 + g_m R_{ac}) C_\mu) V \tag{6.19}$$

Observe that the effect of the capacitance C_μ between the input and output nodes is multiplied by the factor $(1 + g_m R_{ac})$ in the same manner that g_μ was multiplied in Example 6.1. This effective multiplication of capacitance between the input and output nodes of an amplifier is proportional to the voltage gain between these nodes. This effect is known as the *Miller effect* because Miller first discovered it as the cause of *detuning* in early radio sets whenever the voltage gain of an RF amplifier changed.

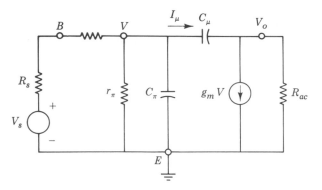

Fig. 6.7. Transistor capacitances added to Fig. 6.6.

The effect of the C_μ on the input node voltage V is not changed if we connect an effective capacitance $(1 + g_m R_{ac}) C_\mu$ between node V and the common node as shown in Fig. 6.8. Increased accuracy would be achieved if we were to also connect a capacitance C_μ between the output node V_o and the common node, but the improvement would be negligible because of the generally small value of C_μ.

Again, the upper cutoff, or half-power, frequency f_h occurs when the reactance of the effective capacitance between the input and common nodes is equal to the resistance between these nodes. Therefore,

$$\omega_h = \frac{G_s'' + g_\pi}{C_\pi + (1 + g_m R_{ac}) C_\mu} \tag{6.20}$$

where $G_s'' = 1/(R_s' + r_x)$ as before.

Let us assume that the transistor manufacturer lists $f_T = 300$ MHz and $C_{ob} = 3$ pF for the transistor used in Example 3.1 at the specified q-point. We will determine the high-frequency cutoff of the amplifier used in that example. From Eq. 6.5 we find $C_\pi = g_m/2\pi f_T - 3$ pF =

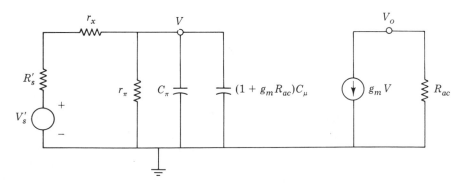

Fig. 6.8. Simplified high-frequency model.

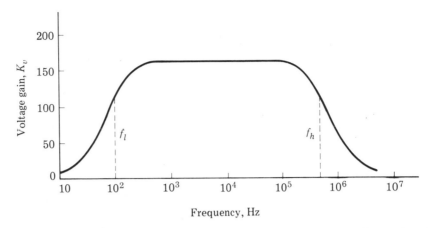

Fig. 6.9. Sketch of the frequency response for the amplifier of Example 6.1.

$.2 / 6.28 \times 3 \times 10^{8} - 3 \times 10^{-12} \simeq 10^{-10}$ F. Then, since $G_{s}'' = 1 / 1,155 = 8.66 \times 10^{-4}$ S and $g_{\pi} = 1 / 695 = 1.44 \times 10^{-3}$,

$$\omega_h = \frac{(8.66 + 14.4) \, 10^{-4}}{10^{-10} + (1 + 200) \, 3 \times 10^{-12}} = \frac{2.304 \times 10^{-3}}{7 \times 10^{-10}} - 3.2 \times 10^{6} \, r / s$$

and $f_h = \omega_h / 2\pi = 5.24 \times 10^{5}$ Hz $= 524$ kHz. Observe that the effective capacitance of C_μ was six times as great as C_π because of the high voltage gain V_o / V.

A sketch of the frequency response for the amplifier of the preceding example is given in Fig. 6.9. The frequency scale is logarithmic so the detail of both the high-frequency and the low-frequency ends of the spectrum can be viewed on the sketch. The low-frequency cutoff or half-power frequency, f_ℓ, is determined by the coupling capacitors C_1 and C_2 of Fig. 6. The determination of f_ℓ is more complicated than that of f_h because there are two coupling capacitors and therefore two time constants involved. The determination of f_ℓ is delayed until Chapter 7, but is given in Fig. 6.9 as 100 Hz.

Although the bandwidth is $f_h - f_\ell$, f_ℓ is normally small in comparison with f_h so the bandwidth is essentially f_h. The mid-frequency range is considered to extend from $10 \, f_\ell$ to $0.1 \, f_h$.

6.2
THE
COMMON-
BASE
AMPLIFIER

The common-base amplifier was introduced in Chapter 4 to illustrate the basic operating principles of the junction transistor. The characteristics of this amplifier configuration are investigated further here. A basic circuit diagram is given in Fig. 6.10 for an ac amplifier. Although two bias supplies are required in this circuit, a single-battery, or power-supply, arrangement is developed in Chapter 7. Load lines could be drawn on the collector characteristics as they were in Fig. 6.2 for the

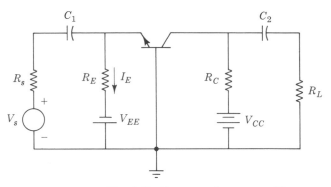

Fig. 6.10. Common-base amplifier.

amplifier of Fig. 6.1. However, once the load-line technique is mastered the load lines may be mentally visualized without actually drawing them. Let us assume that the same transistor is used with the same bias voltages and currents as were used for the common-emitter amplifier. The load resistances are also considered to be the same as in the preceding section. The resistance R_E may be determined from the relationship $I_E R_E = V_{EE} - V_{BE}$, where V_{BE} may be asumed to be about 0.6 V for a silicon transistor, as before.

Let us use the hybrid-π model to analyze the common-base (CB) amplifier so we may determine the frequency characteristics of this amplifier. We will first determine the mid-frequency characteristics using the model of Fig. 6.11. This is a three-node circuit, but it may be reduced to a two-node circuit for all calculations except output resistance if we observe that r_μ is very large compared with all the other

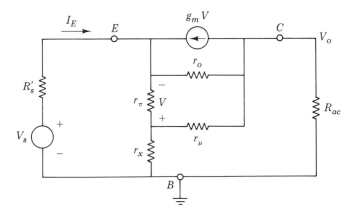

Fig. 6.11. Hybrid-π model of the CB amplifier.

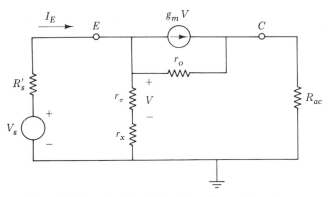

Fig. 6.12. The circuit of Fig. 6.11 with r_μ deleted.

resistances in the circuit and may be neglected as shown in Fig. 6.12. R_E has been combined with R_s to yield R_s' in both circuits. Also the polarity of V is reversed in Fig. 6.12 to make it consistent with the direction of I_E. Writing the nodal equations for Fig. 6.12,

$$I_E = \left(\frac{1}{r_\pi + r_x} + g_o\right) V_E + g_m V - g_o V_C$$

$$g_m V = -g_o V_E + (g_o + G_{ac}) V_C \tag{6.21}$$

Since $V = V_E r_\pi / (r_\pi + r_x)$ we may write Eqs. 6.21 as

$$I_E = \left(\frac{1}{r_x + r_\pi} + g_o + \frac{g_m r_\pi}{r_x + r_\pi}\right) V_E - g_o V_C$$

$$0 = -\left(\frac{g_m r_\pi}{r_x + r_\pi} + g_o\right) V_E + (g_o + G_{ac}) V_C \tag{6.22}$$

Solving Eqs. 6.22 for V_E, and using $g_m r_\pi = \beta_o$ and $g_o \ll (g_o + G_{ac})$

$$V_E \simeq \frac{r_\pi + r_x}{\beta_o + 1} I_E \tag{6.23}$$

Therefore,

$$R_i = \frac{V_E}{I_E} = \frac{r_\pi + r_x}{\beta_o + 1} \tag{6.24}$$

Solving Eqs. 6.22 for V_C and then dividing V_C by V_E we have

$$K_v = \frac{V_C}{V_E} = \frac{\dfrac{\beta_o}{r_x + r_\pi}}{g_o + G_{ac}} \simeq \frac{\beta_o R_{ac}}{r_x + r_\pi} \tag{6.25}$$

Observe that this is the same voltage gain magnitude as obtained for the CE amplifier in Eq. 6.16. However, there is no polarity reversal across the CB amplifier. The current gain from emitter to collector is

$$\frac{I_c}{I_e} \simeq \alpha_o \tag{6.26}$$

However, I_c divides between R_C and R_L, so $I_L = I_C R_C / (R_C + R_L)$ and

$$K_i = \frac{I_L}{I_e} = \frac{\alpha_o R_C}{R_C + R_L} \tag{6.27}$$

In order to find the output resistance of the common-base transistor, let us simplify the circuit of Fig. 6.11 by moving the left end of r_μ from the junction of r_x and r_π to the common, or ground, as shown in Fig. 6.13. This move causes little error because the voltage across r_x is very small compared with the voltage across r_μ. We will first find the relationship between I_C' and V_C. Writing nodal equations,

$$I_C' = g_o V_C - g_m V - g_o V_E$$
$$0 = -g_o V_C + \left(\frac{1}{r_\pi + r_x} + G_s' + g_o\right) V_E + g_m V \tag{6.28}$$

Since $V = V_E r_\pi / (r_\pi + r_x)$ we may write Eqs. 6.28 as

$$I_C' = g_o V_C - \left(g_o + \frac{g_m r_\pi}{r_\pi + r_x}\right) V_E$$
$$0 = -g_o V_C + \left(\frac{g_m r_\pi + 1}{r_\pi + r_x} + G_s' + g_o\right) V_E \tag{6.29}$$

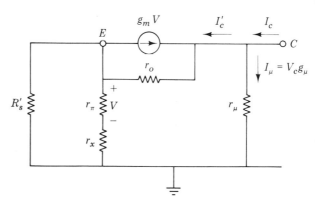

Fig. 6.13. Circuit model for determining the output resistance of the CB transistor.

Using $g_m r_\pi = \beta_o$ and solving Eqs. 6.29 for V_C,

$$V_C = \frac{\left(\dfrac{\beta_o + 1}{r_\pi + r_x} + G_s' + g_o\right) I_C'}{g_o\left(\dfrac{\beta_o + 1}{r_\pi + r_x} + G_s' + g_o\right) - g_o\left(g_o + \dfrac{\beta_o}{r_\pi + r_x}\right)} \tag{6.30}$$

Since $g_o \ll \beta_o/(r_\pi + r_x)$ we may simplify Eq. 6.30 to

$$V_C = \frac{I_C'}{g_o\left(1 - \dfrac{\beta_o}{\beta_o + 1 + G_s'(r_\pi + r_x)}\right)} \tag{6.31}$$

$$G_o' = \frac{I_C'}{V_C} = g_o\left(1 - \frac{\beta_o}{\beta_o + 1 + G_s'(r_\pi + r_x)}\right) \tag{6.32}$$

Adding g_μ, the total output conductance is

$$G_o = G_o' + g_\mu = g_\mu + g_o\left(1 - \frac{\beta_o}{\beta_o + 1 + G_s'(r_\pi + r_x)}\right) \tag{6.33}$$

When the input is open, G_s' approaches zero and the output conductance is

$$G_o\Big|_{G_s' = 0} = \frac{g_\mu + g_o}{\beta_o + 1} = 2\,g_\mu \tag{6.34}$$

When R_s' approaches zero,

$$G_o\Big|_{G_s \to \infty} = g_\mu + g_o \simeq g_o \tag{6.35}$$

Example 6.2

Let us now determine the mid-frequency values of K_v, K_i, K_p, R_i, and R_o for a CB amplifier that uses the same transistor, q-point, load resistances, and source resistance as the CE amplifier of Example 6.1. $K_v = V_o/V_i$ is the same for the CB amplifier as for the CE amplifier = 164 with no polarity reversal. Using Eq. 6.27, $\alpha_o = 139/140 = .993$ so $K_i = 0.496$ and $K_p = 164(.496) = 81$. The input resistance $R_i = (r_x + r_\pi)/(\beta_o + 1) = 850/140 = 6.1$ Ω. Note that this very low input resistance causes V_i to be small compared with V_s since $R_s = 1000$ Ω in this example. Using Eq. 6.33, $G_o = 3.6 \times 10^{-8} + 5 \times 10^{-6}[1 - 139/(140 - .85)] = 3.6 \times 10^{-8} + 5 \times 10^{-6}(1 - .987) = 10^{-7}$ mhos so $R_o = 10$ MΩ.

Finally, let us develop an expression for the high-frequency cutoff ω_h of the CB amplifier and compare it with that of the CE amplifier. The

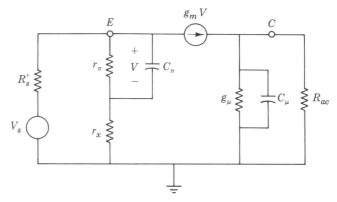

Fig. 6.14. Model for determining the frequency characteristics of a CB amplifier.

circuit model of Fig. 6.13 with r_o deleted and the transistor capacitances added is shown in Fig. 6.14. The driving source is changed to a current source, and $y_\pi = g_\pi + j\omega C_\pi$ and $y_\mu = g_\mu + j\omega C_\mu$ have been used in Fig. 6.15 to facilitate the writing of nodal equations. Writing these equations,

$$I_s' = \left(G_s' + \frac{1}{z_\pi + r_x} + \frac{g_m z_\pi}{z_\pi + r_x} \right) V_i$$

$$0 = -\frac{g_m z_\pi}{r_x + z_\pi} V_i + (y_\mu + G_{ac}) V_o \tag{6.36}$$

Solving Eqs. 6.36 for V_o

$$V_o = \frac{g_m z_\pi I_s'}{[(r_x + z_\pi) G_s' + 1 + g_m z_\pi] (y_\mu + G_{ac})} \tag{6.37}$$

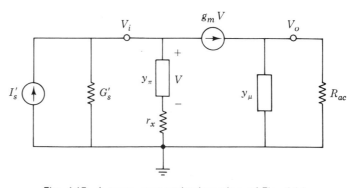

Fig. 6.15. A more convenient version of Fig. 6.14.

Simplifying,

$$\frac{V_o}{I_s'} = \frac{g_m/(g_\mu + G_{ac} + j\omega C_\mu)}{(r_x G_s' + 1)(g_\pi + j\omega C_\pi) + G_s' + g_m} \tag{6.38}$$

Since $g_\mu \ll G_{ac}$ we may rearrange Eq. 6.38 to the following form:

$$\frac{V_o'}{I_s} = \frac{g_m}{[(r_x G_s' + \beta_o + 1)g_\pi + G_s' + j(r_x G_s' + 1)\omega C_\pi](G_{ac} + j\omega C_\mu)} \tag{6.39}$$

We have found that in the case of a single pole, or time constant T, the upper half-power frequency is $\omega_h = 1/T$ or $\omega_h C = G$. In Eq. 6.39 we have two poles or two time constants T_1 and T_2, where

$$T_1 = \frac{(r_x G_s' + 1)C_\pi}{(r_x G_s' + \beta_o + 1)g_\pi + G_s'} \quad \text{and} \quad T_2 = \frac{C_\mu}{G_{ac}}$$

It may be shown that[1] $\omega_h \simeq 1/(T_1 + T_2)$. Therefore,

$$\omega_h = \frac{1}{\dfrac{(r_x G_s' + 1)C_\pi}{(r_x G_s' + \beta_o + 1)g_\pi + G_s'} + \dfrac{C_\mu}{G_{ac}}} \tag{6.40}$$

Example 6.3

We will use the transistor and circuit element values of Examples 6.1 and 6.2 to find ω_h and f_h of the CB amplifier in this example and compare its bandwidth to that of the CE amplifier. Using Eq. 6.40 with $G_s' = .001$, $\beta_o = 139$, $r_x - 155$, $g_\pi = 1.44 \times 10^{-3}$, $G_{ac} = 10^{-3}$, $C_\pi = 10^{-10}$ F and $C_\mu = 3 \times 10^{-12}$ F.

$$\omega_h = \frac{1}{\dfrac{(.155 + 1)10^{-10}}{(140.155)1.44 \times 10^{-3} + 10^{-3}} + \dfrac{3 \times 10^{-12}}{10^{-3}}} = 2.8 \times 10^8 \, r/s$$

$f_h = 2.8 \times 10^8 / 6.28 = 44.6$ MHz. This CB bandwidth is $44.6/.524 = 85$ times as wide as the CE bandwidth.

6.3 THE COMMON-COLLECTOR AMPLIFIER

The collector may also be used as the common element in a transistor amplifier. A typical common-collector (CC) amplifier circuit is shown in Fig. 6.16. It may not be obvious that the collector is at the common, or ground, potential since it is connected to V_{CC}. However, V_{CC} is at *signal ground* potential, so in the incremental circuit model the collector is at common, or ground, potential. The base terminal is the input and the emitter terminal is the output.

---

[1] See *Electronic Circuits* by Gray and Searle, McGraw-Hill.

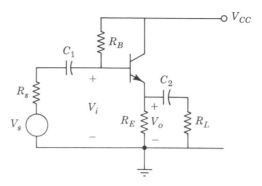

Fig. 6.16. Common-collector amplifier.

You may observe from Fig. 6.16 that

$$V_i = V_o + V_{be} \tag{6.41}$$

Since V_{be} is normally small compared with V_o, V_o is nearly equal to V_i and the voltage gain $V_o/V_i \simeq 1$. Therefore, the CC amplifier is commonly known as an *emitter follower* because the emitter, or output, voltage *follows* the base, or input voltage.

The emitter current I_e is the output current and the base current I_b is the input current, so the current gain through the transistor is approximately $I_e/I_b = (\beta + 1)$. However, the emitter signal current divides between R_E and R_L so I_L/I_b is less than $\beta + 1$.

The hybrid-π model is used in Fig. 6.17 to obtain relationships between the circuit parameters and the voltage gain K_v, the current gain K_i, the input resistance R_i, and the output resistance R_o at midfrequencies. We will neglect the current through r_μ initially and write

$$V_i = I_b\,(r_x + r_\pi) + \frac{(I_b + g_m V)}{g_o + G_{ac}} \tag{6.42}$$

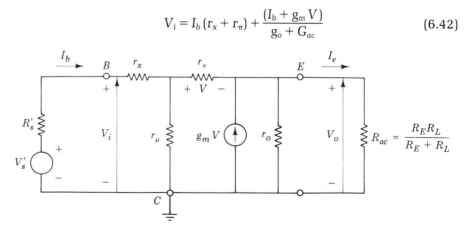

Fig. 6.17. Hybrid-π model for the CC amplifier.

Usually $G_{ac} \gg g_o$. Using this approximation, the relationship $g_m = \beta_o g_\pi$ and $V = I_b r_\pi$, Eq. 6.42 becomes

$$V_i = I_b \left[r_x + r_\pi + (\beta_o + 1) R_{ac} \right] \tag{6.43}$$

From Eq. 6.43, the input impedance is

$$R_i = \frac{V_i}{I_b} = r_x + r_\pi + (\beta_o + 1) R_{ac} \tag{6.44}$$

For the usual values of R_{ac}, $(\beta_o + 1) R_{ac} \gg (r_x + r_\pi)$ and $R_i \simeq (\beta_o + 1) R_{ac}$. Since $\beta_o r_o = r_\mu$, the maximum input resistance is $r_\mu / 2$.

Again, with normal values of R_{ac}, using Fig. 6.17

$$K_v = \frac{V_o}{V_i} = \frac{(\beta_o + 1) R_{ac}}{r_x + r_\pi + (\beta_o + 1) R_{ac}} \tag{6.45}$$

$$K_i = \frac{I_L}{I_b} = \frac{I_e}{I_b} \frac{R_E}{R_E + R_L} = \frac{(\beta_o + 1) R_E}{R_E + R_L} \tag{6.46}$$

The output resistance of the transistor emitter terminal may be determined from Fig. 6.18 where we can see that

$$I_e = \frac{V_o}{R_s' + r_x + r_\pi} + g_m V \tag{6.47}$$

However,

$$V = \frac{V_o r_\pi}{R_s' + r_x + r_\pi}$$

and

$$g_m = \beta_o r_\pi$$

so

$$I_e = V_o \left(\frac{1}{R_s' + r_x + r_\pi} + \frac{\beta_o}{R_s' + r_x + r_\pi} \right) = V_o \left(\frac{\beta_o + 1}{R_s' + r_x + r_\pi} \right) \tag{6.48}$$

and

$$R_o = \frac{V_o}{I_e} = \frac{R_s' + r_x + r_\pi}{\beta_o + 1} \tag{6.49}$$

Observe from Eqs. 6.49 and 6.44 that the CC transistor behaves as an impedance transformer with its output impedance being the impedance of the base circuit divided by $(\beta_o + 1)$ and its input impedance being the impedance of the emitter circuit multiplied by $(\beta_o + 1)$. This impedance

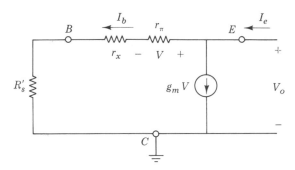

Fig. 6.18. Circuit model for determining the output resistance R_o with r_μ and r_o neglected.

multiplication resulting from a current multiplication occurs frequently in electronic circuits.

The power gain of the CC amplifier is essentially equal to $K_p = (\beta + 1) R_C / (R_C + R_L)$.

Let us use the circuit model of Fig. 6.19 to develop an expression for the bandwidth of a CC amplifier. In this circuit $G_s'' = 1/(r_x + R_s')$ and $y_\mu = j\omega C_\mu$ since g_μ is negligibly small. Writing nodal equations,

$$I_s' = (G_s'' + y_\mu + y_\pi) V_i - y_\pi V_o$$

$$g_m V = -y_\pi V_i + (y_\pi + G_{ac}) V_o \tag{6.50}$$

However, we see from Fig. 6.19 that $V = V_i - V_o$ so Eq. 6.50 becomes

$$I_s' = (G_s'' + y_\mu + y_\pi) V_i - y_\pi V_o$$

$$0 = -(g_m + y_\pi) V_i + (g_m + y_\pi + G_{ac}) V_o \tag{6.51}$$

Solving Eqs. 6.51 for V_o

$$V_o = \frac{(g_m + y_\pi) I_s'}{(G_s'' + y_\mu + y_\pi)(g_m + y_\pi + G_{ac}) - y_\pi (g_m + y_\pi)} \tag{6.52}$$

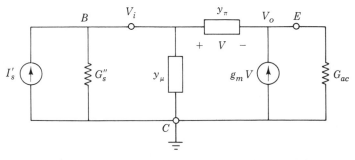

Fig. 6.19. Model for determining the bandwidth of a CC amplifier.

Dividing through by $(g_m + y_\pi)$,

$$V_o = \frac{I_s'}{(G_s'' + y_\mu + y_\pi)\left(1 + \dfrac{G_{ac}}{g_m + y_\pi}\right) - y_\pi} \tag{6.53}$$

Rearranging Eq. 6.53 and assuming $y_\pi \ll g_m$, since $y_\pi = g_m/\beta$,

$$V_o = \frac{I_s'}{(G_s'' + y_\mu)\left(1 + \dfrac{G_{ac}}{g_m}\right) + \dfrac{G_{ac}}{g_m} y_\pi} \tag{6.54}$$

Recognizing $G_{ac}/g_m \ll 1$ and letting $y_\pi = g_\pi + j\omega C_\pi$ and $y_\mu = j\omega C_\mu$,

$$V_o = \frac{I_s'}{G_s'' + \dfrac{G_{ac}}{g_m} g_\pi + j\omega\left(C_\mu + \dfrac{G_{ac}}{g_m} C_\pi\right)} \tag{6.55}$$

Therefore,

$$\omega_h = \frac{G}{C} = \frac{G_s'' + \dfrac{G_{ac}}{g_m} g_\pi}{C_\mu + \dfrac{G_{ac}}{g_m} C_\pi} \tag{6.56}$$

Example 6.4

Let us determine the bandwidth of the CC amplifier using the transistor and components of the preceding amplifiers with $G_{ac} = 1$ mS, $g_m = 0.2$ S, $g_\pi = 1.44$ mS, $G_s'' = 1/(1155) = 0.866$ mS, $C_\pi = 10^{-10}$ F and $C_\mu = 3 \times 10^{-12}$ F.

$$\omega_h = \frac{0.866 \times 10^{-3} + (5 \times 10^{-3})\, 1.44 \times 10^{-3}}{3 \times 10^{-12} + (5 \times 10^{-3})\, 10^{-7}}$$

$$= \frac{8.7 \times 10^{-4}}{3.5 \times 10^{-12}} = 2.49 \times 10^8 \, r/s$$

$$f_h = \frac{2.49 \times 10^8}{6.28} = 39.6 \times 10^6 \, \text{Hz}$$

This bandwidth is about the same as that of the CB amplifier and $39.6/.524 = 76$ times as broad as that of the CE amplifier. Note that the power-gain bandwidth product is nearly the same for all three configurations:

$$\text{CE}\,(K_p f_h) = 11{,}316 \times 5.24 \times 10^5 = 5.93 \times 10^9$$

$$\text{CB}\,(K_p f_h) = \quad\; 82 \times 44.6 \times 10^6 = 3.66 \times 10^9$$

$$\text{CC}\,(K_p f_h) = \quad\; 70 \times 39.6 \times 10^6 = 2.77 \times 10^9$$

6.4 COMPARISON OF CE, CB, AND CC AMPLIFIERS

Some conclusions were drawn and comparisons made among the three basic amplifier configurations as they were analyzed in the preceding sections. These comparisons are summarized in Table 6.1, assuming typical values of load and source resistances. The values of K_i are the ratio (current delivered to total ac load/input current). We also found that a signal polarity reversal occurs from the input to the output of the CE amplifier but *not* for the CB or CC amplifiers.

The variations of K_v, K_i, Z_i, and Z_o are sketched as functions of ac load resistance R_{ac} and driving source resistance R_s' in Fig. 6.20. The numerical values are for the transistor used in the preceding examples of this chapter at the given q-point where

$$r_x = 155\ \Omega \qquad\qquad \beta_o = 139 \qquad\qquad r_o = 200\ \text{k}\Omega$$

$$r_\pi = 695\ \Omega \qquad\qquad g_m = 0.2\ \text{mho} \qquad\qquad r_\mu = 28\ \text{M}\Omega$$

Again, the current gains are to the total R_{ac} load, not just R_L. Although the very large values of load resistance R_{ac} above about 10^5 ohms are impractical for use in transistor amplifiers, they are included on the R_{ac} axes in order to show the limiting values of the functions.

6.5 TRANSDUCER GAIN

The voltage and power gains up to this point have been determined from the input of the amplifier to its load which may be a following amplifier. However, as previously mentioned, the first, or input, stage of an amplifier is normally driven by a transducer of some type having a source resistance R_s. In this case, the power available from the amplifier depends to a large extent upon the relationship between the source resistance R_s and the input resistance R_i of the amplifier. For example, we used a given transistor and similar circuit components and biases in the three different configurations in order to compare their characteristics and found that the voltage gains of the CE and CB amplifiers were the same. However, with the same value of source voltage V_s the output power from the CE amplifier is much higher than that of the CB because of the better *impedance match* between the CE input resistance and the driving-source resistance, as illustrated in Fig. 6.21. Observe that the input voltage, and hence the output voltage, of the CE amplifier is 77

Table 6.1. Comparison of the characteristics of CE, CB, and CC amplifiers with typical load and source resistances

Type	K_v	K_i	Z_i	Z_o	f_h
CE	high	β	medium	medium	medium
CB	high	α	low	high	high
CC	1	$\beta + 1$	high	low	high

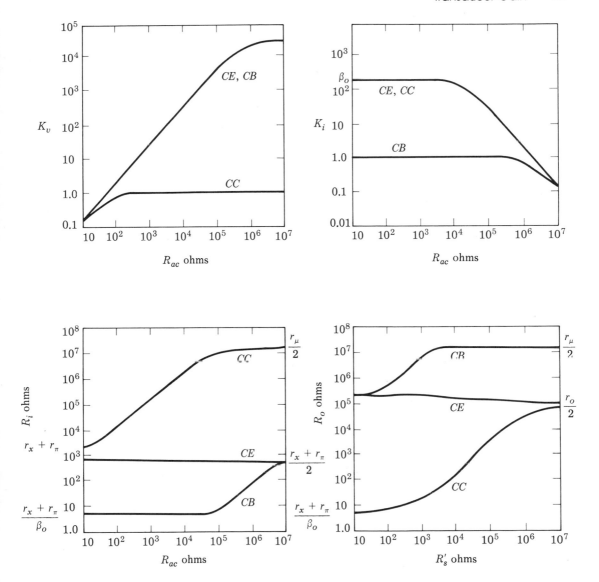

Fig. 6.20. Variations of the amplifier characteristics as functions of R_{ac} and R_s' for CE, CB, and CC configurations.

times as large as that of the CB amplifier. Since the voltage gain of the CB amplifier is 164, its output voltage is 0.98 V_s, so the CB amplifier is of little value in this application.

The true value of an amplifier is measured by the ratio of its load power to the power available from the driving source. As may be seen from Fig. 6.21, the power delivered by a power source to a resistive load

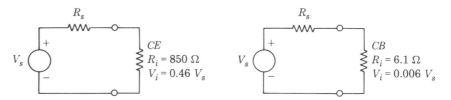

Fig. 6.21. Comparison of the input voltages of the CE and CB amplifiers
when $R_s = 1000 \ \Omega$.

$R_L = I^2 R_L$ where I is the rms current. But $I = V_s / (R_s + R_L)$, so the load power P_L is

$$P_L = \frac{V_s^2 R_L}{(R_L + R_s)^2} \qquad (6.57)$$

The *maximum power transfer theorem* states that maximum power is delivered to the load from a given source when $R_L = R_s$ as shown in Fig. 6.22. This theorem may be proved by either differentiating P_L with respect to R_L (Eq. 6.57) and setting the derivative to zero or by plotting the curve of Fig. 6.22. If either the source or load (or both) are complex, a reactance of proper type is added to the load to reduce the reactance to zero.

The maximum power available, P_a, from any given source is obtained from Eq. 6.57 by letting $R_L = R_s$.

$$P_a = \frac{V_s^2}{4R_s} \qquad (6.58)$$

Since a transformer may be used to match impedances, the available power may be obtained from any source with the aid of a lossless

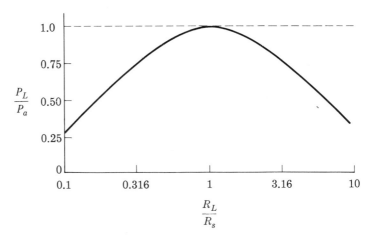

Fig. 6.22. Load power as a function of normalized load resistance R_L / R_S.

transformer. Therefore, a realistic measure of the value of an amplifier is the ratio of the power output of the amplifier to the available power P_a from the source. This ratio is defined as the *transducer gain* G_T, where

$$G_T = \frac{P_o}{P_a} \qquad (6.59)$$

This G_T takes into account the effect of impedance mismatch between the source and the amplifier. Since amplifier gains are normally calculated at mid frequencies, all impedances are normally resistive and Eq. 6.59 may be written, using Eq. 6.58,

$$G_T = \frac{I_o^2 R_L}{V_s^2 / 4R_s} \qquad (6.60)$$

However, the amplifier input current $I_i = V_s/(R_s + R_i)$, so letting $V_s = I_i (R_s + R_i)$ in Eq. 6.60 we have

$$G_T = \frac{I_o^2 \, 4R_s R_L}{I_i^2 (R_s + R_i)^2} = \frac{4R_s R_L K_i^2}{(R_s + R_i)^2} \qquad (6.61)$$

For example, the transducer gain of the CB amplifier having $R_i = 6.1 \ \Omega$ and $K_i - 0.496$ to the load R_L where $R_L = 2 \ k\Omega$, when connected to the source with $R_s = 1 \ k\Omega$, is

$$G_T = \frac{4(10^3)(2 \times 10^3)}{(1,006)^2} (0.496)^2 = 1.94$$

This transducer gain for the CB amplifier is hardly worthwhile because of the poor impedance match between the driving source and the amplifier.

Transducer and amplifier gains are usually expressed in decibels (dB) where

$$dB = 10 \log \frac{P_o}{P_{ref}} \qquad (6.62)$$

Then the total dB gain of an amplifier may be obtained by adding the dB gains of each stage rather than multiplying the individual stage gains to obtain the total numerical gain. The dB gain of the CB amplifier in the example above is $dB = 10 \log (1.94) = 2.9 \ dB$.

It is clear from observing Fig. 6.20 that the CE amplifier is highly superior to the CB and CC amplifiers for cascading two or more similar stages because R_i and R_o are the same order of magnitude in the CE amplifier. Thus, a good impedance match may be achieved between stages. On the other hand, CB and CC amplifiers have vastly different R_i and R_o so they have nearly 0 dB transducer gains when cascaded. However, the output resistance of a CC amplifier matches the input resistance of a CB amplifier, so a combination CC-CB amplifier known as an *emitter-coupled* amplifier is very popular because of its desirable characteristics, including broad bandwidth. The emitter-coupled amplifier is discussed in Chapter 10.

The maximum possible gain would be obtained from an amplifier if the CE configuration were used and the load resistance R_L were matched to the transistor output resistance $\simeq r_o$ by using a transformer with turns ratio $= \sqrt{r_o / R_L}$. This technique is not practical, however, because of the size and cost of a transformer that would provide such a high primary resistance and a suitable bandwidth.

The practical way to achieve maximum power gain in an amplifier stage is to use the CE configuration with $R_C \simeq R_L$ in the capacitively coupled amplifier and then use a driving source that provides a fairly good impedance match for R_i. This technique matches the source to the load and, as previously mentioned, the output, or source, resistance of an amplifier stage is approximately equal to R_C, since normally $r_o \gg R_C$.

Two questions remain. First, will a CB amplifier provide higher transducer gain than a CE amplifier if the driving-source resistance R_S is very low and thus provides a good match for R_i of the common base amplifier? Second, will the CC amplifier provide higher transducer gain if R_s is very high and nearly matches the R_i of the CC amplifier? In order to answer these questions, the transducer power gain G_T for all three configurations is plotted as a function of R_L in Fig. 6.23 for each of three values of R_s: one to match the R_i of the CB amplifier, another to match the R_i of the CE, and the third to match the R_i of the CC amplifier. The transistor and circuit components used in the preceding examples are used to generate these plots. Therefore, R_i for CB = 6.1 Ω, R_i for CE = 850 Ω, R_i for CC = 140 kΩ, R_o for CB and CE = 2 kΩ, and R_o for CC = $(R_s + 850)/140$.

Observe from Fig. 6.23 that the CE amplifier provides higher transducer gain than either the CB or CC for all load resistances above 10 Ω. The CC amplifier provides the highest gain for the lower source resistances and $R_L < 10$ Ω because the impedance match between R_o and R_L is quite good at these values of R_L and the match between R_s and R_i is also fairly good. With $R_s = 140$ kΩ the match between R_s and R_i for the CB amplifier is so poor the CB curve is below -10 dB for all values of R_L and therefore does not appear in the figure. With $R_s = 6$ Ω the CB gain was still about 5 dB less than that of the CE amplifier in spite of the mismatch at the CE input.

Problems   *Section 6.1*

6.1   The following characteristics are given by the manufacturer for a specific n-p-n transistor at the q-point $I_C = 2$ mA, $V_{CE} = 10$ V, and $T = 25°C$.

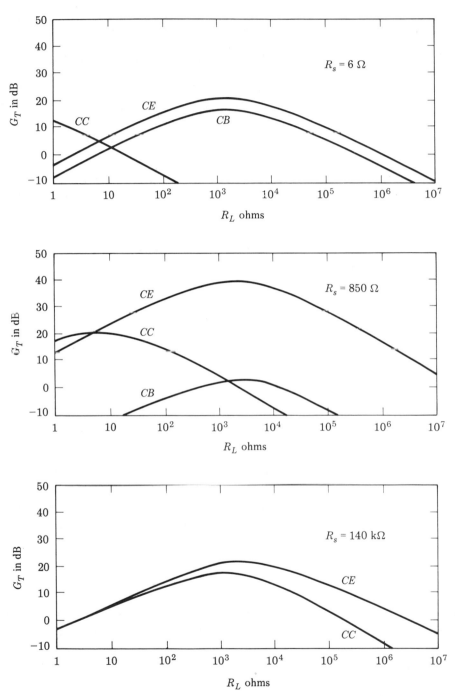

Fig. 6.23. *Transducer gain G_T versus R_L for three values of source resistance R_S.*

$$h_{fe} = 200 \qquad h_{oe} = 1.6 \times 10^{-5} \qquad f_T = 300 \text{ MHz}$$
$$h_{ie} = 2.7 \text{ k}\Omega \qquad h_{re} = 10^{-4} \qquad C_{ob} = 4 \text{ pF}$$

This transistor is used as an ac CE amplifier at the q-point given with $V_{CC} = 20$ V, $R_C = 5$ kΩ, and $R_L = 5$ kΩ. The driving-source resistance is 3 kΩ.

Determine the fixed-bias resistance, the mid-frequency voltage gains V_o/V_i and V_o/V_s, the current gain I_L/I_i, the input resistance, and the output resistance of the transistor. Draw the circuit diagram.

6.2   What percent error is made in the voltage and current gains of the amplifier of Problem 6.1 if r_μ is neglected by assuming that $g_\mu = 0$?

6.3   What percent error is made in the transistor output resistance of the amplifier of Problem 6.1 by neglecting r_μ?

6.4   What is the upper half-power frequency of the amplifier of Problem 6.1?

Section 6.2

6.5   The transistor specified in Problem 6.1 is used in a common-base amplifier circuit with R_C and $R_L = 5$ kΩ. The driving source resistance is 50 Ω.

   a.   Draw the circuit diagram and determine suitable values for R_E and V_{EE} (Fig. 6.10).
   b.   Calculate the voltage gains V_o/V_i and V_o/V_s, the current gain I_L/I_e, the input resistance, and the output resistance of the transistor.

6.6   Determine the upper half-power frequency of the CB amplifier of Problem 6.5 and compare it with that of the CE amplifier of Problem 6.1.

Section 6.3

6.7   The transistor with characteristics given in Problem 6.1 is used as an emitter follower with $R_E = 5$ kΩ and $R_L = 5$ kΩ. Draw the circuit diagram and determine the value of R_B, using the q-point and V_{CC} given in Problem 6.1.

6.8   With $R_S = 10$ kΩ, calculate the input resistance, the voltage gains V_o/V_i and V_o/V_s, the current gain I_L/I_i, and the output resistance of the transistor.

6.9   Determine the upper half-power frequency f_h of the emitter follower of Problems 6.7 and 6.8. Compare this f_h with those of the CE and CB amplifiers.

Section 6.4

6.10 Verify the limiting values or expressions of K_v, K_i, R_i, and R_o given in Fig. 6.20 for the CE, CB, and CC configurations.

6.11 Calculate the power-gain bandwidth products for the CE amplifier of Problems 6.1–3, the CB amplifier of Problems 6.4–6, and the CC amplifier of Problems 6.7–9. Compare them.

Section 6.5

6.12 Prove the maximum power transfer theorem.

6.13 a. Is the CE amplifier of 6.1 well matched to its source and load?
      b. Calculate the transducer gain of this amplifier in dB.
      c. Change R_C to 10 kΩ and change the q-point collector current to maintain the q-point at the center of the dc load line. Calculate the transducer gain of the amplifier in dB and compare it with part *a*.
      d. Change the driving source resistance to match the new input resistance of the amplifier in part *b* and again calculate the transducer gain and compare it with parts *c* and *b*.
      e. Reduce R_C to 2 kΩ with R_L remaining at 5 kΩ and $R_S = 2.7$ kΩ. Change I_C so the q-point remains in the center of the dc load line and calculate the transducer gain in dB. Compare it with that of part *b*.
      f. Change the driving source resistance to match the new input resistance of the amplifier of part *e*. Again, calculate the transducer gain and compare it with those of parts *e* and *b*.
      g. What are your conclusions?

7

Transistor Amplifier Design

The major emphasis in the preceding chapters was on analyzing electronic devices in order to gain an understanding of their characteristics and circuit models. With that understanding you are now prepared to begin selecting the electronic devices, circuit configurations and components that will provide the required power to a specified load over a given frequency range and temperature range. The driving source characteristics must be specified before the design may be completed. The design process normally begins at the load and proceeds toward the source.

**7.1
TRANSISTOR
SELECTION**

After the requirements of the load are established, the design of the amplifier may proceed. For example, the user of an audio music system may determine that the electronic amplifier should be capable of providing 100 watts of audio power to an 8 Ω loudspeaker system over the frequency range 20 Hz to 20,000 Hz within 1 dB and have no more than 1% harmonic distortion. From this specification, the peak load current and voltage is easily established and the catalogs provided by the semiconductor manufacturers must be studied to find a transistor that will deliver the required current and withstand the maximum voltage with adequate safety factors. Then a check must be made to determine whether or not the f_T of the transistor you selected is adequate to provide the required bandwidth, which must be about 40 kHz in order to meet the 1 dB maximum deviation from flat response requirement.

At this point, you are not prepared to design the audio-frequency amplifier specified. Chapter 15 will qualify you for that task and will allow you the option of using integrated circuits to simplify the design task. Negative feedback techniques also will be available to you at that point to limit the distortion to the desired level. This chapter will prepare you for the design of low power amplifiers that might drive such things as power amplifiers, strip chart recorders and TV picture tubes. The amplifiers for visual displays such as picture tubes are known as *video amplifiers* and require bandwidths of several MHz.

After the amplifier that is capable of driving the load has been designed and its input resistance and either current or voltage requirements are known, the amplifier that drives the output stage may be designed, keeping in mind the maximum-power transfer theorem and the bandwidth requirements. Distortion is not usually a problem in a properly-designed lower-level stage because the signals are small and the linear models are highly accurate.

The design process continues toward the input transducer until this transducer has adequate power output capability to drive this *input* stage. Many transducers, such as high-quality microphones, magnetic-type phonograph pickups, tape playback heads and strain gauges, have source voltages of the order of 1 mV or less and source resistances of the order of thousands of ohms, so their output powers are of the order of nanowatts or picowatts. For *low-level* sources, the *noise* generated in

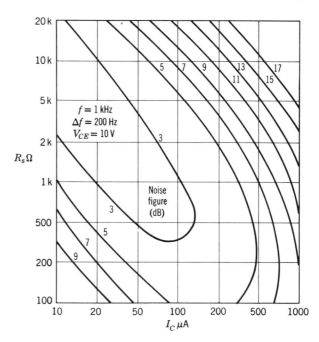

Fig. 7.1. *Constant noise-figure contours for a typical low-noise transistor (2N2443).*

the input amplifier is of great importance because the *noise figure* of this amplifier determines, to a large extent, the noise in the output of the loudspeaker in an audio system, the *snow* on the picture tube or display of a video system, or the fidelity of strip-chart recorders or any other output device. The calculation of noise figures and low-noise design techniques are included in Chapter 11. The manufacturers data sheets include noise-figure data in dB for the transistors most suitable for low-level input stages. The *signal-to-noise ratio (SNR)* is the measure of the quality of a signal with respect to noise. The SNR in the output of an electronic system is essentially the SNR in dB from the input transducer minus the noise figure in dB of the input stage. A typical family of noise-figure contours provided on the transistor data sheet is given in Fig. 7.1.

This figure shows that the noise-figure of the input amplifier will be less than 3 dB if the q-point collector I_C is about 100 μA or less and the driving source resistance R_s is 500 Ω or more, depending upon I_C. Actually I_C is chosen in accordance with R_s. If the 3 dB noise figure is achieved, the output SNR (dB) will be essentially the transducer SNR (dB) $-$ 3 dB, providing the input amplifier is well designed and there are no faulty components in the amplifier.

7.2
VARIATIONS
OF
TRANSISTOR
PARAMETERS

One of the most challenging problems of circuit design is the variation of transistor parameters. We have already seen how the parameters of a given transistor vary with the q-point. This presents no problem, however, because one of the first decisions the designer must make after selecting the transistor is the selection of the q-point. The parameters for this transistor are then easily established, as we have seen, after obtaining h_{fe}, h_{ie}, and either h_{oe} or h_{re} from the transistor curve tracer.

The problems we have not yet faced result from the variations of transistor parameters with temperature and manufacturing tolerances. For example, the variations of h_{FE} with collector current and temperature are taken from the data sheet of a typical transistor and given in Fig. 7.2a. The normalized h_{FE} values listed on the ordinate show that the normalized value of $h_{FE} = 1$ occurs at $I_C = 2$ mA and $T = 25°C$. The data sheet also shows in Fig. 7.2b that h_{FE} at $I_C = 2$ mA and $T = 25°C$ may have values between 20 and 120 because of uncertainties in the manufacturing process. This variation is often known as *beta spread*.

With a large uncertainty in the value of h_{FE}, we could not use fixed bias for a CE amplifier without first obtaining the actual value of h_{FE} by testing each individual transistor and then calculating the bias resistance for each individual transistor. Even then, the variation of h_{FE} with temperature could move the q-point into either the saturation region or too near the cutoff region for satisfactory operation if the

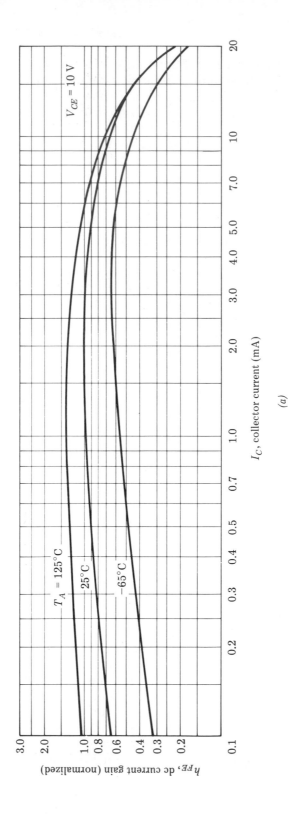

I_C, collector current (mA)

h_{FE}, dc current gain (normalized)

$T_A = 125°C$

$25°C$

$-65°C$

$V_{CE} = 10$ V

(a)

ELECTRICAL CHARACTERISTICS ($T_A = 25°C$ unless otherwise noted)

Characteristic	Symbol	Min.	Max.	Unit
dc current gain ($I_C = 2.0$ mA dc, $V_{CE} = 10$ V dc)	h_{FE}	20	120	—
Current-gain-bandwidth product ($I_C = 2.0$ mA dc, $V_{CE} = 10$ V dc, $f = 100$ MHz)	i_T	300	1200	MHz

(b)

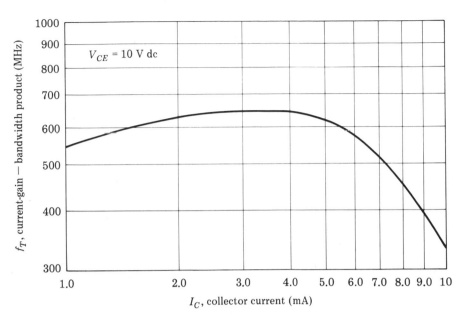

Fig. 7.3. *Typical f_T as a function of I_C for a high-frequency transistor.*

ambient temperature were either increased or decreased too greatly from the temperature at which the transistor was tested. Therefore, fixed bias is not suitable for either assembly line or automated construction, or for a fluctuating environment.

Another parameter that varies considerably with both I_C and manufacturing inaccuracies is the current-gain bandwidth product, or transition frequency f_T. As previously discussed, f_T increases with I_C for values of I_C within the low-level injection range because g_π is proportional to I_C. C_π increases less rapidly than I_C because of its fixed-capacity component C_{je}. However, as I_C is increased into the high-injection level range, both β_o and f_T decrease. The variation of f_T with I_C for one type of high-frequency transistor is given in Fig. 7.3 as extracted from the data sheet.

7.3
STABILIZING-
BIAS CIRCUITS

The variation of transistor parameters with temperature and with manufacturing tolerances essentially eliminates the simple fixed-biased circuit as a satisfactory method for biasing the CE transistor. Therefore, let

Fig. 7.2. *Variations of transistor parameters for a typical high-frequency transistor: (a) variations of h_{FE} with collector current and temperature; (b) variations of h_{FE} and f_T due to manufacturing tolerances.*

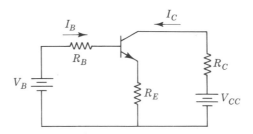

Fig. 7.4. Two-battery bias system with resistance R_E in the emitter circuit.

us momentarily return to the two-battery bias system and investigate the effect of placing a resistance R_E in the emitter circuit as shown in Fig. 7.4. Then, Kirchoff's voltage law may be written for the base loop as

$$V_B = I_B R_B + V_{BE} + (I_B + I_C) R_E \qquad (7.1)$$

Since β_F is the dc value of β or h_{FE}, we can use the relationship $i_C = \beta_F i_B + (\beta_F + 1) I_{CO}$, found in Eq. 4.17, to obtain I_B in terms of I_C and I_{CO}, as follows:

$$I_B = \frac{I_C}{\beta_F} - \frac{\beta_F + 1}{\beta_F} I_{CO} \qquad (7.2)$$

Thus we may use Eq. 7.2 to eliminate I_B in Eq. 7.1 and obtain

$$V_B = \left(\frac{I_C}{\beta_F} - \frac{\beta_F + 1}{\beta_F} I_{CO} \right) R_B + V_{BE} + \left(I_C + \frac{I_C}{\beta_F} - \frac{\beta_F + 1}{\beta_F} I_{CO} \right) R_E \qquad (7.3)$$

Recognizing that $(\beta_F + 1) / \beta_F \simeq 1$ and $I_C / \beta_f \ll I_C$, Eq. 7.3 simplifies to

$$V_B = \left(\frac{I_C}{\beta_F} - I_{CO} \right) R_B + V_{BE} + (I_C - I_{CO}) R_E \qquad (7.4)$$

We may solve Eq. 7.4 explicitly for I_C to obtain

$$I_C = \frac{V_B - V_{BE} + I_{CO} (R_B + R_E)}{R_E + R_B / \beta_F} \qquad (7.5)$$

From Eq. 7.5 we can see that I_C may be relatively constant over a wide range of β_F if $R_E \gg R_B / \beta_F$. Also, if $I_E R_E$ is as large as 1 or 2 volts, the bias supply V_B is large in comparison with the variations of V_{BE} so I_C is almost independent of these variations. In addition, if silicon or GaAs transistors are used over a moderate temperature range, the term $I_{CO} (R_B + R_E)$ will be negligible in comparison with V_B. Therefore, we can see that the emitter resistance R_E stabilizes the q-point I_C. The physical reason for this stability is that an increase in I_C increases the voltage across R_E and thus reduces the forward bais V_{BE}.

The effect of an increase in I_C causing a decrease in forward bias voltage is a large loss of signal gain and is known as *degeneration*. This degeneration may be eliminated for ac signals by connecting a capacitor in parallel with R_E to hold the voltage across R_E constant over the signal-frequency range. This capacitor C_E is known as a *bypass capacitor* and its reactance should be small in comparison with the resistance it is bypassing over the desired frequency range. Its required capacitance will be discussed later.

In our circuit analysis we have used n-p-n transistors as examples for two reasons. First, most of the voltages and currents are positive in the n-p-n circuits so less thought is required to keep the polarities. Second, the available number of n-p-n silicon transistor types greatly exceeds the number of p-n-p silicon types; therefore, n-p-n transistors are used most frequently in design. The use of p-n-p transistors is essenital in some circuits but no problem is encountered in design because all the equations developed with n-p-n transistors are applicable to p-n-p circuits if the voltages and currents carry their own sign.

The two batteries shown in Fig. 7.4 are not required for a stabilized bias system. A single battery system is shown in Fig. 7.5a. Observe that a dc Thevenin's equivalent circuit looking into the base-bias circuit to the left of the points A and B is as shown in Fig. 7.5b, where R_B is the parallel combination of R_1 and R_2, or

$$R_B = \frac{R_1 R_2}{R_1 + R_2} \tag{7.6}$$

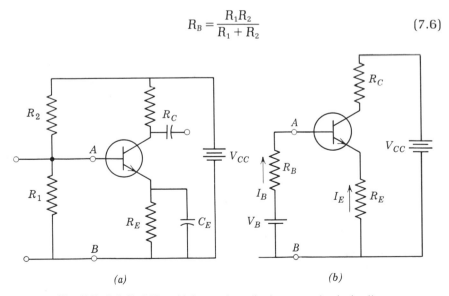

(a)   *(b)*

Fig. 7.5. (a) Stabilized-bias system that uses a single battery.
(b) dc Thevenin's equivalent of this circuit.

and the open-circuit voltage, or voltage between points A and B with the base disconnected, is

$$V_B = \frac{R_1}{R_1 + R_2} V_{CC} \qquad (7.7)$$

Note that this equivalent circuit is the same as the circuit of Fig. 7.4. The value of V_B can be found by adding the voltage drops around the base circuit at the q-point. Thus,

$$V_B = I_B R_B + V_{BE} - I_E R_E \qquad (7.8)$$

Observe that I_E is a negative current since it flows out of the transistor. Therefore, the term $-I_E R_E$ will be positive for an n-p-n transistor. Input charcteristics may not be available, in which case V_{BE} may be assumed to be about 0.2 V for a germanium transistor, 0.6 V for a silicon transistor, or 1.0 V for a GaAs transistor, providing the temperature is near normal.

Since V_{CC} and V_B are now known, there are only two unknowns (R_1 and R_2) in Eqs. 7.6 and 7.7. Therefore, these equations can be solved simultaneously for R_1 and R_2 in terms of V_{CC}, V_B, and R_B. This solution yields

$$R_1 = \frac{V_{CC}}{V_{CC} - V_B} R_B \qquad (7.9)$$

$$R_2 = \frac{V_{CC}}{V_B} R_B \qquad (7.10)$$

A rather simple procedure may now be used to design the base-bias circuit for a CE amplifier that is intended for use over a limited range of ambient temperatures such as for home or laboratory use. After selecting R_C and I_C the base-bias design might proceed as follows:

1. Choose a value of R_E that will provide $I_E R_E$ (or $I_C R_E$) $\simeq 1$ or 2 volts, or perhaps $V_{CC}/10$.

2. Let $R_B \simeq 0.3 \, \beta_{Fmin} R_E$.

3. Calculate V_B from Eq. 7.8.

4. Calculate R_1 and R_2 from Eqs. 7.9 and 7.10.

The choice of R_B is always a compromise. The smaller the value of R_B the better the q-point stability but also the greater the signal loss due to signal current through R_B. For small signal loss R_B should be large in comparison to $r_x + r_\pi$.

**7.4
STABILIZED
BIAS DESIGN**

A more precise bias design is required for reliable operation of mass-produced circuits that are required to operate properly over a wide range of temperatures. Examples of such circuits are automobile radios that must operate properly in Alaska during the winter and also in the Southwestern deserts during the summer. The first step in reliable bias circuit design is the determination of the permissible q-point excursion as illustrated in Fig. 7.6. The dc load line is first drawn on the collector characteristics. However, the total dc resistance between the collector and the emitter is now $R_C + R_E$ so a value of R_E must be selected before the dc load line is drawn. Strictly speaking, the dc load line should have slope $= -1/(R_C + R_E/\alpha_F)$ to account for I_E flowing through R_E rather than I_C. However, the assumption that $\alpha_F \simeq 1$ is adequately accurate. R_E does not influence the ac load line slope $= -1/R_{ac}$ if R_E is properly bypassed.

In order to avoid the distortion caused by the flattening of the input signal as the transistor operation enters either the saturation region or the low-β region as i_C approaches zero, v_{CE} is not allowed to become less than V_o and i_C is not allowed to become less than I_o, as shown in Fig. 7.6. Suitable values for V_o and I_o might be 1 V and 0.1 mA, respectively, for a transistor operating in the range of a few mA. Then, since the maximum signal voltage V_m and the maximum signal-component of collector current V_m/R_{ac} are known from the load-power requirements we may easily find the minimum and maximum permissible values of I_{C1} and I_{C2}. As seen in Fig. 7.6,

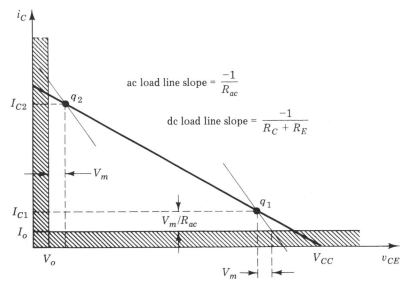

Fig. 7.6. Determination of the permissible q-point excursion.

$$I_{C1} \geq I_o + \frac{V_m}{R_{ac}} \tag{7.11}$$

$$I_{C2} \leq \frac{V_{CC} - (V_o + V_m)}{R_C + R_E} \tag{7.12}$$

With the maximum and minimum q-point collector currents now established, we may write the base-bias voltage-law Eq. 7.4 for each of these q-points.

$$V_B = \left(\frac{I_{C2}}{\beta_{F2}} - I_{CO2}\right) R_B + V_{BE2} + (I_{C2} - I_{CO2}) R_E \tag{7.13}$$

$$V_B = \left(\frac{I_{C1}}{\beta_{F1}} - I_{CO1}\right) R_B + V_{BE1} + (I_{C1} - I_{CO1}) R_E \tag{7.14}$$

Subtracting Eq. 7.14 from Eq. 7.13 we have

$$0 = \left(\frac{I_{C2}}{\beta_{F2}} - \frac{I_{C1}}{\beta_{F1}} - \Delta I_{CO}\right) R_B - \Delta V_{BE} + (\Delta I_C - \Delta I_{CO}) R_E \tag{7.15}$$

where $\Delta I_C = I_{C2} - I_{C1}$, $\Delta I_{CO} = I_{CO2} - I_{CO1}$ and $\Delta V_{BE} = V_{BE1} - V_{BE2}$. This form is necessary since V_{BE1} is larger than V_{BE2} because the larger q-point current occurs at the higher temperature T_2.

Solving Eq. 7.15 explicitly for R_B,

$$R_B = \frac{(\Delta I_C - \Delta I_{CO}) R_E - \Delta V_{BE}}{\dfrac{I_{C1}}{\beta_{F1}} - \dfrac{I_{C2}}{\beta_{F2}} + \Delta I_{CO}} \tag{7.16}$$

After we have determined R_B from Eq. 7.16 we may use Eqs. 7.9 and 7.10 to find R_1 and R_2 (Fig. 7.5), and the bias design is complete. However, we must first find a method for determining the minimum and maximum transistor junction temperatures in order to determine ΔV_{BE} and ΔI_{CO}. The transistor chip temperature is higher than the ambient temperature because of the power dissipation $V_{CE} I_C$ at the collector junction. The relationship between the temperature rise and the power dissipation may be obtained from the derating factor of the transistor. The relationship between ambient temperature, power dissipation capability, and derating factor is shown in Fig. 7.7. From this figure it is seen that the derating factor (the slope of the T_{jmax} curve) is

$$\text{derating factor (Watts /°C)} = -\frac{P_{dmax}}{T_{jmax} - 25°C} \tag{7.17}$$

For example, if a transistor has a maximum power dissipation capability of 300 mW at ambient temperature $T_a = 25°C$ (the usual reference)

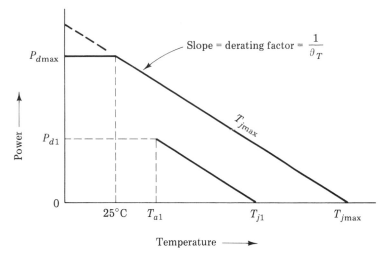

Fig. 7.7. Relationship among maximum power dissipation capability P_{dmax}, maximum permissible junction temperature T_{jmax}, and the derating factor.

and has a maximum safe junction temperature rating of 175°C, then its derating factor is 300 mW/(175 − 25)°C = 2 mW/°C. Then, rearranging Eq. 7.17, $P_{dmax} = (T_{jmax} − 25°C)$ (derating factor). For a general junction dissipation P_d less than P_{dmax} we could similarly write

$$P_d = (T_j − T_a) \text{ (derating factor)} \qquad (7.18)$$

Because of the analogy between heat flow and electric current flow with the temperature difference $(T_j − T_a)$ being the forcing function, a thermal resistance θ_T has been defined such that

$$P_d = \frac{T_j − T_a}{\theta_T} \qquad (7.19)$$

where we can see by comparing Eqs. 7.18 and 7.19 that θ_T is the reciprocal of the derating factor. Our interest is in T_j which may be obtained explicitly from Eq. 7.19.

$$T_j = T_a + \theta_T T_d \qquad (7.20)$$

We may now find $T_{j2} = T_{amax} + \theta_T (V_{CE2} I_{C2})$ and $T_{j1} = T_{amin}$. The rise in junction temperature due to power dissipation $V_{CE1} I_{C1}$ should not be included because we want the equipment to operate properly from the instant the switch is turned on, without waiting for it to *warm up*.

The value of I_{CO} typically doubles for each 10°C increase in junction temperature. Also, the maximum expected value of I_{CO} at 25°C is normally given in the data sheet of a given transistor. Therefore, the value of I_{CO2} at the maximum junction temperature T_{j2} may be obtained from the relationship

$$I_{CO2} = I_{CO}2^{(T_{j2}-25°C)/10}$$

where I_{CO} is given at 25°C and T_{j2} is in °C. At T_{amin}, I_{CO} is normally so small in a silicon or GaAs transistor that it may be neglected and $\Delta I_{CO} \simeq I_{CO2}$.

We found in Chapter 3 (See Fig. 3.7a) that the voltage across a forward-biased diode, with current held constant, decreases about 2 mV/°C. Therefore, we have all the information needed to design the bias circuit for a CE amplifier.

Example 7.1

Let us assume that our assignment is to design a CE amplifier that will provide 0.5 mA max to a 3 kΩ load resistance. The available V_{CC} is 15 V. The transistor with characteristics given in Fig. 7.2 is selected, with $R_C = 3$ kΩ to match the load. The amplifier will be mass-produced and it must operate satisfactorily over the ambient temperature range from $-40°C$ to 75°C.

We have no collector characteristics but may assume that $V_o = 1$ V will avoid the saturation region. Also, Fig. 7.2a indicates that β_F decreases only by 20% at $I_C = 0.3$ mA, so we will select $I_o = 0.3$ mA. Then, $I_{C1} = (0.3 + 0.5)$ mA $= 0.8$ mA. Let us choose $R_E = 0.2 R_C = 600$ Ω. Since $V_m - I_m R_{ac} = 0.5$ mA $(1.5$ kΩ$) = 0.75$ V, $I_{C2} = (15 - 1.75)$ V$/3.6$ kΩ $= 3.68$ mA and $\Delta I_C = 3.68 - .8 = 2.88$ mA. The maximum I_{CO} for any unit of this transistor type at 25°C is 1 μA. The derating factor is 2mW/°C, so $\theta_T = 0.5°C/mW$. Then, $T_{j2} = 75°C + 0.5 (1.75 \times 3.68)°C = 78.22°C$ and $I_{CO2} = 1$ μA $(2^{5.322}) = 40$ μA. Also $\Delta V_{BE} = (78.22 - (-40)) (2$ mV$) = 0.236$ V.

Substituting these values into Eq. 7.16 we have

$$R_B = \frac{((2.88 - .04) \times 10^{-3} \times 600 - 0.236) \text{ V}}{\dfrac{0.8 \text{ mA}}{.5 \times 20} - \dfrac{3.68 \text{ mA}}{1.1 \times 120} + .04 \text{ mA}} = \frac{1.458 \text{ V}}{.092 \text{ mA}} = 15.85 \text{ k}\Omega$$

The value of V_B must be determined from Eq. 7.8. The set of values determined at either the low temperature q_1-point or the high temperature q_2-point could be substituted into Eq. 7.8 but I_{CO2} must be included in Eq. 7.2 in order to determine I_{B2} if the values at q_2 are used. However, if the values at q_1 are used, I_{CO1} is negligible so we will use q_1. Then, $I_{B1} = I_{C1}/\beta_{F1} = .8$ mA$/10 = .08$ mA and $I_{E1} = -(I_{C1} + I_{B1}) = -.88$ mA. Also $V_{BE1} = .6$ V $+ (25° - (-40)) 2 \times 10^{-3} = .73$ V. Therefore

$$V_B = .08 \times 15.85 + 0.73 + .88 \times 0.6 = 2.526 \text{ V}.$$

From Eqs. 7.9 and 7.10

$$R_1 = 15\,(15.85\ \mathrm{k})/(15 - 2.526) = 19\ \mathrm{k\Omega}$$

$$R_2 = 15\,(15.85\ \mathrm{k})/2.526 = 94\ \mathrm{k\Omega}$$

The next smaller stock size resistors would be used. This completes the design since $r_i \simeq 300 + r_x$, which is small in comparison with R_B, and only about 5% of the signal current is lost at the low-temperature limit worst case. At the high-temperature, maximum β case, $r_\pi \simeq 7{,}500\ \Omega$ so about 33% of the signal current will be lost when the highest β transistor is at the maximum temperature. This is not bad because the gain of all units will be more nearly equal due to the reduction of the gain of the higher-gain units.

When Eq. 7.16 yields values of R_B which are too low, and therefore seriously shunt the input currents, either a larger value of R_E is needed or a larger value of ΔI_C should be permitted. Also, negative values of R_B may be obtained from Eq. 7.16. The causes and possible solutions to this problem follow:

1. The numerator may become negative because ΔV_{BE} is negative and $(\Delta I_C - \Delta I_{CO})\,R_E$ is either: (a) negative, or (b) positive, but smaller in magnitude than ΔV_{BE}. In case (a), ΔI_{CO} is larger than ΔI_C, and if ΔI_C cannot be increased considerably, a different transistor with lower ΔI_{CO} must be selected. In case (b), the same solutions as listed for case (a) are applicable. In addition, R_E may be increased.

2. The denominator may be negative because $I_{C1}/\beta_1 - I_{C2}/\beta_2$ is negative and has a greater magnitude than ΔI_{CO}. This indicates that for the given temperature range, you have chosen ΔI_C larger than can be obtained with any positive value of R_B. Decreasing ΔI_C will solve the problem. Also, if your value of ΔI_C gives positive but very large values of R_B so that V_B is greater than V_{CC}, R_1 will be negative. The solution to this problem is also to either reduce ΔI_C or arbitrarily reduce R_B to a practical value.

Single-battery bias may also be used in the CB configuration as shown in Fig. 7.8. The circuit components and q-point stability may be determined in precisely the same manner as for the CE circuit. However, R_E and not R_B shunts the input so R_B, and thus R_1 and R_2, may have comparatively low resistance. The lower limit of R_1 and R_2 is determined by their power dissipation capabilities and their permissible drain on the power supply V_{CC}. Therefore, the CB circuit may be very thermally stable. The capacitor C_B holds the base at signal-ground potential.

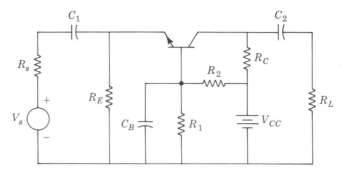

Fig. 7.8. Single battery bias for the CB circuit.

The stabilized bias circuit may be precisely the same for the CC amplifier as for the CE amplifier. The CC amplifier is inherently more thermally stable, however, because R_E is the dc load resistance and may be larger than the permissible value of R_E in the CE amplifier. The design procedure is the same, however. If the stability requirements are not too stringent, R_B may be large enough that $V_B = V_{CC}$ and R_1 is not required. Even then, the bias circuit may shunt much of the input current and greatly reduce the input resistance of the emitter follower.

7.5 BOOT-STRAPPING FOR THE CC AMPLIFIER

A technique which may be used to increase the input impedance of a CC amplifier and to greatly reduce the loading of the bias is known as *bootstrapping*. This word comes from the phrase "lifting one's self by his own bootstraps." The bootstrap circuit is shown in Fig. 7.9. In this circuit, the capacitor C_3 is a dc blocking capacitor and has negligible reactance at the lowest signal frequency. Therefore, the output voltage v_o is applied at the junction of resistors R_1 and R_2. These resistors provide forward bias as they do in the CE and CB amplifiers. However, the ac voltage across R_3 is the difference between the input voltage v_i and the output voltage v_o, or $v_{R3} = (v_i - v_o)$. Since v_o is nearly equal to v_i in an emitter follower, there is very little signal voltage across R_3 and, hence, very little signal current through R_3. This current can be determined by dividing the voltage by the resistance, or $i_{R3} = (v_i - v_o)/R_3$. The effective resistance of the bias circuit, as seen by the input voltage, is the ratio of the input voltage to the current i_{R3}. then

$$R_{eff} = \frac{v_i}{i_{R3}} = \frac{v_i}{(v_i - v_o)/R_3} = \frac{R_3 v_i}{v_i - v_o} \tag{7.21}$$

If both the numerator and the denominator of Eq. 7.21 are divided by v_i, and it is recognized that v_o/v_i is the voltage gain K_v, Eq. 7.21 may be rewritten as

$$R_{eff} = \frac{R_3}{1 - K_v} \tag{7.22}$$

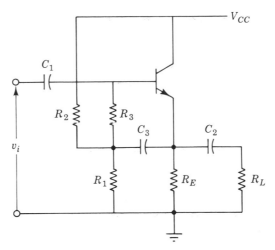

Fig. 7.9. A bootstrapped bias circuit.

For example, if the CC amplifier of Fig. 7.9 has $R_3 = 5$ kΩ and $K_v = 0.996$, the effective resistance of the bias circuit is $5 \times 10^3/.004 = 1.25 \times 10^6$ Ω.

This high effective resistance is encouraging for the attainment of a high input resistance. The fact that the resistance of R_1 and R_2 did not appear directly in the effective resistance formula may give the impression that the values of these resistors are unimportant. However, R_1 and R_2 are in parallel with R_{ac}, so far as the signal is concerned, and should be large in comparison with R_{ac}. Otherwise, the voltage gain will be reduced, with a resulting reduction of effective resistance.

If the $I_B R_3$ drop is not negligible in comparison with V_B then V_B should be increased sufficiently to compensate for the $I_B R_3$ drop. Otherwise, the bias design of the CC amplifier may proceed in the same manner as for the CE amplifier.

**7.6
THE DETERM-
INATION
OF COUPLING
AND BYPASS
CAPACITORS**

We learned in the preceding work that when a series RC circuit is driven, or excited, by a sinusoidal source of varying frequency, the current is reduced to 0.707 of its mid-frequency value and the lower half-power frequency f_ℓ is reached when $1/\omega_\ell C = R_{\text{series}}$. Then,

$$\omega_\ell = \frac{1}{CR_{\text{series}}} \tag{7.23}$$

Note that ω_ℓ is the reciprocal of the time constant of the series RC circuit. Therefore, the low-frequency cutoff is easily determined when a single series capacitor appears in a circuit. However, an RC coupled stage normally includes two or three capacitors as needed for coupling and bypass, as shown in the hybrid-π model of Fig. 7.10. This circuit is a

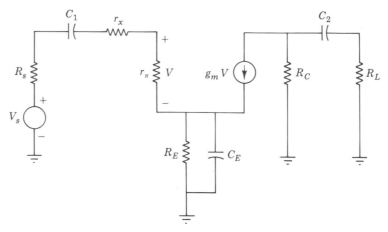

Fig. 7.10. Hybrid-π model for a CE amplifier including the coupling and bypass capacitors.

low-frequency model, as opposed to the mid-frequency or high-frequency models we have used in the past.

Let us first derive an expression for the low-frequency cutoff ω_ℓ for the amplifier, assuming that C_1 only determines ω_ℓ and both C_2 and C_E are very large and therefore have negligible reactance at ω_ℓ. Then, since R_E is perfectly bypassed, $R_{\text{series}} = R_s + r_x + r_\pi = R_s + h_{ie}$ and

$$\omega_{\ell 1} = \frac{1}{(R_s + h_{ie})\,C_1} \tag{7.24}$$

Similarly, if C_1 and C_E are both very large, the resistance in series with, or *facing*, C_2 is $R_C + R_L$ and

$$\omega_{\ell 2} = \frac{1}{(R_C + R_L)\,C_2} \tag{7.25}$$

However, if we assume that both C_1 and C_2 are very large and their reactances are negligible at ω_1, the resistance facing C_E is R_E in parallel with $(r_\pi + r_x + R_s)/(\beta_o + 1)$. This is the output resistance of the CC amplifier (Eq. 6.49). Therefore,

$$\omega_{\ell E} = \frac{1}{\left(R_E \middle\| \dfrac{R_s + h_{ie}}{\beta_o + 1}\right) C_E} \tag{7.26}$$

Since R_E is generally large in comparison with $(R_s + h_{ie})/(\beta_o + 1)$, $\omega_{\ell E} \simeq (\beta_o + 1)/(R_s + h_{ie})\,C_E$. A comparison of Eqs. 7.24, 7.25, and 7.26

reveals that for the same values of ω_ℓ, the capacitor C_E must be approximately $(\beta_o + 1)\,C_1$, and C_2 would be of the same order of magnitude as C_1.

A rigorous solution of the circuit of Fig. 7.10 using either loop or nodal equations would reveal the pole and zero locations of the entire circuit, from which the low frequency response could be predicted. However, this technique would be a tedious task and does not yield a direct approach to circuit design. Therefore, we are encouraged to find a simpler, more direct approach.

One simple but somewhat pessimistic method is to use the approximation that the total time constant τ_T of the circuit is the reciprocal of the sum of the reciprocals of the individual time constants considered separately as above. This amounts to adding the individual half power frequencies caused by the individual capacitors taken separately. This approach leads to a viable design procedure where each capacitance could be calculated to produce individually the desired ω_ℓ for the circuit. Then, each capacitance would be multiplied by the number of capacitors in the circuit to obtain the required capacitance for each capacitor. For example, the capacitances that would be obtained from Eqs. 7.24, 7.25, and 7.26 would each be multiplied by 3 to obtain their required capacitance. However, this technique would result in a physically large and costly capacitance for C_E.

A better design approach for the CE amplifier is to make the capacitances of C_1 and C_2 large enough that their reactances are negligible at ω_ℓ, and let C_E determine ω_ℓ as we assumed it would in writing Eq. 7.26. This method may be achieved by making the capacitances of C_1 and C_2 ten times as large as the relationships of Eqs. 7.24 and 7.25 would normally yield. Then we may write the design equations for the CE amplifier, using Eqs. 7.24, 7.25, and 7.26.

$$C_1 = \frac{10}{(R_s + h_{ie})\,\omega_\ell} \tag{7.27}$$

$$C_2 = \frac{10}{(R_C + R_L)\,\omega_\ell} \tag{7.28}$$

$$C_E = \frac{1}{\left(R_E \middle\| \dfrac{R_s + h_{ie}}{\beta_o - 1}\right)\omega_\ell} \tag{7.29}$$

Generally this combination yields the minimum total size and cost for the CE amplifier. The foregoing principles may be used to determine the coupling and bypass capacitances for the other configurations. A typical frequency-response curve for a high-quality audio-frequency (af) amplifier is given in Fig. 7.11.

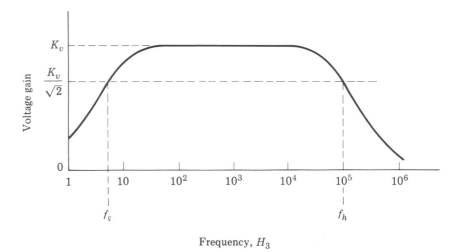

Fig. 7.11. Typical frequency-response curve for a high-quality audio-frequency amplifier.

7.7
TIME
RESPONSE

In the preceding sections the input voltage was assumed to be a steady state sinusoid and the frequency response of the amplifier was of primary concern. However, when the ultimate load is a visual display rather than an audio device, the *time response* is of primary concern. For example, the intensity of light on a TV picture tube must be capable of changing from black to white in less than 0.1 μs in order to provide good picture resolution. In other words, the eye sees time response and the ear hears frequency response.

In order to test the response time of an amplifier circuit, we will apply a step voltage V to the input and determine the effect of the shunt capacitances on the output waveform. Let us use the CE hybrid-π model given in Fig. 7.12. The capacitance $C_{sh} = C_\pi + C_\mu (1 + g_m R_{sh})$. Since V_i is a function of time (a step voltage), we must use either differential equa-

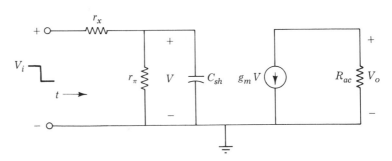

Fig. 7.12. CE hybrid-π model with step-input voltage V_s.

tions or Laplace Transform techniques to obtain V_i and V_o as functions of time. Since the Laplace transform method is easier and more direct, we will use it and write $V_o(s)$ in terms of $V_i(s)$, as though the input voltage were sinusoidal. However, in this case, $j\omega$ is replaced by s. Then, from Fig. 7.12,

$$V(s) = \frac{V_i(s)\left(\dfrac{1}{g_\pi + sC_{sh}}\right)}{r_x + \dfrac{1}{g_\pi + sC_{sh}}} = \frac{V_i(s)}{r_x(g_\pi + sC_{sh}) + 1} \tag{7.30}$$

$$V_o(s) = -\frac{g_m V_i(s) R_{ac}}{r_x(g_\pi + sC_{sh}) + 1} = -\frac{g_m V_i(s) R_{ac}}{C_{sh} r_x\left(s + \dfrac{g_\pi + g_x}{C_{sh}}\right)} \tag{7.31}$$

The voltage transfer function $G_v(s)$ is defined as

$$G_v(s) = \frac{V_o(s)}{V_i(s)} = -\frac{g_m R_{ac}}{C_{sh} r_x\left(s + \dfrac{g_\pi + g_x}{C_{sh}}\right)} \tag{7.32}$$

We may obtain an expression for $G_v(s)$ in terms of the mid-frequency voltage gain K_v, since from Eq. 6.16,

$$K_v = -\frac{g_m r_\pi R_{ac}}{r_x + r_\pi} \tag{7.33}$$

Then,

$$g_m R_{ac} = -\frac{K_v(r_x + r_\pi)}{r_\pi} \tag{7.34}$$

Substituting Eqs. 7.34 into Eq. 7.32

$$G_v = \frac{K_v(r_x + r_\pi)}{r_\pi r_x C_{sh}\left(s + \dfrac{g_\pi + g_x}{C_{sh}}\right)} = \frac{K_v}{\dfrac{r_\pi r_x}{r_\pi + r_x} C_{sh}\left(s + \dfrac{g_\pi + g_x}{C_{sh}}\right)} \tag{7.35}$$

From the relationship of Eq. 6.20 we may see that $\omega_h = 1/R_{sh}C_{sh} = 1/\tau_h$ and write Eq. 7.35 as

$$G_v = K_v \frac{\omega_h}{s + \omega_h} \tag{7.36}$$

Then the s-domain output voltage $V_o(s) = G_v V_i(s)$. If V_i is assumed to be a negative step voltage, $V_i(s) = -V/s$, where V is the magnitude of the voltage step. Then,

$$V_o(s) = -\frac{K_v V \omega_h}{s(s + \omega_h)} \tag{7.37}$$

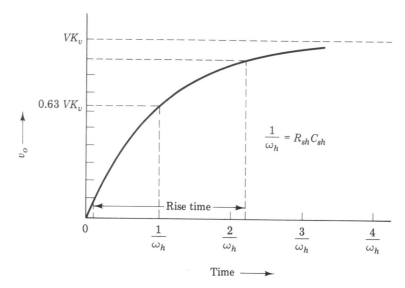

Fig. 7.13. Plot of v_o as a function of time when a negative step voltage input is applied to an amplifier whose upper half-power frequency is ω_h.

Equation 7.37 may be expanded by partial fractions to obtain

$$V_o(s) = -K_v V \left(\frac{1}{s} - \frac{1}{s + \omega_h}\right) \tag{7.38}$$

The inverse transform of Eq. 7.38 is

$$v_o(t) = -K_v V \left(1 - e^{-\omega_h t}\right) \tag{7.39}$$

The plot of v_o as a function of time is given in Fig. 7.13.

The *rise time* of an amplifier is defined as the time required for the output voltage to rise from 10 percent to 90 percent of its final value when a step voltage is applied to the input. Fig. 7.13 shows that the rise time t_r is about 2.2 time constants, or

$$t_r = \frac{2.2}{\omega_h} = 2.2 \, R_{sh} C_{sh} \tag{7.40}$$

When a dc voltage is suddenly applied to an RC coupled amplifier, the output voltage will rise quickly, as discussed above, but must eventually return to zero because of the blocking capacitors. Let us now consider the effect of the coupling capacitance on the time response of the amplifier. For simplicity, let us assume that the input coupling capacitance is the only significant capacitance in the CE model given in

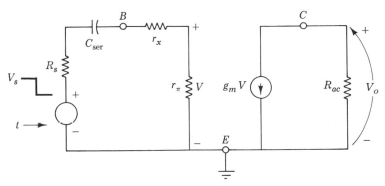

Fig. 7.14. Hybrid-π model for determining the time response resulting from an input blocking capacitor in a CE amplifier.

Fig. 7.14. Then, using s in place of $j\omega$ to obtain an expression for V_o in terms of V_i,

$$V(s) = \frac{V_i(s)\, r_\pi}{R_s + r_x + r_\pi + \dfrac{1}{sC_{ser}}} = \frac{r_\pi C_{ser}\, sV_i(s)}{sC_{ser}(R_s + r_x + r_\pi) + 1} \tag{7.41}$$

The output voltage is

$$V_o(s) = -\frac{r_\pi C_{ser}\, sV_i(s)\, g_m R_{ac}}{sC_{ser}(R_s + r_x + r_\pi) + 1} \tag{7.42}$$

The transfer function is

$$G(s) = \frac{V_o(s)}{V_i(s)} = -\frac{r_\pi C_{ser}\, s g_m R_{ac}}{sC_{ser}(R_s + r_x + r_\pi) + 1} \tag{7.43}$$

Since the mid-frequency transducer gain is

$$K_v = -\frac{V_o}{V_s} = -\frac{r_\pi g_m R_{ac}}{R_s + r_x + r_\pi} \tag{7.44}$$

We may substitute $(R_s + r_x + r_\pi)K_v = -r_\pi g_m R_{ac}$ in Eq. 7.43 to obtain

$$G(s) = \frac{K_v C_{ser}\, s}{sC_{ser} + \dfrac{1}{R_s + r_x + r_\pi}} = \frac{K_v\, s}{s + \dfrac{1}{R_{ser} C_{ser}}} \tag{7.45}$$

where $R_{ser} = R_s + r_x + r_\pi$. Then, if $V_s(s) = -V/s$

$$V_o = G(s)\, V_s(s) = -\frac{K_v V}{s + \omega_\ell} \tag{7.46}$$

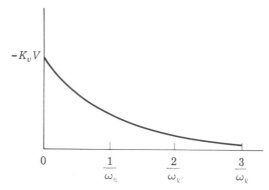

Fig. 7.15. Sketch of output voltage as a function of time if the input signal is a negative step voltage.

where $\omega_\ell = 1/R_{ser} C_{ser}$ as we found in Eq. 7.23. The inverse transform of Eq. 7.46 is

$$v_o(t) = -VK_v e^{-\omega_\ell t} \qquad (7.47)$$

The output voltage v_o is plotted as a function of time in Fig. 7.15. The signal steps (at time $t = 0$) from zero to $-VK_v$ volts as expected. However, since the coupling capacitor blocks the dc, the output voltage exponentially drops to zero with a time constant $\tau_1 = 1/\omega_1$. Of course, the output signal is inverted with respect to the input signal because we have assumed that the common-emitter configuration, with its inherent polarity reversal, is used.

Although the RC-coupled amplifier is not suitable for amplifying dc signals, the information gained from the step function analysis is very useful in predicting the behavior of the amplifier when it is called on to amplify rectangular pulses. A series of rectangular pulses is shown in Fig. 7.16. These pulses have amplitude V and pulse duration d. We may express the time-domain input voltage as the sum of a series of step voltages.

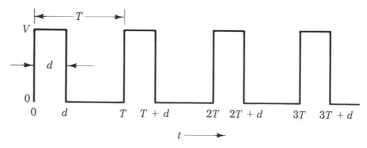

Fig. 7.16. A series of rectangular pulses.

$$v_i = V[u(t) - u(t - d) + u(t - T) - u(t - (T + d)) + u(t - 2T)\dots] \quad (7.48)$$

The $u(t - a)$ symbolizes a step voltage that occurs at $t = a$ and so on. The Laplace transform of this excitation voltage is

$$V_i = V\left(\frac{1}{s} - \frac{e^{-ds}}{s} + \frac{e^{-Ts}}{s} = \frac{e^{-(T+d)s}}{s} + \frac{e^{-2Ts}}{s}\cdots\right) \quad (7.49)$$

The s domain output voltage is $V_o = V_i G_v$. Therefore,

$$V_o = -VK_v\left(\frac{1}{s + \omega_1} - \frac{e^{-ds}}{s + \omega_1} + \frac{e^{-Ts}}{s + \omega_1} - \frac{e^{-(T+d)s}}{s + \omega_1} + \frac{e^{-2Ts}}{s + \omega_1}\cdots\right) \quad (7.50)$$

The time-domain response is obtained from the inverse transform[1]

$$v_o = VK_v(e^{-\omega_1 t} - u(t - d)e^{-\omega_1(t-d)} + u(t - T)e^{-\omega_1(t-T)}$$
$$- u(t - T - d)e^{-\omega_1(t-T-d)} + u(t - 2T)e^{-\omega_1(t-2T)}\dots) \quad (7.51)$$

The time response may be plotted by adding the exponential terms of Eq. 7.51, as illustrated in Fig. 7.17. Since the average steady-state current through the coupling capacitor must be zero, the cross-hatched area labeled A below the reference axis must approach the corresponding area labeled B above the axis after several periods T have elapsed.

Ideally, the output voltage should remain constant during the pulse. However, as shown, the output current and voltage decay exponentially because of the charging of the coupling capacitor. The departure of the actual output from the ideal flat response at the termination of the pulse is called *sag*. From a consideration of the first pulse as shown in Fig. 7.17, the sag can be determined by solving Eq. 7.51 at $t = d$. Then

$$\text{Sag} = VK_v - VK_v e^{-\omega_1 d} = K_v V(1 - e^{-\omega_1 d}) \quad (7.52)$$

The fractional sag, or ratio of sag to the output at the beginning of the pulse, is of greater significance than the amount of sag.

$$\text{Fractional sag} = \frac{\text{Sag}}{K_v V} = 1 - e^{-\omega_1 d} \quad (7.53)$$

If the pulse duration d is small in comparison with $1/\omega_1 = (R_o + R_i)C_C$, the fractional sag can be determined more easily than by the use of Eq. 7.53. The initial slope of the exponential decay is, using Eq. 7.51

$$\text{Initial slope} = \frac{d(-K_v V e^{-\omega_1 t})}{dt}\bigg|_{t=0} = -K_v V \omega_1 \quad (7.54)$$

---

[1] This time transformation is treated in *Network Analysis*, Second Edition by M. E. Van Valkenburg, Prentice-Hall, Englewood Cliffs, New Jersey, 1964.

Function   Plot

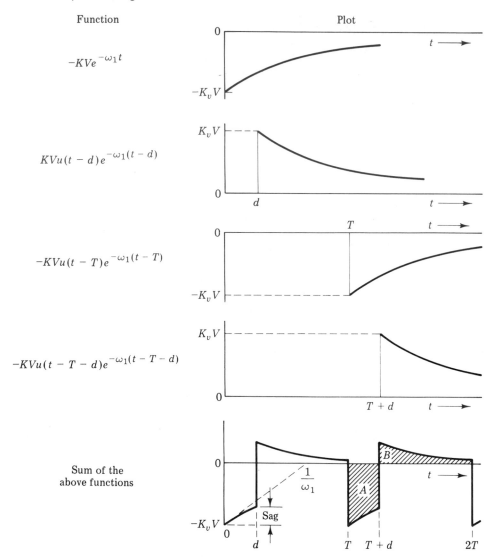

Fig. 7.17. Time response of an RC-coupled amplifier to a periodic rectangular pulse.

The exponential sag closely follows the initial slope as long as the pulse duration is small (0.1 or less) in comparison with the time constant $1/\omega_1$. The total sag is approximately equal to the pulse duration times the magnitude of the initial slope

$$\text{Sag} \simeq K_v V \omega_1 d \tag{7.55}$$

and

$$\text{Fractional sag} \simeq \omega_1 d \tag{7.56}$$

Equation 7.55 points up the desirability of a small value of ω_1 when rectangular pulses are to be amplified.

Figure 7.17 shows that the first pulse after $t = 0$ has greater sag than the succeeding pulses. The sag decreases somewhat as the transient dies out. However, the fractional sag remains constant if the pulse amplitude is measured from the base line.

We may conclude from the foregoing relationships that both the time response and the frequency response may be obtained from either a frequency response measurement or a time response measurement.

Problems   *Section 7.1*

7.1   A given microphone has an internal resistance of 500 Ω. The signal-to-noise ratio in the output of the microphone is 55 dB when it is picking up a typical level of program material. As a designer of an amplifier for the microphone, what value of q-point collector current I_C would you use for the 2N 2443 transistor that has been chosen for the amplifier? What SNR ratio would you expect in the output of the amplifier?

Section 7.2

7.2   A mass-produced amplifier is guaranteed to operate over the ambient temperature range $-65°$ to $125°C$. If you have chosen the transistor with characteristics given in Figs. 7.2 and 7.3 for the amplifier, what are the maximum and minimum values of h_{FE} and f_T expected over this temperature range if I_C is allowed to vary from 1 mA to 4 mA?

Section 7.3

7.3   Design a stabilized bias circuit for an amplifier intended for indoor use if the $h_{FE} = 100$ at the selected $I_C = 2$ mA and $V_{CC} = 20$ V.

7.4   What fraction of the input signal current will be lost in the bias circuit of Problem 7.3 if $r_x = 200$ Ω in the transistor?

Section 7.4

7.5   Show that Eq. 7.16 is obtained from Eq. 7.15.

7.6   A given transistor has a maximum power dissipation capability of 400 mW at 25°C ambient temperature and a derating factor of 3 mW/°C. Determine the transistor junction temperature when $V_{CE} = 8$ V, $I_C = 5$ mA and the ambient temperature is 50°C. What is T_{jmax} for this transistor?

7.7   Assume that you have been assigned the task of designing an audio-frequency amplifier for an automobile radio that must operate over the ambient temperature range $-60°C$ to $60°C$. The silicon transistor you have chosen has the following characteristics:

h_{FE} min at $-60°C = 40$; h_{FE} max at $60°C = 250$.
I_{CO} (max) $= 0.1$ µA at 25°C.
Maximum power dissipation $= 400$ mW at $T_a = 25°C$.
Derating factor $= 3$ mW/°C.

The maximum output signal amplitude is 0.5 V into a load resistance $R_L = 2$ kΩ. The nominal power source is 13 V. Design the amplifier except for capacitor values. Make reasonable assumptions and approximations when needed.

Section 7.5

7.8   A transistor having h_{fe} (min) $= 100$, $h_{ie} = 1.5$ k, $h_{oe} = 10^{-5}$ S and $h_{re} = 10^{-4}$ at $I_C = 2$ mA and $V_{CE} = 10$ V is used as an emitter follower with $R_E = 5$ kΩ and $R_L = 10$ kΩ. The maximum permissible base-circuit resistance $R_B = 50$ kΩ because of thermal stability requirements. Design a bootstrapping circuit that will provide high (near maximum) input resistance to ac signals. Determine the input resistance of your circuit. What is a suitable value for the capacitor C_3 (Fig. 7.9) if the lowest frequency of interest is 20 Hz?

Section 7.6

7.9   A given CE amplifier uses a transistor having $h_{ie} = 2700$ Ω, $h_{fe} = 200$ and $h_{oe} = 2 \times 10^{-5}$ S at the q-point $I_C = 2$ ma, $V_{CE} = 10$ V. If $R_s = 3$ kΩ, $R_B = 35$ kΩ, $R_C = 4.7$ kΩ, $R_L = 6$ kΩ and $R_E = 1$ kΩ. Determine suitable values for the input and output dc blocking capacitors and the emitter bypass capacitor for $f_\ell = 20$ Hz.

7.10   Two amplifier stages, each identical to the amplifier of Problem 7.9, are connected in cascade. What are suitable values for the coupling and bypass capacitors if $f_\ell = 20$ Hz for the two-stage amplifier. Of course, R_L of the input amplifier is R_i of the second, or output, amplifier.

7.11 Determine suitable values for the input and output dc blocking (or coupling) capacitors for the emitter follower circuit of Problem 7.8.

Section 7.7

7.12 A 50 Hz square wave is applied to the input of an amplifier that has $f_\ell = 20$ Hz and $f_h = 100$ kHz.

  a. Sketch the output voltage as a function of time for 60 ms, using an appropriate time scale.
  b. Sketch the output voltage as a function of time for 10 μs, using an appropriate time scale.
  c. What is the rise time of the amplifier?

7.13 What must be the bandwidth of an amplifier if the rise time is to be no greater than 1 μs?

7.14 What must be the low-frequency cutoff f_ℓ if the sag is to be 10% when a 50 Hz square wave is applied to the input?

8

Field-Effect Transistors

The *field-effect transistor* (FET) is a voltage-controlled semiconductor that has a very high input impedance, particularly at low frequencies where the junction capacitance is ineffective in shunting the signals. The FET has only one p-n junction and is therefore sometimes known as a *unipolar* field-effect transistor (UNIFET). In order to unambiguously distinguish the two-junction transistor, with which we are familiar, from the FET, the two-junction transistor is often known as a bipolar junction transistor (BJT). In the preceding work, however, we have followed the traditional custom of referring to the BJT simply as a transistor, since the bipolar type was the only transistor in the field for many years.

Both the FET and the BJT have their specific advantages. The FET is especially adaptable to high-density, low-power-consumption digital integrated circuits which are used in digital watches, hand held calculators, and for many other applications. The FET may also provide lower noise and better linearity in amplifier applications than the BJT. On the other hand, the BJT provides higher gain and wider bandwidths and therefore is used extensively in amplifiers and high-speed digital circuits.

There are several types of FETs. We will begin with the junction FET (JFET) because it is most like the familiar BJT.

8.1
THE JUNCTION
FET

A typical JFET structure is given in Fig. 8.1. The schematic representation in Fig. 8.2 will be used to explain the principles of operation of the junction FET. Figure 8.2a shows that a narrow semiconductor channel

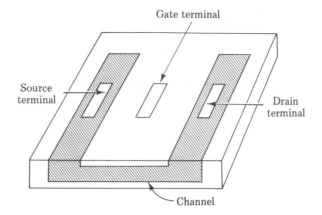

Fig. 8.1. *Typical field-effect transistor structure.*

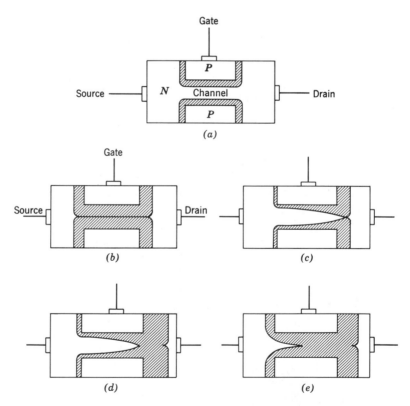

Fig. 8.2. *Schematic representation of the junction FET: (a)* $v_{GS} = 0$, $v_{DS} = 0$;
(b) $v_{GS} = V_p$ *(Pinch off)*, $v_{DS} = 0$; *(c)* $v_{GS} = 0$, $v_{DS} = V_p$; *(d)* $v_{GS} = 0$, $v_{DS} > V_p$;
(e) $v_{GS} < 0$, $v_{DS} > V_p$.

provides a conducting path between the *source* and the *drain*. This channel may be either an *n*- or *p*-type crystal. The *n*-type is used in this discussion. With no biases applied to the transistor, the channel conductance $G_c = \sigma (wt / \ell)$, where σ is the conductivity of the crystal and w, t, and ℓ are the width, thickness, and length of the channel, respectively. For example, if $\sigma = 100$ S/m, $w = 10^{-3}$ m, $t = 10^{-6}$ m, and $\ell = 10^{-5}$ m, the channel $G_c = .01$ S and the channel resistance $r_c = 100\ \Omega$.

If reverse bias is applied between the gate and the source, the depletion region width is increased, the thickness of the channel is decreased, and therefore the conductivity of the channel is decreased. The gate bias required to just reduce the channel thickness to zero, as shown in Fig. 8.2b, is called the *pinch-off* voltage, V_p.

When the gate-source voltage v_{GS} is zero and the drain is made positive with respect to the source, electrons drift through the channel because of the electric field. The drain current i_D is equal to the drain-source voltage v_{DS} times the channel conductance G_c, providing v_{DS} is very small. However, the positive drain voltage reverse biases the *p-n* junction near the drain end of the channel, and when the drain voltage is increased to the pinch-off voltage, the channel thickness is reduced to zero at a point near the drain end of the channel, as shown in Fig. 8.2c. The drain current does not stop when the drain voltage reaches pinch-off because a voltage equal to V_p still exists between the pinch-off point and the source, and the resulting electric field along the channel causes the free carriers in the channel to drift from the source to the drain.

As the drain voltage is increased above V_p, the depletion region thickness is increased between the drain and the gate, as shown in Fig. 8.2d. In fact, the additional drain voltage is absorbed by the increased field in the wider pinched-off region, and the electric field between the original pinch-off point and the source remains essentially unchanged. Therefore, the channel current, and hence the drain current, remains essentially unchanged. The carriers which arrive at the pinch-off point are swept through the depletion region in the same manner that carriers which arrive at the collector junction are swept from the base into the collector region in a conventional or *bipolar* transistor. Thus, whenever v_{DS} is higher than the pinch-off voltage V_p, the drain current is essentially independent of the drain voltage.

The field-effect transistor normally operates with the drain voltage v_{DS} beyond the pinch-off voltage V_p and with reverse bias applied between the gate and the source. The electric field, and thus the drain current in the channel, is then controlled by the gate voltage v_{GS}. The effect of the gate voltage v_{GS} on the channel conductance is shown in Fig. 8.2e. The channel thickness is reduced as a result of the reverse gate bias. The drain current is essentially independent of the drain voltage

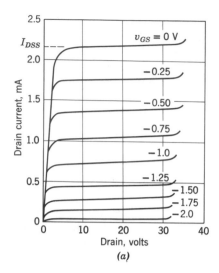

Fig. 8.3. Drain characteristics of a typical n-channel FET.

whenever the sum of the magnitudes of the drain voltage and the reverse-bias gate voltage exceeds the pinch-off voltage.

The drain characteristics of a typical n-channel FET are shown in Fig. 8.3. The drain current which flows when $v_{GS} = 0$ and $v_{DS} = V_p$ is known as I_{DSS}, which means saturated drain current with input shorted. The transistor of Fig. 8.3 has $I_{DSS} = 2.2$ mA. Observe that the slope of the drain characteristic curves are quite flat in the normal operating range. This means the output resistance (equivalent to $1/h_{oe}$ in a bipolar transistor) is very high. Avalanche breakdown occurs at the junction whenever the drain-gate voltage exceeds a given value.

The input characteristics of a typical n-channel FET are shown in Fig. 8.4. The gate current is the saturation current of the reverse-biased junction. This current is of the order of 10^{-9} A or 1 nanoamp for a low-power silicon FET at 25°C. The dynamic input conductance g_g is the slope of the input characteristics. The dynamic input conductance of the FET of Fig. 8.4 is $\Delta i_G / \Delta v_{GS} = 1$ nA/5 V $= 4 \times 10^{-10}$ S, and the dynamic or incremental input resistance is 2.5×10^9 Ω when the junction is reverse biased. Although the input resistance decreases rapidly as the junction becomes forward biased, Fig. 8.4a shows that it remains quite high (a megohm or more) in a silicon FET so long as the forward bias does not exceed about 0.25 V at 25°C. Therefore, $v_{GS} = 0$ may be a suitable q-point for a small-signal amplifier. Note that the input resistance drops abruptly when avalanche breakdown occurs. Avalanche breakdown occurs whenever the algebraic difference between the gate and drain potentials exceed the avalanche breakdown voltage.

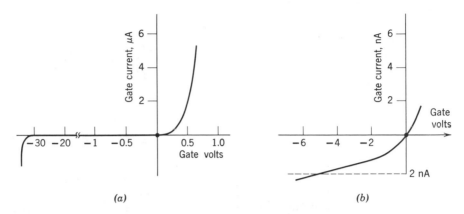

Fig. 8.4. Input characteristics of a typical n-channel FET: (a) expanded
voltage scale; (b) expanded current scale.

The symbols for both n-channel and p-channel JFETs are given in
Fig. 8.5a. The use of the familiar diode symbol is evident with the
extra-long bar representing the channel. The arrow which indicates the
direction of forward-bias current flow is sometimes located near the
source end of the channel to identify that terminal, although some FETs
perform with drain and source terminals interchanged.

The JFET is often used as a current source to provide an essentially-
constant current to a load with varying voltage. This is accomplished by
shorting the gate to the source, as shown in Fig. 8.5b. Then the current
through this *diode current regulator* or *current source* is essentially
equal to I_{DSS} so long as v_{DS} is greater than the pinch-off voltage V_p. This
device has many uses. One common use is as the dc load resistance in

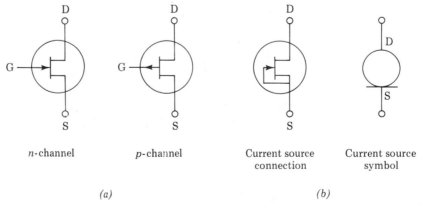

Fig. 8.5. (a) JFET transistor symbols; (b) current source connection
and symbol.

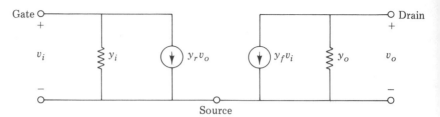

Fig. 8.6. Y-parameter equivalent circuit of a FET.

a transistor amplifier, because it may allow relatively large q-point currents that provide high values of g_m, and also since it may provide very large values of load resistance which result in very high gain.

**8.2
AN FET
CIRCUIT
MODEL**

In order to analyze the FET amplifier, we need to develop a circuit model for the FET device. The basic y-parameter circuit shown in Fig. 8.6 is the most useful configuration. At low frequencies, the depletion-region capacitances can be neglected and the admittances become conductances. The admittance y_i is the gate-to-source conductance and is normally given the symbol g_g for *gate input conductance*. The gate conductance is found from the relationship

$$g_g = \frac{\Delta i_G}{\Delta v_{GS}}\bigg|_{V_{DS}} \tag{8.1}$$

We have already noted this conductance is very low (about 10^{-9} S or so). In fact, for many applications, g_g can be considered essentially equal to zero, which simplifies the calculations a bit.

The parameter y_r is given by the relationship

$$y_r = \frac{\Delta i_G}{\Delta v_{DS}}\bigg|_{V_{GS}} \tag{8.2}$$

However, at low frequencies, the drain-source voltage has essentially no effect on the gate current, so y_r is normally assumed to be zero.

The parameter y_f is called the *forward transconductance* or *mutual conductance* of the transistor and is normally given the symbol g_m. The mutual conductance is found from the relationship

$$g_m = \frac{\Delta i_D}{\Delta v_{GS}}\bigg|_{V_{DS}} \tag{8.3}$$

If the transfer characteristics of the transistor are provided, as in Fig. 8.7, the value of g_m is the slope of this transfer curve at the required q point. In addition, g_m can be obtained with less accuracy from the drain char-

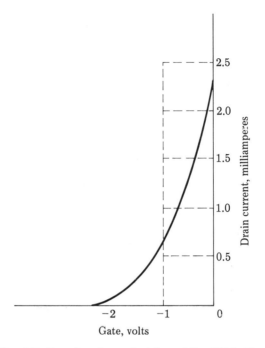

Fig. 8.7. Transfer characteristics of the FET in Fig. 8.3.

acteristics of Fig. 8.3 by using the same technique as was used to obtain h_{fe} or h_{re} in the bipolar transistors.

Finally, the parameter y_o is the output or *drain admittance* with the input terminals shorted. This drain admittance is usually given the symbol g_d and is found from the following relationship.

$$g_d = \frac{\Delta i_D}{\Delta v_{DS}}\bigg|V_{GS} \qquad (8.4)$$

Therefore, g_d is the slope of the drain characteristic (Fig. 8.3) at the desired q-point. From this analysis, a low-frequency equivalent circuit for the FET may have the form shown in Fig. 8.8.

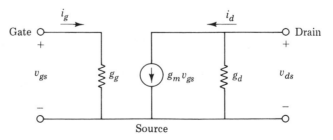

Fig. 8.8. Equivalent circuit for a FET.

A typical amplifier configuration (corresponding to the common-emitter configuration of a bipolar, n-p-n or p-n-p transistor) is shown in Fig. 8.9a. In Fig. 8.9b, the FET of Fig. 8.9 has been replaced by its circuit model. We have assumed C_S is effectively a short circuit for the ac signals in developing the circuit model. Now if C_{C1} is sufficiently large, $v_{gs} \simeq v_i$, and if C_{C2} is also large, $v_d \simeq v_o$. Then, if $G_D = 1/R_D$,

$$v_o = \frac{-g_m v_i}{g_d + G_D + G_L} \tag{8.5}$$

The voltage gain is

$$K_v = \frac{v_o}{v_i} = -\frac{g_m}{g_d + G_D + G_L} \tag{8.6}$$

In many amplifiers, r_d is much greater than R_{ac}. Then, letting $R_{ac} = 1/(G_D + G_L)$, the voltage gain is

$$K_v \simeq -g_m R_{ac} \tag{8.7}$$

Of course, Eqs. 8.6 and 8.7 are only valid if the transistor operates on the linear portion of its characteristic curves.

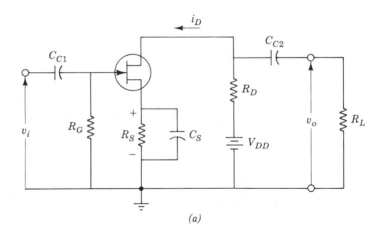

(a)

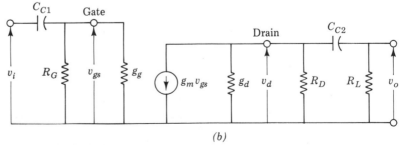

(b)

Fig. 8.9. FET amplifier circuit: (a) actual circuit; (b) equivalent circuit model.

Since y_r in Fig. 8.5 is zero, there is no feedback from the output to the input loop in the equivalent circuit. (This condition is similar to $h_{re} = 0$ for the bipolar transistor.) Then, the input impedance of the transistor is $R_{it} = 1/g_g = r_g$, and the output impedance of the transistor is $R_{ot} = 1/g_d = r_d$. Since the signal input current to a FET is so small, the current gain is very high but of little importance. However, the current gain can be readily found from the circuit model if it is required.

The resistor R_G in Fig. 8.9a permits the saturation current, which flows across the junction, to flow to ground and to the negative terminal of V_{DD}. Thus, the gate is maintained at approximately ground potential, for dc, which is negative with respect to the source. Usually, no more than 0.1 V or so should be dropped across R_B. If bias at $v_{GS} = 0$ is desired, the source would be returned to ground. However, as already noted, the gate signal should be restricted to a few tenths of a volt in magnitude if $V_{GS} = 0$.

Reverse bias is usually desired for the gate of the FET with respect to the source. Under these conditions, the resistor R_S of Fig. 8.9 is used to obtain the proper value of bias potentials for the FET. The drain current then flows through this resistor in the direction which will provide reverse bias. If the gate circuit resistance R_G is low enough so that the drain-gate saturation current I_{DO} produces negligible voltage across R_G, the gate is maintained at dc ground potential and the bias voltage is $I_D R_S$. Thus, for a desired gate bias voltage V_{GS}, R_S can be obtained from the relationship

$$R_S = \frac{V_{GS}}{I_D} \tag{8.8}$$

A bypass capacitor C_S across R_S will prevent degeneration and loss of gain for ac signals. This capacitor must bypass the impedance looking in at the FET source terminal in parallel with R_S. This source impedance may be obtained by making a small change in the source voltage Δv_S and noting the change in source current Δi_S, as shown in Fig. 8.10. However, the gate is maintained at ground potential, so the source-to-ground voltage v_S is equal to the source-to-gate voltage v_{SG}. Also, $\Delta i_S = \Delta i_D = g_m v_{GS}$, providing g_d is small compared with G_{ac}, which is the usual situation. Therefore, the resistance into the source is

$$r_s = \frac{\Delta v_S}{\Delta i_S} = \frac{\Delta v_{GS}}{\Delta i_D} = \frac{1}{g_m} \tag{8.9}$$

If the reactance of the bypass capacitor C_S is to be equal to the magnitude of the resistance being bypassed at some low-frequency cutoff ω_ℓ, then

$$\frac{1}{\omega_\ell C_S} = \frac{1}{(g_m + G_S)} \tag{8.10}$$

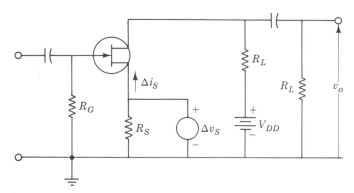

Fig. 8.10. Configuration used to determine the impedance into the source of a FET.

where $G_S = 1/R_S$, and

$$C_S = \frac{g_m + G_S}{\omega_\ell} \tag{8.11}$$

The drain current in a FET varies approximately as the square of the gate-source voltage and can be expressed in the pinch-off (saturation) region by the following relationship:

$$i_D \simeq I_{DSS}\left(1 - \frac{v_{GS}}{V_p}\right)^2 \tag{8.12}$$

Since $g_m = di_D/dv_{GS}$, g_m can be obtained by differentiating Eq. 8.12 to obtain

$$g_m \simeq 2I_{DSS}\left(1 - \frac{v_{GS}}{V_p}\right)\left(-\frac{1}{V_p}\right) \tag{8.13}$$

$$g_m \simeq -\frac{2I_{DSS}}{V_P}\left(1 - \frac{v_{GS}}{V_p}\right) \tag{8.14}$$

The value of g_m obtained at zero bias, or $v_{GS} = 0$, is usually given as g_{mo} and can be readily obtained from Eq. 8.14 as

$$g_{mo} = -\frac{2I_{DSS}}{V_P} \tag{8.15}$$

Observe that g_m or g_{mo} will always be positive, since either V_P or I_{DSS} will be a negative quantity. Substituting Eq. 8.15 into Eq. 8.14,

$$g_m \simeq g_{mo}\left(1 - \frac{v_{GS}}{V_P}\right) \tag{8.16}$$

Thus, g_m can be obtained analytically from the data normally given in the manufacturer's data sheets.

The variation of drain current with temperature in a field-effect transistor is determined by two factors. One factor is the temperature variation of depletion-region width, as discussed in Chapter 3, which results from the temperature variation of barrier height $(V_{ho} - V)$, where V_{ho} is the zero-bias barrier height. As previously discussed, the temperature coefficient of this voltage is about -2.0 mV/°C, which results in an increased drain current with increased temperature. The other factor is the variation of majority carrier mobility with temperature. This mobility influences the transconductance g_m. As the temperature increases, the carrier mobility, and hence g_m, decreases and tends to compensate for the variation of V_{ho} with temperature. In fact, the proper choice of q-point will give essentially zero temperature coefficient of drain current from about -50 to 100°C. The temperature coefficient due to mobility change is about 0.7 percent/°C. Therefore, the condition for zero temperature coefficient is

$$0.007\,(-i_D)/°C = g_m\,(-0.002)/°C \tag{8.17}$$

$$\frac{i_D}{g_m} = +0.286 \text{ V} \tag{8.18}$$

Substituting the expression for i_D (Eq. 8.12) and g_m (Eq. 8.13) into Eq. 8.18, the following relationship results:

$$-\frac{V_P}{2}\left(1 - \frac{v_{GS}}{V_P}\right) = +0.286 \text{ V} \tag{8.19}$$

or

$$v_{GS} = V_P + 0.572 \text{ V} \tag{8.20}$$

Equation 8.20 shows that zero thermal drift may be achieved if the FET is biased 0.572 V above the pinch-off voltage V_P. Note that V_P and v_{GS} are negative for an n-channel FET. All signs should be reversed for a p-channel FET. The constant in Eq. 8.20 may vary somewhat from FET to FET.

8.3
THE
DEPLETION-
MODE MOSFET

If extremely high-input resistance is desired, insulated gate FETs (IGFETs) may be used. These devices are more commonly known as metal-oxide semiconductor (MOS) field-effect transistors (FET). Usually, they are referred to simply as MOSFETs. There are two basic types of MOSFET's: *depletion mode* and *enhancement mode*.

A depletion-mode MOSFET is one in which an applied voltage v_{DS} causes an appreciable current flow with $v_{GS} = 0$. The JFET is also a

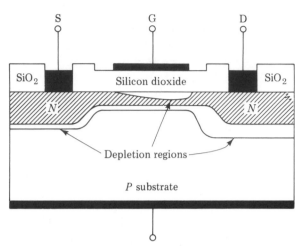

Fig. 8.11. Construction features of a depletion-mode n-channel MOSFET.

depletion-mode device, so we will consider the depletion-mode MOSFET first because of its similarity to the JFET. The physical construction of a typical depletion-mode MOSFET is illustrated in Fig. 8.11. The channel between the source and the drain is thin and lightly n-doped. The silicon dioxide insulating layer between the channel and the metal gate electrode, usually aluminum, is very thin, so a potential difference, or voltage, between the gate electrode and the channel causes a rather strong electric field in the SiO_2 layer. This field causes the negative carriers in the channel to move away from the SiO_2 layer and to form a depletion region between the insulating layer and the channel when the gate electrode becomes negative with respect to the source. This depletion region thickens, and thus the channel becomes thinner as distance is increased from the source toward the drain when a positive v_{DS} is applied to the drain because of the increasing electric field. If the drain voltage is sufficiently positive, the channel thickness reduces to zero toward the drain end of the channel. Also, when v_{GS} becomes sufficiently negative, the channel thickness is reduced to zero throughout the channel and i_D is reduced to zero. This cutoff value of v_{GS} is usually called the pinch-off voltage V_P, as in the JFET, but is sometimes known as the threshold voltage V_T. The substrate must have either zero bias or reverse bias with respect to the channel. It is usually connected to the source either externally or internally. A typical set of drain characteristics is given in Fig. 8.12.

The most evident difference between the JFET and the enhancement-mode MOSFET is that v_{GS} may be positive as well as negative. The gate current is essentially zero for all values of v_{GS} because of the SiO_2 insulating layer between the gate and the channel. Of course, the tran-

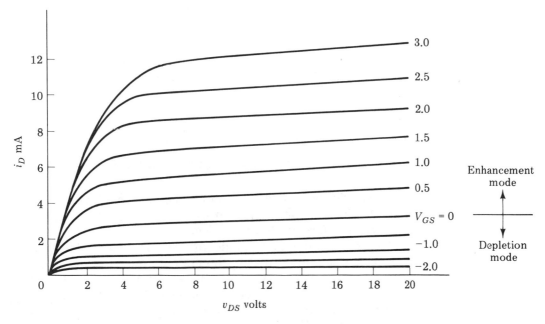

Fig. 8.12. Drain characteristics of an n-channel depletion-mode MOSFET.

sistor will be destroyed if v_{GS} becomes too high and this insulating layer breaks down. As v_{GS} becomes positive, more electrons are attracted into the channel, thus enhancing its conductance. Therefore, the MOSFET is said to be operating in the *enhancement mode* when v_{GS} is positive, even though it is a depletion-mode MOSFET. Also, the saturation of i_D may occur at a higher voltage than $(v_{DS} - V_P)$ because of the increased channel length and the consequent increased iR drop between the gate electrode and the drain region, as shown in Fig. 8.11. This extra channel length is allowed in order to reduce the capacitance C_{gd} between the gate and drain and thus improve the high-frequency characteristics.

The SiO_2 insulating layer normally contains some fixed positive ions which attract electrons into the n-channel, thus increasing the conduction. In fact, these ions may form an n-channel in the lightly-doped p-substrate and thus eliminate the need for n-doping the channel in the manufacturing process, therefore reducing the cost. In addition to being easier to manufacture than a p-channel depletion-mode MOSFET, the n-channel has better high-frequency characteristics because of the higher mobility of electrons. The carriers must traverse the channel during a small part of the gate signal V_{gs}, or gain will be sacrificed. For these reasons, depletion-mode MOSFETs are usually n-channel, although p-channel types are available.

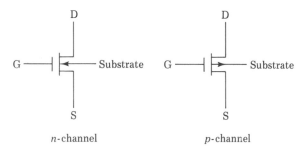

Fig. 8.13. Symbols for depletion-mode MOSFETs.

The symbols for both n-channel and p-channel MOSFETs are given in Fig. 8.13. The substrate is shown as a separate connection, but is often connected internally to the source.

Like the JFET, the depletion-mode MOSFET is a square-law device. That is, over the saturation region, the drain current is proportional to the square of the gate-source voltage with reference to pinch-off. Therefore, all of the relationships developed for the JFET are applicable to the depletion-mode MOSFET, and the same circuit models may be used. The input of the MOSFET is essentially an open circuit, however, as indicated in Fig. 8.14. The gate-to-source capacitance C_{GS} and the gate-to-drain capacitance C_{gd} are included in this model, so it is suitable for the entire frequency range. These capacitances would also be included in the JFET model when high-frequency performance is of importance.

The circuit diagram for a typical n-channel MOSFET amplifier is given in Fig. 8.15. When capacitor C_1 is used, R_G is required to maintain the gate at dc ground potential. Otherwise, the gate potential is unpredictable. The resistance of R_G may be many megohms. If $V_{GS} = 0$ is a suitable bias condition, no additional bias components are needed.

8.4 THE ENHANCEMENTMODE MOSFET

Enhancement-mode MOSFETs are typically constructed as shown in Fig. 8.16. As compared with the depletion-mode MOSFET, observe that it is a p-channel type, although no channel is shown in Fig. 8.16, and the metal gate electrode is large enough to overlap the edges of the

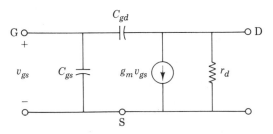

Fig. 8.14. High-frequency circuit model for the depletion-mode MOSFET.

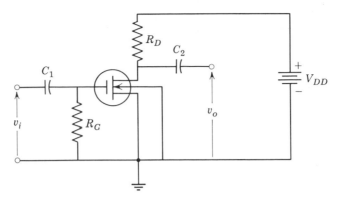

Fig. 8.15. N-channel depletion-mode MOSFET amplifier.

source and drain regions. All dimensions are small, being either microns or fractions of a micron.

As in the depletion-mode MOSFET, the trapped positive ions in the oxide insulating layer attract free electrons from the lightly-doped n substrate (n^- or p^- mean lightly, while n^+ or p^+ mean heavily doped). Thus, the free-electron density is greater near the oxide layer than the doping density or average density in the substrate, as illustrated in Fig. 8.17a. This increased density is known as an *accumulation layer*. As a negative potential is applied to the gate electrode, the resulting electric field tends to cancel the field produced by the positive ions in the oxide layer, and the negative carrier density near the oxide layer decreases. The net carrier density at the surface of the oxide layer becomes zero when v_{GS} is equal to the *threshold voltage* V_T, as illustrated in Fig. 8.17b. A depletion region then exists adjacent to the oxide insulator.

As the magnitude of v_{GS} becomes greater (more negative) than V_T, a layer of positive carriers forms adjacent to the oxide layer, thus forming a p-channel through which drain current flows when a negative voltage v_{DS} is applied to the drain electrode with respect to the source. This induced channel is called an *inversion layer*, as shown in Fig. 8.17c.

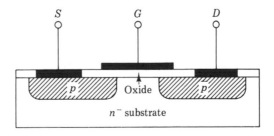

Fig. 8.16. Typical construction of a p-channel enhancement-mode MOSFET.

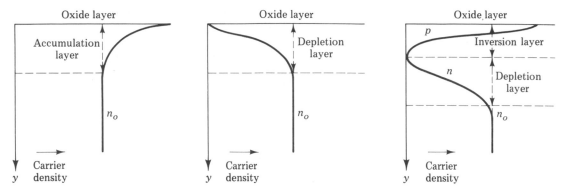

Fig. 8.17. Effect of v_{GS} on the carrier density near the oxide insulator in an enhancement-mode MOSFET: (a) $v_{GS} = 0$; (b) $v_{GS} = V_T$; (c) $|v_{GS}| > |V_T|$.

The gate electrode overlaps both the source and the drain regions to insure that the induced channel extends into both of these regions.

Like the JFET and the depletion-mode MOSFET, the drain current saturates when the magnitude of v_{DS} exceeds $|v_{GS} - V_T|$. This is because the channel thickness is zero at the drain end, and an increased magnitude (more negative) of v_{DS} merely increases the depletion-region thickness around the drain area, as shown in Fig. 8.18. This increased depletion-region thickness does shorten the channel slightly, reducing its resistance and causing the drain characteristics to have a small slope in the saturation region.

A set of drain characteristics for a typical enhancement-mode

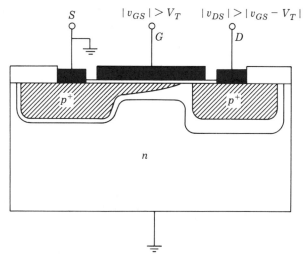

Fig. 8.18. Enhancement-mode MOSFET with $|v_{GS}| > |V_T|$ and $|v_{DS}| > |v_{GS} - V_T|$.

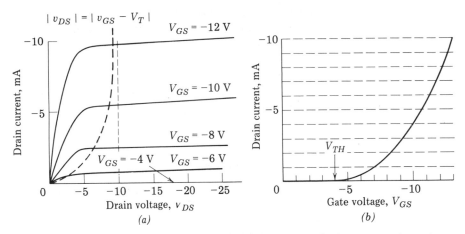

Fig. 8.19. Characteristics of a typical p-channel enhancement-mode MOSFET: (a) drain characteristics; (b) transfer curves.

MOSFET is given in Fig. 8.19a. You may observe that the threshold voltage V_T is 4 V and a dashed line has been drawn through the points where $|v_{DS}| = |v_{GS} - V_T|$, marking the boundary of the saturation region. Also, a transfer curve for this MOSFET with $v_{DS} = -10$ V is presented in Fig. 8.19b.

Although the preceding discussion has focused on the p-channel, it is possible to make n-channel enhancement-mode MOSFETs. One technique is to implant negative ions in the oxide layer to offset the effect of the positive ions naturally trapped there. Another technique is to use special gate-electrode materials, other than the usual aluminum, that provide the desired threshold voltage. These special processes increase the cost of the n-channel enhancement-mode MOSFET over the p-channel, but the availability of an n-channel is very important for use in complementary (CMOS) circuits that are discussed later in the text. The symbols for both p-channel and n-channel enhancement-mode MOSFETs are given in Fig. 8.20. Observe that these symbols differ from

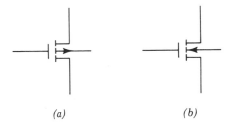

Fig. 8.20. Symbols for the enhancement-mode MOSFET: (a) p-channel; (b) n-channel.

those of the depletion-mode MOSFET's in that the line representing the channel is dashed in the enhancement-mode symbols, indicating the channel does not always exist.

The enhancement-mode MOSFET, like the other types, exhibits the square-law relationship between i_D and v_{GS}. However, since I_{DSS} does not exist for the enhancement mode, this relationship is expressed in the saturation region, for $|v_{GS}| \geq |V_T|$, as

$$i_D = k \, (v_{GS} - V_T)^2 \tag{8.21}$$

where, for the p-channel,

$$k = -\left(\frac{\mu_h \, \epsilon_{ox}}{2t_{ox}}\right)\left(\frac{w}{\ell}\right) \tag{8.22}$$

and μ_h = hole mobility, ϵ_{ox} = permittivity, t_{ox} = thickness of the oxide insulating layer, and w/ℓ is the width-to-length ratio of the channel. The permittivity of silicon dioxide is about $4\epsilon_o$, with $\epsilon_o = 8.854 \times 10^{-12}$ F/m, and t_{ox} is usually between .08 and .02 μm.

Since the gate bias of the enhancement-mode MOSFET has the same polarity as the drain bias, the gate bias may be simply obtained by a voltage divider, as shown in Fig. 8.21a, or by direct coupling, as shown in Fig. 8.21b. The direct coupling may be a distinct advantage of the enhancement-mode MOSFET. A disadvantage may be the larger capacitances between the gate and both the source and the drain which result from the overlapping of the gate electrode with these elements as compared with JFETs and depletion-mode MOSFETs.

The characteristics of a MOSFET may be controlled by adjusting the bias voltage V_{BS} between the substrate or *bulk* material and the source, as shown in Fig. 8.22. However, the substrate must never be forward biased with respect to the source, channel, or drain.

Because of the extremely high input impedance of MOSFETs, one

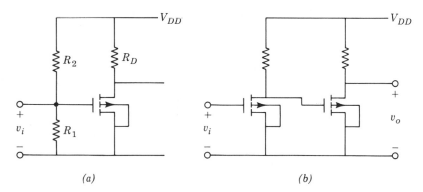

Fig. 8.21. Simple bias circuits for the enhancement-mode MOSFET: (a) voltage-divider bias; (b) direct coupling.

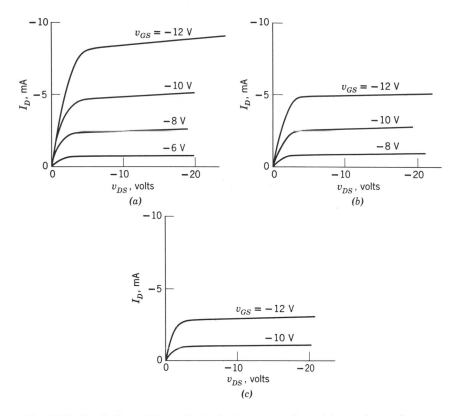

Fig. 8.22. *Illustration of the effect of substrate voltage V_{BS} on the collector characteristics of an enhancement-mode MOSFET: (a) $V_{BS} = 0$ V; (b) $V_{BS} = +4$ V; (c) $V_{BS} = +10$ V.*

must be very careful when installing or handling them. To more fully understand this statement, consider the situation illustrated in Fig. 8.23. In this figure, a positive charge $(+q)$ is placed near the gate lead of a MOSFET. The positive charge attracts electrons to the end of the gate lead. The loss of these electrons creates a positive charge on the gate; a potential is developed across the insulation between the gate and substrate. In fact, the gate, insulation, and substrate form a capacitor, and the voltage across the insulation in a capacitor is

$$V = \frac{Q}{C} \qquad (8.23)$$

As mentioned previously, the input capacitance of the gate in a MOSFET is typically in the order of one picofarad. Thus, the voltage across the insulation is

$$V \simeq 10^{12} \, Q \qquad (8.24)$$

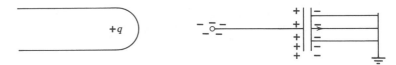

Fig. 8.23. Effect of a static charge near the gate of a MOSFET.

Hence, a small charge can produce a large potential across the insulation.

The gate insulating material in a MOSFET is typically 10^{-5} to 2×10^{-5} cm thick. Consequently, gate-to-substrate voltages in the order of 50 V or so will cause breakdown of the insulation. Once breakdown has occurred, the insulating qualities of the insulator are destroyed and the MOSFET is ruined.

The charge shown in Fig. 8.23 can be a static charge on a person's finger or some tool that is being used. Thus, *one can destroy a MOSFET without even touching it.* To prevent destruction, MOSFETs are shipped with the leads shorted by special clips or packing cases. Some manufacturers insert the leads into small pieces of a conducting foam plastic material. Manufacturers recommend a grounded wrist strap be worn by those handling MOSFETs, and grounding leads should be connected to the soldering irons used.

**8.5
DUAL
INSULATED-
GATE
FIELD-EFFECT
TRANSISTORS**

The dual-gate MOSFET has two control gates, as shown in Fig. 8.24. With this configuration, the output can be controlled by two different signals while the substrate remains grounded. Even if only one signal is to be used, the dual-gate MOSFET has characteristics which make it superior to a single-gate MOSFET for some applications.

As shown in Fig. 8.24, the first gate (gate No. 1) controls the channel near the source, while the second gate (gate No. 2) controls the channel near the drain. The characteristics of a 3N200 (an n-channel depletion-mode dual-gate MOSFET) are presented in Fig. 8.25. Notice that a pair of diodes is internally connected back-to-back from each gate to the

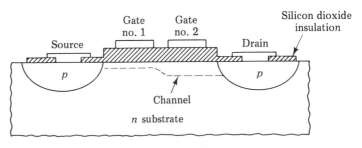

Fig. 8.24. Dual-gate MOSFET configuration.

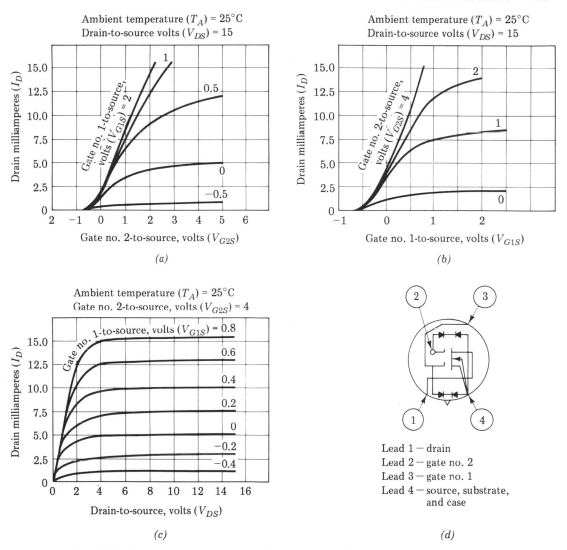

Fig. 8.25. *Characteristics of a 3N200 transistor: (a) i_D versus v_{G2S}; (b) i_D versus v_{G1S}; (c) i_D versus v_{DS}; (d) terminal diagram. (Courtesy of RCA Corporation, New Jersey.)*

source and substrate. These diodes break down in the avalanche mode when the reverse voltage across any of them exceeds approximately 10 V. Consequently, the internal impedance of each gate is very high until the input signal exceeds +10 V or drops below −10 V. Then the diodes break down to protect the thin gate region insulators. The addition of this protection has reduced the input impedance of each gate so that up to 5 μA of current can flow (at 100°C) when 6 V is applied to the gate

Electrical characteristics at $T_A = 25°C$ unless otherwise specified	Symbols	Test conditions		Limits Min.	Limits Typ.	Limits Max.	Units
Gate no. 1-to-source cutoff voltage	$V_{G1S(off)}$	$V_{DS} = +15$ V, $I_D = 50$ μA, $V_{G2S} = +4$ V		-0.1	-1	-3	V
Gate no. 2-to-source cutoff voltage	$V_{G2S(off)}$	$V_{DS} = +15$ V, $I_D = 50$ μA, $V_{G1S} = 0$		-0.1	-1	-3	V
Gate no. 1-terminal forward current	I_{G1SSF}	$V_{G1S} = +1$ V, $V_{G2S} = V_{DS} = 0$	$T_A = 25°C$ $T_A = 100°C$	— —	— —	50 5	mA μA
Gate no. 1-terminal reverse current	I_{G1SSR}	$V_{G1S} = -6$ V, $V_{G2S} = V_{DS} = 0$	$T_A = 25°C$ $T_A = 100°C$	— —	— —	50 5	mA μA
Gate no. 2-terminal forward current	I_{G2SSF}	$V_{G2S} = +6$ V, $V_{G1S} = V_{DS} = 0$	$T_A = 25°C$ $T_A = 100°C$	— —	— —	50 5	mA μA
Gate no. 2-terminal reverse current	I_{G2SSR}	$V_{G2S} = -6$ V, $V_{G1S} = V_{DS} = 0$	$T_A = 25°C$ $T_A = 100°C$	— —	— —	50 5	mA μA
Zero-bias drain current	I_{DS}	$V_{DS} = +15$ V, $V_{G1S} = 0$, $V_{G2S} = +4$ V		0.5	5.0	12	mA
Forward transconductance (gate no. 1-to-drain)	g_{fs}	$f = 1$ kHz		10,000	15,000	20,000	μmho
Small-signal, short-circuit input capacitance	C_{iss}	$V_{DS} = +15$ V, $I_D = 10$ mA, $V_{G2S} = +4$ V, $f = 1$ MHz		4.0	6.0	8.5	pF
Small-signal, short-circuit, reverse transfer capacitance (drain-to-gate-no. 1)	C_{rss}			0.005	0.02	0.03	pF
Small-signal, short-circuit output capacitance	C_{oss}			—	2.0	—	pF

Fig. 8.25. (Continued)

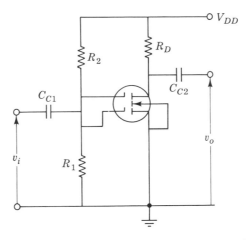

Fig. 8.26. Dual-gate MOSFET with both gates connected in parallel.

(see the electrical characteristics in Fig. 8.25). Thus, the input imped-ance to each gate may be as low as 6 V / 5 × 10⁻⁶ A = 1.2 MΩ.

The simplest dual-gate MOSFET configuration is shown in Fig. 8.26. In this configuration, both gates are connected in parallel and the dual-gate MOSFET acts the same as a single-gate MOSFET. An example will be used to illustrate the use of this configuration.

Example 8.1

A 3N200 transistor is connected, as shown in Fig. 8.26. Determine the characteristics of this amplifier at $V_{G1} = V_{G2} = 1$ V. Let $V_{DD} = 12$ V.

From both Fig. 8.25a and Fig. 8.25b, we note $I_D = 7.5$ mA when $V_{G1} = V_{G2} = 1$ V. Also, notice in Fig. 8.25c that I_D is almost indepen-dent of V_{DS} in the range of $V_{DS} = 4$ V to $V_{DS} = 15$ V. If we assume $I_D = 7.5$ mA when $V_{G1} = V_{G2} = 1$ V and $V_{DS} = 6$ V, the q-point can be located at $I_D = 7.5$ mA and $V_{DS} = 6$ V. Then, $R_D = \Delta V_{DS}/\Delta I_C = 6$ V/7.5 mA = 800 Ω. The minimum input impedance to each gate at 25°C is 6 V / 50 nA = 120 MΩ. Let $R_1 = 1$ MΩ and $R_2 = 11$ MΩ to obtain the proper bias point.

The curves given in Fig. 8.25 show the effect of electrode poten-tials on drain current. The manufacturer also provides a set of curves which shows the effect of electrode potentials on g_m, as shown in Fig. 8.27. A plot of g_m versus V_G when $V_G = V_{G1} = V_{G2}$ can be made as shown in Fig. 8.27b. In this configuration, points of $V_{G1} = V_{G2}$ are connected to produce the new g_m versus V_G curve. Then, notice that g_m at $V_G = 1$ V is equal to $1.5 × 10^3$ μS.

The value of r_d (see Fig. 8.25c) is much greater than 800 Ω. Then, $K_v = -g_m R_D = -1.5 × 10^{-3} × 800 = -1.2$. The output impedance of the amplifier is about 800 Ω and $R_i = R_1 \| R_2 \| R_{inG1} \| R_{inG2} = 900$ kΩ.

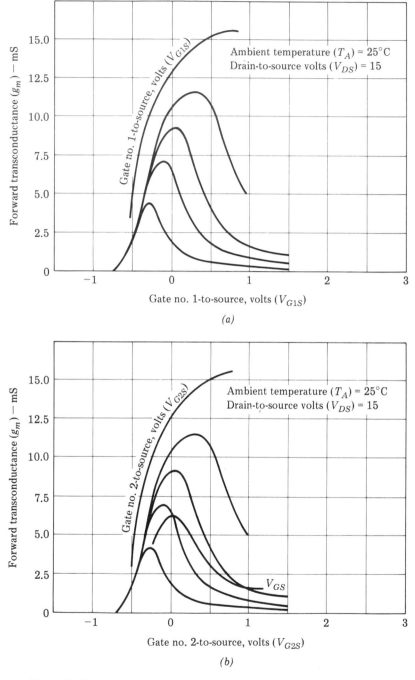

Fig. 8.27. Plots of g_m versus gate voltages for a 3N200: (a) the original curves; (b) modified for $V_{GS} = V_{G1S} = V_{G2S}$.

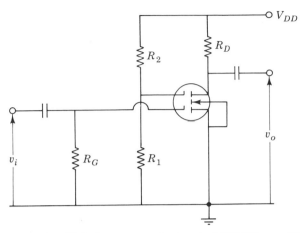

Fig. 8.28. High-frequency dual-gate MOSFET amplifier.

The capacitance from the output of an amplifier to the input of the amplifier can cause stability problems at high frequencies. Part of the output signal is coupled back into the input circuit. The dual-gate MOSFET can be used to minimize this stability problem. When the transistor is connected as shown in Fig. 8.28, the input signal is connected to gate No. 1, and gate No. 2 is maintained at a constant potential. Then, the capacitance between the input (gate No. 1) and the output (drain) terminals can be much lower than with a single gate MOSFET. The capacitance between gate No. 1 and the drain of a 3N200 MOSFET is typically only 0.02 pF. If gate No. 2 is maintained constant, the equivalent circuit for a single gate MOSFET (Fig. 8.8) can be used to represent the dual-gate MOSFET.

An example may help clarify the design of a high-frequency MOSFET amplifier.

Example 8.2 | A 3N200 is connected, as shown in Fig. 8.28. Design the circuit if $V_{DD} = 12$ V. Then find the characteristics of the amplifier.

If $V_{G2S} = 4$ V, the characteristics of Fig. 8.25c will apply. As we found in Example 8.1, the input impedance of each gate at 25°C is at least 120 MΩ. Therefore, let $R_1 = 1$ MΩ and $R_2 = 2$ MΩ, so V_{G2S} will be 4 V. Also, let $R_G = 1$ MΩ, or less, and V_{G1S} will be 0 V. From Fig. 8.25c, the knee of the curves extends up to about $V_{DS} = 4$ V. Therefore, our signal swing can be from $v_{DS} = 4$ V to $v_{DS} = 12$ V. The center of this range will be chosen as our q-point. Then, the q-point occurs at $V_{DS} = 8$ V and $I_D = 5.2$ mA. $R_D = \Delta V_{DS}/I_D = 4$ V / 5.2 mA ≈ 770 Ω.

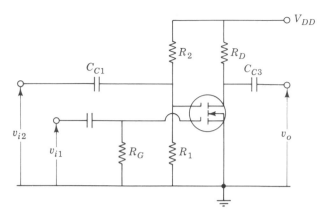

Fig. 8.29. Dual-gate MOSFET signal mixer circuit.

From Fig. 8.25a, the value of g_m is 12.5 mS. Again, $r_d \gg R_D$ so $K_v \simeq -g_m R_D = -12.5 \times 10^{-3} \times 770\ \Omega = -9.6$. The output impedance of the amplifier is $R_o \simeq 770\ \Omega$. The input impedance of the amplifier is $R_i = R_G \| R_{it} = 0.99$ MΩ. For high-frequency circuits, R_D may be replaced by a tuned circuit.

The dual-gate MOSFET is used extensively in automatic gain control circuits and in signal mixer circuits. The basic circuit for these applications is shown in Fig. 8.29. Of course, if v_{i2} is a dc or slow varying voltage, the three circuit elements, C_{C1}, R_1, and R_2 would be removed and v_{i2} would be applied directly to gate No. 2.

The drain current in Fig. 8.29 is controlled by both v_{G1S} and v_{G2S}. In fact, as Fig. 8.27 shows, the g_m of the transistor is controlled by v_{G2S}. Using this concept, the equivalent circuit for a dual-gate MOSFET can be drawn, as shown in Fig. 8.30. The effective g_m of the transistor is given by the relationship

$$g_m = g_{m1} + D_{m2}\, v_{G2S} \tag{8.25}$$

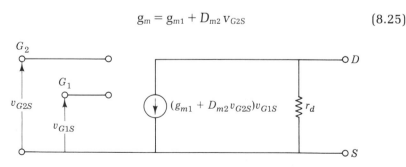

Fig. 8.30. Equivalent circuit for a dual-gate MOSFET.

An example will be used to illustrate how to find the values of the parameters in this equivalent circuit.

Example 8.3

A 3N200 is connected, as shown in Fig. 8.29. Design the circuit if $V_{DD} = 12$ V. Then, find the parameters of the equivalent circuit in Fig. 8.30.

If $V_{G2S} = 4$ V, the characteristics of Fig. 8.25c will apply. Therefore, as in Example 8.2, let $R_1 = 1$ MΩ, $R_2 = 2$ MΩ, $R_G = 1$ MΩ, and $R_D = 770$ Ω.

The value of r_d is found from the inverse of the slope of $v_{G1S} = 0$ V in Fig. 8.25c. Then, $r_d = \Delta v_{DS} / \Delta i_D = (14 - 4)$ V$/ (5.1 - 5.0)$ mA $= 100$ kΩ. The values of g_{m1} and D_{m2} can be found from Fig. 8.27a. A plot of g_m versus v_{G2S} is made with v_{G1S} held constant at 0 V (the q-point value of v_{G1S}). If $v_{G1S} = 0$ V and $v_{G2S} = 0$ V, the value of g_m is 6.5×10^{-3} S. Then, with $v_{G1S} = 0$ V and $v_{G2S} = 1$ V, the value of g_m is 9×10^{-3} S. This process is repeated until enough points are determined to plot the curve of Fig. 8.31. The actual curve is approximated by the dashed straight-line curve in Fig. 8.31. The value of g_{m1} is the value of g_m when v_{G2S} is equal to zero. In this example, $g_{m1} = 7 \times 10^{-3}$ S. The value of D_{m2} is the rate at which g_m changes with v_{G2S}. In this example, $D_{m2} = \Delta g_m / \Delta v_{G2S} = (13 - 7)10^{-3}/4 = 1.5 \times 10^{-3}$ S/V.

The voltage gain of this amplifier is $K_v \simeq -g_m R_D$. Then, $v_o = K_v v_i = -g_m R_D v_{G1S} = -[7 \times 10^{-3} + (1.5 \times 10^{-3})v_{G2S}]770 \times v_{G1S}$. Finally, $v_o = -5.39 v_{G1S} - 1.16 v_{G2S} v_{G1S}$.

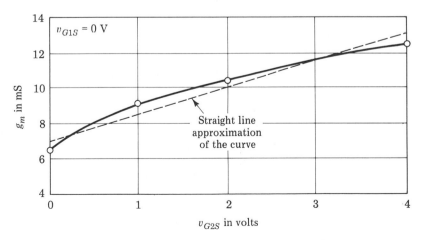

Fig. 8.31. Plot of g_m versus v_{G2s} for a 3N200 MOSFET.

If the straight-line approximation for g_m that was used in Example 7.6 is not accurate enough, a more accurate approximation can be achieved by using the form

$$g_m = g_{m1} + D_{m2} v_{G2S} + D_{m3} (v_{G2S})^2 \qquad (8.26)$$

With this equation for g_m, three points on the approximate curve can be made to coincide with the points on the actual curve. The values of g_{m1}, D_{m2}, and D_{m3} can be found by writing three equations for three points on the curve. Thus, if $g_m = 6.5$ mS when $v_{G2S} = 0$ V, $g_m = 9$ mS when $v_{G2S} = 1$ V, and $g_m = 12.6 \times 10^{-3}$ when $v_{G2S} = 4$ V, the three equations are:

$$6.5 \times 10^{-3} = g_{m1} + 0 + 0 \qquad (8.27)$$

$$9 \times 10^{-3} = g_{m1} + 1 \times D_{m2} + 1^2 \times D_{m3} \qquad (8.28)$$

$$12.6 \times 10^{-3} = g_{m1} + 4 \times D_{m2} + 4^2 \times D_{m3} \qquad (8.29)$$

These equations may be solved simultaneously to yield the three unknowns.

Problems   *Section 8.1*

8.1   The transistor that has the drain characteristics given in Fig. 8.3 is used as a current source. What is the value of current from this source? What is the range of voltage over which it behaves as a current source and what is the approximate source resistance over this range?

Section 8.2

8.2   Assuming the input characteristics of Fig. 8.4 are for the same JFET as the characteristics of Figs. 8.3 and 8.7, determine the low-frequency y-parameters of this transistor at the q-point $V_{GS} = -1$ V, $V_{DS} = 15$ V.

8.3   The JFET with characteristics given in Fig. 8.3 is used in the amplifier circuit of Fig. 8.9a with $V_{DD} = 30$ V and $R_D = R_L = 20$ kΩ. Choose a q-point near the center of the load line and determine:

    a.   Suitable values for R_S and R_G.
    b.   The voltage gain of the amplifier.
    c.   Suitable values for the capacitors if the lower half-power frequency f_ℓ is to be 50 Hz.

8.4 A given FET has $I_{DSS} = 2.2$ mA and $V_P = -2.2$ V. Calculate g_{mo} and the value of g_m at $V_{GS} = -1$ V. Compare these values with those obtained graphically from Figs. 8.3 and 8.7.

8.5 What is the theoretical value of V_{GS} for zero thermal drift for the FET that has characteristics given in Fig. 8.3? For what value of R_D would this be an ideal q-point if $V_{DD} = 30$ V?

Section 8.3

8.6 List the advantages and disadvantages of the depletion-mode MOSFET as compared with the JFET.

Section 8.4

8.7 A given p-channel enhancement-mode MOSFET has an oxide layer thickness $= 0.2$ μm, a channel length $= 5$ μm, and a channel width $= 60$ μm. The threshold voltage is -4 V.

   a. Determine the value of k for this transistor.
   b. Find the value of g_m at $V_{GS} = -8$ V, assuming $|V_{DS}|$ is large enough so the drain current is saturated.

8.8 Calculate the value of k for the MOSFET with characteristics given in Fig. 8.19.

8.9 A MOSFET having the characteristics given in Fig. 8.19 is used in the circuit of Fig. 8.21a. The desired q-point is at $V_{DS} = -10$ V, $V_{GS} = -8$ V. Determine suitable values for the resistors in the circuit, assuming dc blocking capacitors are used in both the input and output leads if needed, and find the voltage gain of the amplifier.

8.10 The amplifier of Fig. 8.21b uses the MOSFET of Fig. 8.19. $V_{DD} = -20$ V and the desired q-point is $V_{GS} = V_{DS} = -10$ V for both stages. Determine the values of the resistors in the circuit and calculate the voltage gain for the two-stage circuit.

Section 8.5

8.11 Modify the circuit of Example 8.2 so the q-point $I_D = 3$ mA and $V_{DS} = 8$ V can be used. Determine the value of all circuit resistors and find K_v for your amplifier. The value of V_{DD} is 12 V.

8.12 Find the voltage gain of the amplifier in Example 8.3 when $v_{G2S} = 0$ V. Repeat when $v_{G2S} = 4$ V.

Answer: $K_v = -5.39$, $K_v = -10$.

8.13 An amplifier is connected as in Example 8.3. Find v_o if $v_{i1} = 0.1 \sin 10^6 t$ and $v_{i2} = 0.1 \sin 10^3 t$.

8.14 Determine the values of g_{m1}, D_{m2}, and D_{m3} in Eqs. 8.27, 8.28, and 8.29. Use Eq. 8.26 for g_m to find the gain of the amplifier in Example 8.3 when $v_{G2S} = 0$ V. Repeat when $v_{G2S} = 4$ V. Compare with the results in Problem 8.12.

9

FET Amplifiers

Some elementary amplifiers were considered in Chapter 8 in order to clarify the characteristics of FETs. However, the effects of the variations of transistor parameters due to manufacturing inaccuracies on amplifier design were not considered. Also, the characteristics of FET amplifier configurations other than common source were not investigated. These topics, in addition to the frequency-response characteristics of the several configurations, are considered in the design of FET amplifiers in this chapter. The use of the FET as a voltage-controlled resistance is also included.

9.1
BIAS CIRCUIT
DESIGN FOR
THE FET

A self-bias circuit using a source resistor R_S was discussed in Sec. 8.2 and shown in Fig. 8.9. This simple circuit provides suitable bias for a JFET or depletion-mode MOSFET when the parameters of the FET are well established, as they are when measured with a curve tracer for an individual transistor. Also, the JFET was shown to have an essentially zero-temperature-coefficient bias point (Eq. 8.20). As temperature increases, the pinch-off voltage V_P increases because of the reduced depletion region thickness, but the channel conductance decreases because of the decreased mobility of the charge carriers, as illustrated by the transfer of Fig. 9.1a. In addition to the variation of transfer characteristics with temperature, the variations due to manufacturing tolerances may be even greater, as illustrated in Fig. 9.1b, where the transfer characteristic of a randomly selected FET may lie anywhere within the cross-hatched area specified by the manufacturer. Therefore, a bias circuit is needed that will limit the q-point variation to a range that will

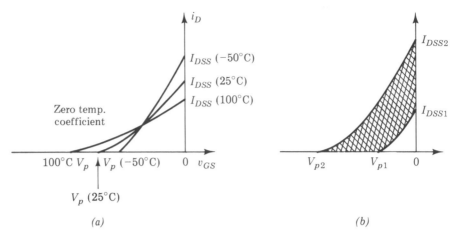

Fig. 9.1. *Variations of FET transfer characteristics with temperature and manufacturing tolerances: (a) variations with temperature; (b) manufacturing tolerances.*

provide linear operation, with adequate power gain over the required ambient temperature range when the transistors are randomly selected. Since this is the same problem we faced with the bipolar transistor, we will seek a similar solution.

The bias circuit of Fig. 9.2a is suitable for JFETs or depletion-mode MOSFETs. This circuit is similar to that of the bipolar junction transistor (BJT) except R_3 has been added to maintain high impedance between

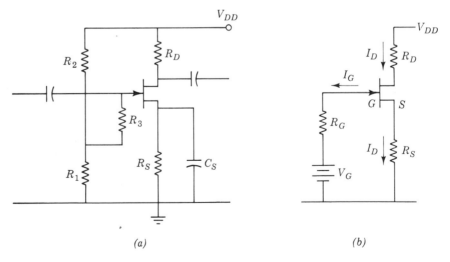

Fig. 9.2. *Stabilized bias circuit for a depletion-mode FET: (a) bias circuit; (b) equivalent circuit.*

the gate and ground. Using Thévenin's theorem to the left of the gate electrode, the equivalent bias circuit of Fig. 9.2b is obtained, where

$$V_G = \frac{R_1}{R_1 + R_2} V_{DD} \qquad (9.1)$$

and

$$R_G = R_3 + \frac{R_1 R_2}{R_1 + R_2} \qquad (9.2)$$

Writing Kirchhoff's voltage equation around the gate-bias circuit in Fig. 9.2b, we obtain

$$V_G = -I_G R_G + V_{GS} + I_D R_S \qquad (9.3)$$

where I_G is the thermal saturation current that flows across the reverse-biased gate-channel junction in a JFET or through the oxide insulator in a depletion-mode MOSFET. Since I_G is very small compared with I_D, even at high operating temperatures, the current through R_S is essentially equal to I_D.

The permissible q-point excursion may be determined with the aid of Fig. 9.3, where the dc and ac load lines have been drawn on the drain characteristics. The lower current limit I_o should be high enough to avoid the low values of g_m, and thus the gain, that occurs at low values of I_D. Perhaps I_o should not be lower than about $I_{DSS1}/10$. Then,

$$I_{D1} = I_o + \frac{V_m}{R_{ac}} \qquad (9.4)$$

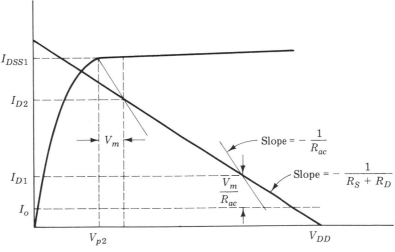

Fig. 9.3. Illustration of the technique for determining the limits for I_{D1} and I_{D2}.

where V_m is the peak value of the output signal voltage. For maximum transducer gain, the capacitively coupled load R_L should be approximately equal to R_D, so a good starting point is to assume $R_{ac} = R_L / 2$.

The upper current limit for the JFET is I_{DSS1}, and the upper q-point I_{D2} may not exceed

$$I_{D2}(\text{max}) = I_{DSS1} - \frac{V_m}{R_{ac}} \tag{9.5}$$

where I_{DSS1} is the minimum I_{DSS} listed on the manufacturer's data sheet for 25°C, corrected *downward* for temperatures significantly above 25°C, as indicated in Fig. 9.1. This limit is not imposed on the depletion-mode MOSFET, however.

The gate current I_{G1} is essentially always negligible at moderate temperatures, being in the nanoampere range. However, a JFET at high temperatures may have a significant saturation current I_{G2}, perhaps in the microampere range. Thus, we may write Eq. 9.3 for each set of currents at the q-point limits as follows:

$$V_G = -|I_{G2}|R_G + V_{GS2} + I_{D2}R_S \tag{9.6}$$

$$V_G = V_{GS1} + I_{D1}R_S \tag{9.7}$$

Subtracting Eq. 9.7 from Eq. 9.6 and solving for R_S,

$$R_S = \frac{|V_{GS2} - V_{GS1}| + |I_{G2}|R_G}{|I_{D2} - I_{D1}|} \tag{9.8}$$

The magnitude signs make Eq. 9.8 applicable for both n- and p-channel FETs. We may find V_{GS1} and V_{GS2} from the relationship $I_D = I_{DSS}(1 - V_{GS}/V_P)^2$, which yields

$$V_{GS1} = V_{P1}\left(1 - \sqrt{\frac{I_{D1}}{I_{DSS1}}}\right) \tag{9.9}$$

$$V_{GS2} = V_{P2}\left(1 - \sqrt{\frac{I_{D2}}{I_{DSS2}}}\right) \tag{9.10}$$

We can see from Fig. 9.3 that

$$I_{D2} = \frac{V_{DD} - V_{P2} - V_m}{R_S + R_D} \tag{9.11}$$

Then the maximum allowable value of R_D is, from Eq. 9.11,

$$R_D(\text{max}) = \frac{V_{DD} - V_{P2} - V_m}{I_{D2}} - R_S \tag{9.12}$$

In the event this value of R_D is much less than R_L, the transducer gain may be unsatisfactory and an FET type with tighter tolerances is needed. On the other hand, if Eq. 9.12 yields a value of R_D much greater than R_L, I_{D1} may be increased to improve the gain characteristics of the amplifier. Please observe from Eq. 9.12 that the maximum permissible value of V_{DD} yields the largest value of R_L, and therefore the highest transducer gain.

A suggested design procedure for the bias circuit of a JFET or depletion-mode MOSFET follows:

1. Select R_D the next stock size above R_L and let $R_S \simeq 0.2\ R_D$.

2. Use V_{DD} as high as practical without exceeding the V_{DS} breakdown rating of the transistor.

3. Choose an FET with sufficiently high current capability $(I_{DSS1} > (V_{DD}/(R_D + R_S) + V_m/R_{ac})$ for a JFET) and $|V_{P2}|$ small (for high gain).

4. Calculate I_{D2} using Eq. 9.11.

5. Let $I_o \simeq I_{DSS1}/10$ and determine I_{D1} from Eq. 9.4.

6. Choose R_G as the minimum desirable input resistance and calculate R_S from Eq. 9.8, assuming I_G doubles for each 10°C increment of temperature.

7. In the event R_S is much larger than the 0.2 R_D value assumed, either find an FET with smaller spread of V_{P1} and V_{P2}, if $(V_{GS1} - V_{GS2}) \gg I_{G2}R_G$, or use a MOSFET instead of a JFET if $I_{G2}R_G$ is too large.

8. In the event R_S is much smaller than 0.2 R_D, increase I_{D1} until $R_S \simeq 0.2\ R_D$.

Example 9.1

Let us assume that we need an amplifier with 5 MΩ input resistance that will provide a 1 V peak signal to a 5 kΩ load over the ambient temperature range $-25°C$ to $75°C$. Then we will choose $R_D \simeq 5$ kΩ and $R_S = 1$ kΩ, so $R_{ac} = 2.5$ kΩ. The amplifier is to be mass-produced. Let us further assume that a suitable n-channel JFET has $I_{DSSmin} = 6$ mA, $I_{DSSmax} = 15$ mA, $V_{Pmin} = -2$ V, $V_{Pmax} = -6$ V, $I_{GSS} = 1$ nA at 25°C, and $V_{DSmax} = 30$ V. Then, using Eq. 9.11 $I_{D2} = (30 - 6 - 1)/6K = 3.8$ mA; $I_{D1} = (.6 + .4)\,mA = 1\ \ mA$. Then, $V_{GS1} = -2(1 - \sqrt{1/6}) = -1.2\ \ $V, $V_{GS2} = -6(1 - \sqrt{3.8/15}) = -3$ V and $I_{G2} = -10^{-9}(2)^5 = -3.2 \times 10^{-8}$ A. Therefore, using Eq. 9.8,

$$R_S = \frac{(3 - 1.2)\,V + .16\ V}{(3.2 - 1.0)\,mA} = \frac{1.96\ V}{2.2\ mA} = 891\ \Omega$$

Since $.2R_D = 1$ kΩ, I_{D1} could be increased slightly, but we will consider that $I_{D1} = -1$ mA is satisfactory. From Eqs. 8.12,

$$I_D = I_{DSS}\left(1 - \frac{V_{GS}}{V_P}\right)^2 \qquad (8.12)$$

and 8.14,

$$g_m = \frac{2I_{DSS}}{V_P}\left(1 - \frac{V_{GS}}{V_P}\right) \qquad (8.14)$$

we can see that $g_m = (2/V_P)\sqrt{I_{DSS}/I_D}$. Therefore, at I_{D1}, $g_{m1} = \frac{2}{2}\sqrt{(6 \text{ mA})(1 \text{ mA})} = 2.45$ mS, and at I_{D2}, $g_{m2} = \frac{2}{6}\sqrt{(15 \text{ mA})(3.8 \text{ mA})} = 2.52$ mS, so $K_v \simeq 2.5 \times 2.5 = 6.25$ over the range of operation. Although this voltage gain is rather low, the power gain is very high because of the large ratio R_{in}/R_L. To complete the design, we find from Eq. 9.7 that $V_G = -1.2 + 1.0 = -.2$ V. However, V_{DD} is positive, so in order to avoid the problem of obtaining a negative voltage, we may either increase I_{D1} to 1.2 mA or increase R_S to 1.2 kΩ. In either case, $V_G = 0$, so R_1 and R_2 are not required, and $R_3 = 5$ MΩ may be connected between the gate and common ground.

Although the simple self-bias circuit proved adequate in the example above, this is not generally the case. If either the maximum ambient temperature had been much higher, thus increasing $I_{G2}R_{G1}$, or V_m had been greater, reducing I_{D2} and increasing I_{D1}, Eq. 9.8 shows that the required R_S would have been increased and V_G would have been positive. Then, R_1 and R_2 in Fig. 9.2a would have been required. However, the values of R_1, R_2, and R_3 cannot be obtained directly from Eqs. 9.1 and 9.2 because there are three unknowns and only two equations. Therefore, we must select one of the resistances from other considerations. For example, we may see from Fig. 9.1a that the series combination of R_1 and R_2 causes a current drain on the power supply equal to $V_{DD}/(R_1 + R_2)$. Therefore, we may arbitrarily choose a value of R_2 large enough to limit this current drain to a fraction of a milliampere. For example, $R_2 = 100$ kΩ would be satisfactory. Then, R_1 and R_3 may be readily determined using Eqs. 9.1 and 9.2.

The direct-coupling bias method for an enhancement-mode MOSFET was briefly discussed in Chapter 8. This bias method may also be used for a single stage, as shown in Fig. 9.4a. The capacitor C_2 is required to prevent signal feedback from drain to gate. Otherwise, degeneration and low gain would result. Since essentially no bias current flows through R_1 and R_2, the gate-bias voltage is

$$V_{GS} = V_{DD} - I_D R_D \qquad (9.13)$$

Observe from Fig. 9.4b that this two-battery bias circuit with $V_G = V_{DD}$ and $R_G = R_1 + R_2$ yields the same gate-bias voltage V_{GS}. Therefore, as

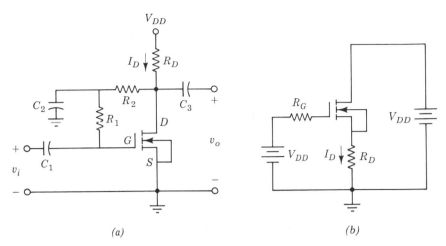

(a)　　　　　　　　*(b)*

Fig. 9.4. Self-bias circuit for an enhancement-mode MOSFET: (a) actual circuit; (b) two-battery model.

may be seen from Fig. 9.4b, this self-bias circuit provides maximum q-point stability because this stability is the same as though the total drain-circuit resistance were between the source and common ground. The q-point is fixed for a given value of R_D, as shown in Fig. 9.5. Since the saturation region, or normal operating region, of the enhancement-

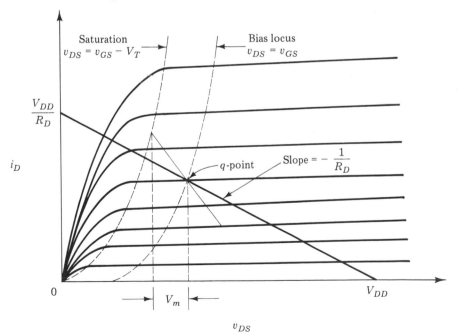

Fig. 9.5. Enhancement-mode self-bias locus.

mode MOSFET is to the right of the dashed line $v_{DS} = v_{GS} = V_T$, as discussed in Chapter 8, and the bias locus is the line $v_{DS} = v_{GS}$, as seen in Fig. 9.4a, the q-point is always located V_T (threshold voltage) volts to the right of the saturation line at the intersection of the load line and the bias-locus line. Therefore, the peak signal voltage V_m must be less than the threshold voltage V_T, depending upon the slope of the ac load line, as shown in Fig. 9.5. Consequently, the MOSFET must have a threshold voltage greater than V_m in order to maintain operation in the linear region.

Since the gate current of a MOSFET is of the order of 1 pA at normal temperatures, $R_G = R_1 + R_2$ (Fig. 9.4a) may be many megohms. However, the input impedance for ac signals is R_1 only, because of the bypass capacitor C_2. Therefore, R_1 must be large enough to provide the desired input impedance. The resistance R_2 need only be large in comparison with R_{ac} to avoid loading the output. The capacitor C_2 must have small reactance compared with R_1 and R_2 in parallel at the lowest signal frequency of interest.

**9.2
COMMON
SOURCE
FREQUENCY
CHARACTER-
ISTICS**

The FET circuits, like the BJT circuits, are limited in their high-frequency response by their interelectrode capacitances. In addition, the upper frequency limit of an FET is limited by the time required for the charge carriers to traverse the length of the channel. This time must be less than about one-fourth of the period of the input signal. Therefore, n-channel GaAS FETs have superior frequency characteristics because of their high-carrier mobility. Channel lengths are very short, of the order of 1 μm, so the transit times may be very short.

A suitable model for determining the high-frequency characteristics of a common source FET amplifier is given in Fig. 9.6. This is the same basic y-parameter model used in Chapter 8, except the interelectrode capacitances are added. Also, y_{fs} (common *source forward* transfer ad-

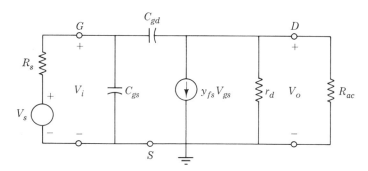

Fig. 9.6. High-frequency model for the CS FET amplifier.

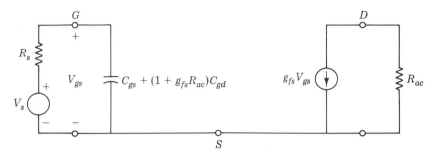

Fig. 9.7. Simplified high-frequency model.

mittance) is used instead of g_m because y_{fs} is complex when the channel transit time is significant, so I_d lags V_{gs}. Over the frequency range in which the transit time is negligible, $y_{fs} = g_{fs} = g_m$.

We may simplify the circuit model of Fig. 9.6 by assuming that y_{fs} is real, the usual case, and placing the effective value of C_{gd} in parallel with C_{gs}, as we did in Section 7.5 for the BJT. This effective or *Miller* capacitance is, assuming $r_d \gg R_{ac}$,

$$C_{gd} \text{(effective)} = C_{gd} (1 + g_{fs} R_{ac}) \tag{9.14}$$

The simplified model is given in Fig. 9.7.

Since the upper edge of the passband in r/s may be obtained from the reciprocal of the sum of the time constants, we can see from Fig. 9.7 that

$$\omega_h = \frac{1}{R_s [G_{gs} + (1 + g_{fs} R_{ac}) C_{gd}]} \tag{9.15}$$

There is also a small capacitance across the output terminals, but this capacitance is usually small in comparison with the input capacitance. Also, R_{ac} is usually small when compared with R_s in an FET amplifier, so the output time constant is normally negligible in comparison with the input time constant.

FET data sheets do not usually give the values for C_{gs} and C_{gd} directly. Instead, they give values for C_{iss} and C_{rss}, where C_{iss} is the *common-source input capacitance with the output shorted*, and C_{rss} is the reverse transfer capacitance in the common-source mode with the input shorted. Figure 9.6 shows that $C_{rss} = C_{gd}$ and $C_{iss} = C_{gs} + C_{gd}$. Therefore,

$$C_{gs} = C_{iss} - C_{gd} = C_{iss} - C_{rss} \tag{9.16}$$

Equation 9.15 may be written in terms of C_{iss} and C_{rss}, using Eq. 9.16,

$$\omega_h = \frac{1}{R_s (C_{iss} + g_{fs} R_{ac} C_{rss})} \tag{9.17}$$

Since the FET capacitances are small, typically a few picofarads, and g_{fs} is small in comparison with the transconductance of a BJT, one would expect the bandwidth of an FET to be much greater than the bandwidth of a BJT. However, the resistance levels in a FET model are usually much higher than in a BJT model. Therefore, their time constants, and thus their bandwidths, are generally of the same order of magnitude.

**9.3
THE SOURCE
FOLLOWER**

You may recall that the emitter-follower configuration of the BJT is useful because of its high-input resistance and low-output resistance in comparison with the common-emitter configuration. Similarly, the *source follower*, or common-drain (CD) configuration, of the FET may be used to obtain high-input impedance and low-output impedance in comparison with the common-source configuration. A simple self-biased JFET source follower circuit is given in Fig. 9.8a, and its circuit model is shown in Fig. 9.8b.

We may avoid writing nodal equations for the solution of the circuit of Fig. 9.8b if we observe from Fig. 9.8a that the self bias is developed across R_{S1}, and $R_{S1} + R_{S2}$ is the dc load resistance. Therefore, R_{S1} is usually small compared with R_{S2} and most of v_o appears across R_{S2}. Thus, the voltage across R_G is small in comparison with v_o, and R_G is very large in order to provide a very high input resistance. The current i_i through R_G is negligible in comparison with $g_{fs}(v_i - v_o)$, and we may write

$$v_o = \frac{g_{fs}(v_i - v_o)}{G_L + g_d + 1/(R_{S1} + R_{S2})} \qquad (9.18)$$

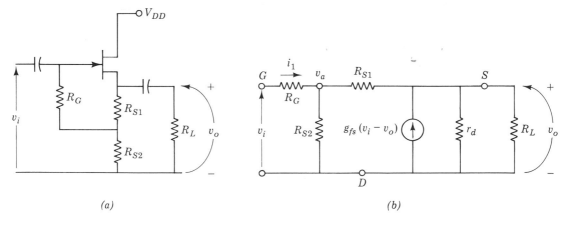

(a)   (b)

Fig. 9.8. Source-follower or CD amplifier: (a) self-bias circuit;
(b) mid-frequency model.

However, the parallel combination of R_L and $(R_{S1} + R_{S2})$ is the ac load resistance R_{ac}, and r_d is normally very large in comparison with R_{ac}, so we may simplify Eq. 9.18 to

$$v_o \simeq g_{fs}(v_i - v_o)R_{ac} \tag{9.19}$$

Solving for the voltage gain $K_v = v_o/v_i$ from Eq. 9.19,

$$K_v = \frac{g_{fs}R_{ac}}{1 + g_{fs}R_{ac}} \tag{9.20}$$

The input resistance of the source follower may be determined by observing from Fig. 9.8b that

$$v_a = \frac{v_o R_{S2}}{R_{S1} + R_{S2}} \tag{9.21}$$

Therefore,

$$i_i = \frac{v_i - v_a}{R_G} = \frac{v_i}{R_G} - \frac{v_o R_{S2}}{(R_{S1} + R_{S2})R_G} \tag{9.22}$$

Using the relationship $v_o = K_v v_i$, we have

$$G_i = \frac{i_i}{v_i} = \frac{1}{R_G} - \frac{K_v R_{S2}}{(R_{S1} + R_{S2})R_G} = \frac{R_{S1} + R_{S2}(1 - K_v)}{R_G(R_{S1} + R_{S2})} \tag{9.23}$$

so

$$R_i = \frac{R_G(R_{S1} + R_{S2})}{R_{S1} + R_{S2}(1 - K_v)} = \frac{R_G}{1 - K_v R_{S2}/(R_{S1} + R_{S2})} \tag{9.24}$$

Observe from Eq. 9.24 that R_i is large in comparison with R_G if R_{S2} is large in comparison with R_{S1}.

Since the voltage gain of the source follower is nearly one, the current gain is approximately R_i/R_{ac}. The power gain is essentially equal to the current gain.

The general stabilized-bias circuit may also be used for the source follower, as shown in Fig. 9.9a. Bootstrapping is also included to provide maximum input impedance. Observation of the circuit model of Fig. 9.9b will verify that this circuit has the voltage gain given by Eq. 9.20. However, R_{ac} now includes R_1 and R_2, so these resistances need to be large in comparison with R_S and R_L in order to avoid degrading the voltage gain.

The input resistance R_i may be readily obtained from Fig. 9.9b if we observe that $i_i = (v_i - v_o)/R_G$. Then,

$$R_i = \frac{v_i}{i_i} = \frac{R_G}{1 - K_v} \tag{9.25}$$

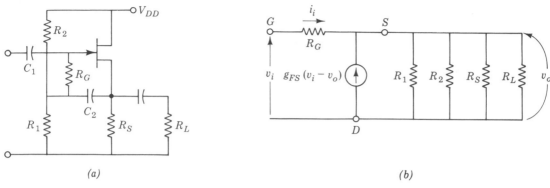

(a)   (b)

Fig. 9.9. Source follower with stabilized bias: (a) circuit diagram;
(b) mid-frequency model.

Since K_v typically ranges between 0.90 and 0.98, the input resistance
may be 10 to 50 times R_G, and since R_G may be several megohms for a
JFET, R_i may be of the order of hundreds of megohms for a JFET or
thousands of megohms for a MOSFET. Although we have used JFETs as
examples, MOSFETs may also be used as source followers.

The stabilized bias circuit design for the source follower may follow
the same general procedure as given for the CS amplifier, except R_S is
established from power-transfer considerations rather than q-point stability. Then Eqs. 9.8, 9.9, and 9.10 may be used to determine I_{D1} and I_{D2}.
Since R_S is the entire dc load resistance, excellent q-point stability may
be expected. Therefore, I_{D1} may be chosen near the center of the dc load
line and I_{D2} calculated from Eq. 9.8.

The high-frequency characteristics of a source follower may be obtained by including the FET capacitances in the circuit model, as shown

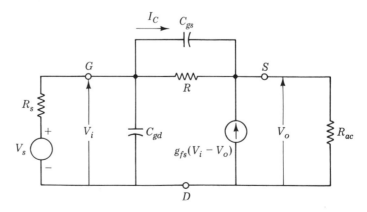

Fig. 9.10. High-frequency model of the source follower.

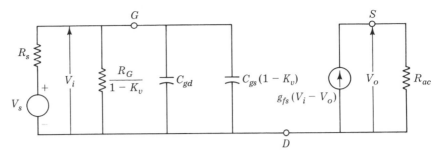

Fig. 9.11. Simplified high-frequency model.

in Fig. 9.10. Observe that the current I_c through C_{gs} is $j\omega C_{gs}(V_i - V_o)$. Therefore, the admittance of C_{gs} as seen from the input node is

$$jB_{gs} = \frac{I_c}{V_i} = \frac{j\omega C_{gs}(V_i - V_o)}{V_i} = j\omega C_{gs}(1 - K_v) \qquad (9.26)$$

Therefore, the Miller effect is working to our benefit in the case of the source follower, and the model of Fig. 9.10 may be simplified, as shown in Fig. 9.11. Then, we may obtain the high-frequency cutoff in r/s as the reciprocal of the time constant. Assuming $R_G/(1 - K_v) \gg R_S$,

$$\omega_h = \frac{1}{R_S[C_{gd} + C_{gs}(1 - K_v)]} = \frac{1}{R_S(C_{iss} - K_v C_{rss})} \qquad (9.27)$$

Thus, for the same magnitude of driving source resistance R_S, the bandwidth of the source follower is much wider than that of the CS amplifier because the effective stray capacitance is reduced by the ratio of their voltage gains, assuming the same transistor with the same R_{ac} is used in both configurations.

9.4 THE COMMON-GATE AMPLIFIER

The common-gate (CG) FET amplifier is comparable with the common-base BJT amplifier. In this configuration, the gate is connected directly to ground, or common; the input signal is applied to the source terminal, and the drain terminal is the output, as shown in the circuit diagram given in Fig. 9.12a. The midfrequency model of the CG amplifier is shown in Fig. 9.12b. Let us use nodal equations to analyze this circuit. Observing that $g_{fs} v_{gs} = -g_{fs} v_i$,

$$i_i = (G_S + g_d + g_{fs}) v_i - g_d v_o$$

$$0 = -(g_{fs} + g_d) v_i + (g_d + G_{ac}) v_o \qquad (9.28)$$

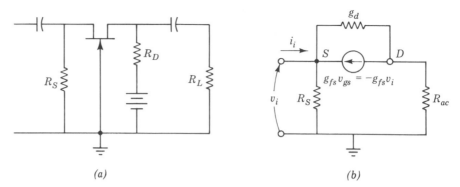

Fig. 9.12. Common-drain amplifier: (a) circuit diagram;
(b) mid-frequency model.

Using Cramer's rule, or determinants,

$$v_i = \frac{(g_d + G_{ac})\, i_i}{(G_S + g_d + g_{fs})\,(g_d + G_{ac}) - g_d\,(g_{fs} + g_d)} \tag{9.29}$$

$$v_o = \frac{(g_{fs} + g_d)\, i_i}{(G_S + g_d + g_{fs})\,(g_d + G_{ac}) - g_d\,(g_{fs} + g_d)} \tag{9.30}$$

Then,

$$K_v = \frac{v_o}{v_i} = \frac{g_f + g_d}{g_d + G_{ac}} \simeq g_{fs}\, R_{ac} \tag{9.31}$$

If $g_d \ll G_{ac}$, as usual, we may neglect the term $g_d\,(g_{fs} + g_d)$ in Eq. 9.29 to obtain

$$R_i = \frac{v_i}{i_i} = \frac{1}{G_S + g_d + g_{fs}} \simeq \frac{1}{G_S + g_{fs}} \tag{9.32}$$

Using this same approximation, and $g_d \ll g_{fs}$,

$$K_i = \frac{v_o}{R_L\, i_i} \simeq \frac{1}{R_L\, G_{ac}\left(1 + \dfrac{G_S}{g_{fs}}\right)} = \frac{1}{(1 + G_S/g_{fs})\,(R_L + R_D)/R_D} \tag{9.33}$$

The output resistance R_o looking into the drain terminal may be obtained from the model of Fig. 9.13. Here the symbol R_s' is used for the driving source resistance to distinguish it from the FET source resistor R_S. Writing nodal equations,

$$I_o = g_d\, v_o - (g_{fs} + g_d)\, v_i$$

$$0 = -g_d\, v_o + (G_s' + G_S + g_d + g_{fs})\, v_i \tag{9.34}$$

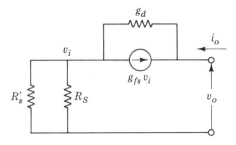

Fig. 9.13. Model for determining R_o.

Since $g_d \ll g_{fs}$, we may use determinants to obtain

$$v_o = \frac{i_o (G_s' + G_S + g_{fs})}{g_d (G_s' + G_S + g_{fs}) - g_d g_{fs}} \qquad (9.35)$$

Therefore,

$$R_o = \frac{V_o}{i_o} = \frac{r_d}{1 - g_{fs}/(G_s' + G_S + g_{fs})} \qquad (9.36)$$

Observe that the voltage gain K_v of the CG amplifier is the same as that of the CS amplifier; the current gain to the total load R_{ac} is unity, but divides to $R_D/(R_D + R_L)$ into the load; the input resistance R_i is low, being R_S in parallel with $1/g_{fs}$; and the output resistance R_o is high, being much greater than r_d if $g_{fs} \gg (G_s' + G_S)$.

The CG amplifier is used primarily in very high-frequency tuned amplifiers because of its high stability, low noise, and good gain in that application. A high-frequency CG model with a tuned LC load is given in Fig. 9.14 to illustrate the superior stability of this amplifier. Instability normally occurs in a high-frequency tuned amplifier because

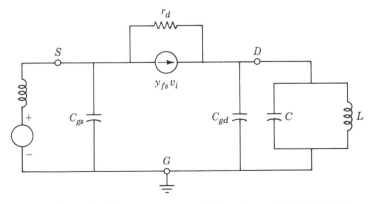

Fig. 9.14. High-frequency model for a tuned CG amplifier.

of the coupling of energy, or feedback, from the output of the amplifier to its input through the interelectrode capacitances or other coupling elements.

Observe that there is essentially no coupling capacitance between the output and the input of a CG amplifier. The input capacitance C_{gs} and the output capacitance C_{gd} just become part of the tuning capacitance of the LC circuits. Although r_d is a coupling element, its resistance is so high that it does not cause instability. One may expect the CB amplifier to have the same stability, but the base-spreading resistance r_x causes appreciable coupling at high frequencies in the CB amplifier.

The voltage gain of the CG amplifier of Fig. 9.14 may be quite high because a fairly large drain curent with its accompanying high value of y_{fs} may flow through the low-resistance coil L, and the ac load resistance $R_{ac} = Q\omega_o L$ at the resonant frequency, where ω_o is the resonant frequency, $Q = \omega_o/B$, and B is the bandwidth in r/s. Therefore, R_{ac} may be quite high, yielding high gain. Also, the tuned coupling-circuits may provide impedance matching and maximum power transfer. The gain is limited at high frequencies by y_{fs}, which becomes complex and decreases in magnitude as the channel transit time becomes a significant part of the signal period, as previously discussed.

9.5
A
COMBINATION
FET-BJT
AMPLIFIER

FETs are often used in conjunction with BJTs in order to utilize the advantages of each type of transistor. An example of such a combination is the microphone amplifier of Fig. 9.15a. Some types of microphones have output voltages of the order of 1 mV or less and have small capacitances as their internal impedances. Therefore, in order to obtain high-quality sound from these microphones, the amplifier must have low-noise and very high-input resistance. The high-input resistance is required because of the small internal capacitance C_m of the microphone, which appears in series with the input resistance R_{ia} of the amplifier, as shown in the model of Fig. 9.15b. This capacitance may be of the order of 100 pF, and the circuit time constant must be long enough to provide good response at low-audio frequencies such as 50 Hz or lower. Chapter 11 treats amplifier noise in considerable detail.

The amplifier of Fig. 9.15a uses direct coupling between the FET and the BJT. Either a bootstrapping circuit or a depletion-mode MOSFET could be used to increase the input impedance above that obtainable from the simple JFET circuit shown. However, microphones are usually used at normal room temperatures, so R_G may be as high as 100 MΩ in the circuit shown if I_{DO} is 1 nA or less at normal room temperatures.

The biases and q-points of the FET and BJT in Fig. 9.15a are interdependent because of the direct coupling. A sensible design approach might be to begin with the BJT (output) stage and select R_C to match, or be compatible with, the intended load on the amplifier, which most

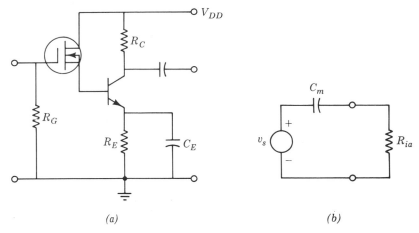

Fig. 9.15. Amplifier for a capacitor-type microphone: (a) amplifier diagram;
(b) capacitor microphone model.

likely would be a power amplifier discussed in Chapter 15. Then a
suitable q-point for the BJT could be selected and the value of R_E determined to provide the desired q-point for the FET. The desired q-point
for the FET is at $I_D = I_B$ and $V_{GS} = V_{BE} + I_E R_E$, of course. Also, $V_{DS} = V_{DD} - (V_{BE} + I_E R_E)$.

<table>
<tr><td>**9.6**
VOLTAGE-
CONTROLLED
RESISTANCE</td><td>

The FET may be used as a voltage-controlled resistance, as may be seen
in Fig. 9.16. At low values of v_{DS} in the nonsaturated region, we may
observe that each drain characteristic has a slope di_D/dv_{DS} that is fairly
constant at low values of v_{DS} and is dependent upon V_{GS}. The re-
lationship between i_D, v_{DS}, and v_{GS} in the nonsaturated region is given
by the expression,

</td></tr>
</table>

$$i_D = \frac{I_{DSS}}{V_P^2}\left[2\left(v_{GS} - V_P\right)v_{DS} - v_{DS}^2\right] \qquad (9.37)$$

Therefore,

$$\frac{di_D}{dv_{DS}} = \frac{2I_{DSS}}{V_P^2}\left[v_{GS} - V_P - v_{DS}\right] \qquad (9.38)$$

This slope is essentially constant and represents a linear conductance,
provided $v_{DS} \ll (v_{GS} - V_P)$. With this restriction,

$$r_{ds} = \frac{V_P^2}{2I_{DSS}\left(v_{GS} - V_P\right)} = \frac{|V_P|}{2I_{DSS}\left(1 - \left|\dfrac{v_{GS}}{V_P}\right|\right)} \qquad (9.39)$$

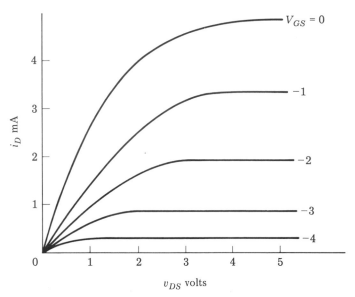

Fig. 9.16. Drain characteristics of a JFET at low values of v_{DS}.

One application of the FET variable resistance is the automatic gain control of an amplifier, as shown in Fig. 9.17. In this circuit, the JFET acts as an unbypassed resistance in the emitter circuit of a BJT. Therefore, the signal voltage across the JFET is $i_e\, r_{ds}$. However, the signal voltage v_{be} across the base-emitter junction is normally only a few millivolts, so the input voltage $v_i \simeq i_e\, r_{ds}$. Then, the voltage gain of the BJT amplifier is

$$|K_v| \simeq \frac{i_c\, R_{ac}}{i_e\, r_{ds}} \tag{9.40}$$

But $i_c = \alpha_f\, i_e \simeq i_e$, so

$$|K_v| \simeq \frac{R_{ac}}{r_{ds}} \tag{9.41}$$

Using Eq. 9.39,

$$|K_v| = \frac{2I_{DSS}\, R_{ac}}{|V_P|}\left(1 - \left|\frac{v_{GS}}{V_P}\right|\right) \tag{9.42}$$

The JFET should be chosen so the $I_{DSS} \gg I_C$; so $v_{DS} \ll V_P$ in order to maintain good linearity in the BJT amplifier and also to provide high gain when $v_{GS} = 0$.

The gain of a BJT could be adjusted remotely by applying a variable

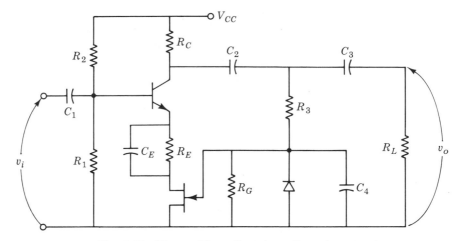

Fig. 9.17. CE amplifier with automatic gain cotnrol.

voltage to the gate of the FET. However, the circuit gain is adjusted automatically in Fig. 9.17, so the ouptut voltage remains relatively constant for varying levels of input voltage. This is accomplished by using a rectifier and filter circuit in the output of the amplifier to provide a negative quasi-dc voltage V_{GS} to the FET gate. Therefore, as the output signal level increases, the gain of the BJT decreases. The resistor R_3 (Fig. 9.17) prevents C_4 from shunting the output currents around R_L. Therefore, R_3 should be large in comparison with R_{ac}, since R_3 is part of R_{ac}, but small in comparison with R_G; so V_{GS} will be nearly equal to the peak value of v_o and thus provide maximum gain control.

Problems *Section 9.1*

9.1 A given type of n-channel JFET has a spread of V_P from -1.5 V to -4 V and a spread of I_{DSS} from 2 mA to 6 mA for random selection over the temperature range $-25°C$ to $60°C$. What is the maximum value of R_D that may be used in a mass-produced amplifier using this transistor type if $V_{DD} = 25$ V, $R_S = 0.2\,R_D$, the capacitively coupled load is 10 kΩ, and the peak value V_m of the output signal is 1 V?

9.2 The JFET having characteristics given in Problem 9.1 is used in a mass-produced amplifier over the temperature range given. The maximum I_{DO} is 2 nA at 25°C and the required input resistance to the amplifier is 10 MΩ. Design the amplifier circuit using $V_{DD} = 25$ V.

9.3   Determine the voltage gain, current gain, and power gain of the amplifier of Problem 9.2 at both the maximum I_D and minimum I_D q-point values. What is the percentage variation of voltage gain?

9.4   The enhancement-mode MOSFET having characteristics given in Fig. 8.19 is used in the amplifier circuit of Fig. 9.4a. The capacitively coupled load is 100 kΩ and $V_{DD} = -25$ V.

    a.   Determine suitable values for all of the circuit components if the desired input resistance is 20 MΩ, $R_D = 10$ kΩ, and $f_\ell = 10$ Hz. Calculate K_v.
    b.   Repeat part (a) if R_D is increased to 100 kΩ.

Section 9.2

9.5   The transistor of Problem 9.4 has $C_{iss} = 10$ pF and $C_{rss} = 5$ pF.

    a.   What is the bandwidth of the amplifier of Problem 9.4a.?
    b.   What is the bandwidth of Problem 9.4b?

Section 9.3

9.6   A given JFET has $I_{DSS} = 2$ mA and $V_P = -2$ V. This transistor is used in the circuit of Fig. 9.8a with $V_{DD} = 25$ V, $I_D = 1$ mA, $R_{S1} + R_{S2} = 12$ kΩ, $R_L = 15$ kΩ, and $R_G = 4.7$ MΩ. Determine R_{S1}, the voltage gain K_v, and the input resistance of the amplifier.

9.7   The transistor specified in Problem 9.6 is used in the circuit of Fig. 9.9a with $V_{DD} = 25$ V, $I_D = 1$ mA, $R_S = 12$ kΩ, $R_L = 15$ kΩ, and $R_G = 4.7$ MΩ. Determine suitable values for R_1 and R_2 and calculate the input resistance R_i.

9.8   The amplifier of Problem 9.6 is driven by a source having an internal resistance $= 1$ MΩ. If the transistor has $C_{iss} = 8$ pF and $C_{rss} = 3$ pF, determine the upper cutoff frequency of the amplifier.

Section 9.4

9.9   A JFET having $I_{DSS} = 10$ mA and $V_P = -6$ V is used in the circuit of Fig. 9.12a with $V_{DD} = 20$ V, $I_D = 5$ mA, and $R_D = R_L = 200$ Ω. The driving source resistance $R_s = 100$ Ω. Determine the Source Circuit Resistance R_S, the voltage gain, the current gain, the input resistance of the transistor, and the output resistance of the transistor.

9.10   The transistor of Problem 9.9 is used in the circuit of Fig. 9.14. The q-point is the same ($I_D = 3$ mA, $V_{DD} = 20$ V). The tuned drain circuit has $L = 1$ mH, $C = 100$ pF, and $Q = 100$. What is the voltage gain of the amplifier?

Section 9.5

9.11   The microphone amplifier (Fig. 9.15) has a capacitively coupled load $R_L = 10$ kΩ. The supply voltage $V_{DD} = 20$ V and the BJT has $B_o = 200$ at $I_C = 1$ mA and $r_x = 100$ Ω. The FET has $I_{DSS} = 2$ mA and $V_P = -2$ V. Select suitable values for R_C and R_E. Calculate the required value of R_G if $C_m = 50$ pF and the low-frequency cutoff $f_\ell = 20$ Hz. What is the voltage gain, current gain, and power gain of the total amplifier?

Section 9.6

9.12   A given JFET having $I_{DSS} - 5$ mA and $V_P = -6$ V is used as a voltage-controlled resistance.

   a.   Sketch r_{ds} as a function of v_{GS} for values of v_{GS} from 0 to 5.5 V.
   b.   Sketch the maximum value of v_{DS} as a function of v_{GS} for values of v_{GS} from 0 to -5.5 V if $v_{DS} \le .05 (v_{GS} - V_P)$.

10

Direct-Coupled Amplifiers

The need frequently arises for an amplifier which will faithfully reproduce slowly varying signals. The very low cutoff frequency required for such an amplifier may eliminate capacitive or transformer coupling from practical consideration and leave only direct coupling as a feasible solution. The main disadvantage of direct coupling is that thermal currents generated in the amplifier are amplified along with the signal currents. Thus, thermal stability problems are increased and thermal currents may mask signal currents. Particular attention must be paid to thermal stability in a direct-coupled or dc amplifier. Five amplifier types will be studied in this chapter. They are the Darlington connection, n-p-n–p-n-p arrangements, differential amplifiers, the cascode configuration, and the complementary-symmetry amplifier.

Stabilizing-bias circuits were discussed in Chapter 7 where it was shown mathematically that the collector current I_C of a transistor may become more stable as the ratio of base-circuit resistance R_B to emitter-circuit resistance R_E is decreased. The reason for this improvement in q-point stability is illustrated in Fig. 10.1, where the thermally generated current I_{CO} is shown to divide between the base circuit and the emitter circuit. Only that portion, KI_{CO}, which flows across the emitter junction is amplified by β; therefore, the collector current which results from the thermal current is $(I_{CO} + K\beta I_{CO}) = (1 + K\beta) I_{CO}$ and K can have values between 1 and 0. Since the shift in q-point is caused primarily by the increase of I_{CO} with temperature, the best stability is obtained when K is 0.

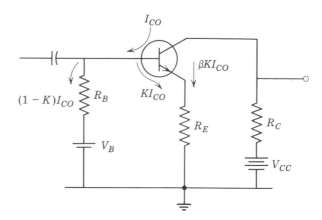

Fig. 10.1. Circuit showing the division of thermal current I_{CO} between the emitter and base currents.

A current stability factor S_I is defined as the ratio of collector current change to the change in I_{CO}, or

$$S_I = \frac{\Delta I_C}{\Delta I_{CO}} \tag{10.1}$$

Thus, if I_{CO} in Fig. 10.1 is increased by ΔI_{CO}, the collector current is increased by $(1 + K\beta)\Delta I_{CO}$, and the current stability factor $S_I = (1 + K\beta)$. Note that a small stability factor results in good q-point stability. Therefore, S_I is actually an *instability* factor. If the base resistor R_B is very large and the emitter resistor R_E is zero, which occurs when fixed bias is used, the value of K is 1 and the value of S_I is $(1 + \beta)$. On the other hand, if R_B is zero and R_E is large, which occurs when the common-base configuration is used, $K = 0$ and $S_I = 1$. Intermediate values of R_E and R_B give values of S_I which lie between one and $(\beta + 1)$. In fact, it can be shown[1] that when S_I is large compared with one, but small compared with $(\beta + 1)$,

$$S_I \simeq R_B / R_E \tag{10.2}$$

A desirable value for S_I can be determined after a transistor has been chosen for a given application.

10.1
THE
DARLINGTON
CONNECTION

One method of direct coupling bipolar transistors, known as the Darlington connection, is shown in Fig. 10.2. In this arrangement, the emitter current of transistor T_1 is the base current of transistor T_2. If R_L is small, $i_{E1} = (\beta_1 + 1)i_{B1}$ and $i_{C2} = \beta_2 i_{B2}$. Then the ratio $i_{C2}/i_{B1} =$

[1] *Electronic Engineering*, Alley and Atwood, Second Edition, John Wiley and Sons Inc., New York, New York, p. 195.

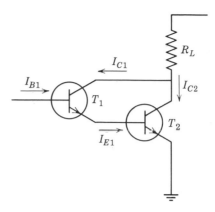

Fig. 10.2. Darlington connection.

$(\beta_1 + 1)\beta_2$. The current i_{C1} adds to i_{C2} in the load resistor, but if β_2 is large, i_{C1} is negligible and the total amplification factor is approximately the product $\beta_1 \beta_2$. Three transistors are sometimes used in the Darlington connection to produce a current gain approximately equal to $\beta_1 \beta_2 \beta_3$.

The thermal currents are also amplified in the Darlington connection as stated previously. The preceding discussion showed that the thermal current amplification is equal to the current stability factor in any given stage. Resistors may be used in the Darlington circuit to reduce the thermal currents as shown in Fig. 10.3. The thermal current $S_{I1} I_{CO1}$ from transistor T_1 is divided between resistor R_{B2} and the input of transistor T_2. Also, part of the thermal current I_{CO2} of transistor T_2 may flow through R_{B2} to reduce the stability factor S_{12}. Note that R_E is not by-passed because the expected signal frequencies are too low for effective

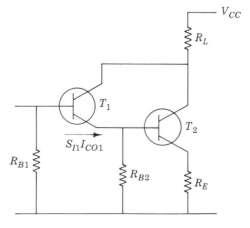

Fig. 10.3. Darlington-connected amplifier with linear thermal stabilization.

bypassing. Therefore, for good voltage gain, R_E must be small in comparison with R_L. The signal gain is also reduced because part of the signal current is shunted through the stabilizing resistors. In fact, a change in thermal current is indistinguishable from a signal current. Therefore, with this linear-type stabilization, high gain with adequate thermal stability can be achieved only by the use of transistors which have very small values of I_{CO}.

We will now consider the design of a Darlington amplifier.

Example 10.1

A circuit is connected as shown in Fig. 10.3. The load resistor R_L is determined by the amplifier application. Let us assume that $R_L = 100\ \Omega$ and $V_{CC} = 20$ V. We will next choose $R_E = 10\ \Omega$. The transistor selected for T_2 has $\beta_2 = 100$, and $r_{x2} = 12\ \Omega$. Since the maximum value of $i_C \simeq 20$ V$/(R_L + R_E) \simeq 180$ mA, the average value of collector current i_{C2} should be about 90 mA. Then, $h_{ie2} = r_{x2} + \beta_2/g_{m2}$. The value of g_{m2} is $\simeq 40\ I_C = 40(.09) = 3.6$, and $h_{FE2}/g_{m2} = 100/3.6 = 28\ \Omega$. Therefore, $h_{ie2} = 40\ \Omega$ at this average collector current. Now, transistor T_2 has an unbypassed resistor in the emitter circuit so the input resistance of transistor T_2 is $h_{ie2} + (h_{FE2} + 1)R_E = 40 + (101)\,10 = 1050\ \Omega$. A sensible choice of value for R_{B2} might be about the same as the input resistance of T_2. Then half of the signal current will be shunted through R_{B2}. We will choose $R_{B2} = 1$ kΩ. Then the total resistance in the emitter circuit of T_1 (Fig. 10.3) is approximately $R_{eq} = (1.05$ k $\times 1$ k$)/2.05$ k $= 512\ \Omega$. We will now select transistor T_1 with $h_{FE1} = 120$ and $r_{b1} = 100\ \Omega$. The average base current of transistor T_2 is $i_{C2}/\beta_2 = 90$ mA$/100 = 0.9$ mA. Assuming v_{BE} of transistor T_2 to be 0.6 V and recognizing that the voltage drop across R_E at the average value of emitter current is $I_{E2}\,R_E \simeq .09$ A $(10\ \Omega) = 0.9$ V, the voltage across R_{B2} is 1.5 V and the current through R_{B2} is 1.5 V$/1$ k $= 1.5$ mA at this average value of current. Then the average emitter current of transistor T_1 is $I_{RB2} + I_{B2} = 1.5 + 0.9 = 2.4$ mA, and the average base current of T_1 is $i_{E1}/h_{FE1} = 2.4$ mA$/120 = 20$ μA.

The average input impedance of transistor T_1 is $(h_{FE1} + 1)(R_{eq}) + h_{ie1} \simeq h_{FE1}\,R_{eq} + h_{FE1}/g_{m1} = 120\,(512) + 120\,/\,(40 \times .0024) = 61.5$k $+ 1.25$k $\simeq 63$ kΩ. Note that this high input impedance is due to the impedance in the emitter circuit, R_{eq}. Let us again sacrifice about one-half of the signal current and select $R_{B1} = 68$ kΩ. Then the total current gain for the Darlington amplifier is $K_i \simeq \beta_1/2$ times $B_2/2$ or $\beta_1\beta_2/4 = 3000$ and the voltage gain $K_v = v_O/v_I = K_i\,R_L/R_i = 3000 \times 100/63$ k $= 4.77$. Note that transistor T_1 acts as an emitter follower and provides current gain but not voltage gain.

Diodes can be used to stabilize the Darlington amplifier as shown in Fig. 10.4. If the reverse saturation current I_{S2} of diode D_2 is equal to the

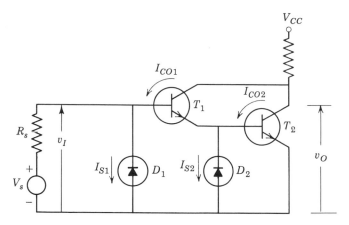

Fig. 10.4. Diode-stabilized Darlington amplifier.

thermal current I_{CO2} of transistor T_2, the current stability factor of transistor T_2 is unity. Similarly, if the saturation current I_{S1} is equal to the thermal current I_{CO1}, the current stability factor of transistor T_1 is one. The only problem arises in finding diodes which match the transistors in thermal currents. Since the resistance of a reverse-biased diode is very high, the approximate current gain of the amplifier of Fig. 10.4, in terms of the transistor betas, is $\beta_1\beta_2$.

Example 10.2

Let us assume that the Darlington amplifier of the preceding example has the stabilizing resistors replaced by diodes. At the average currents previously determined, the total current gain is approximately $K_i = h_{FE2} \times h_{FE1} = 120 \times 100 = 12{,}000$; the approximate input resistance of transistor T_1 is $\beta_1 h_{ie2} + h_{ie1} = 120 \times 40 + 1.25$ $k = 6.05$ kΩ; and the approximate voltage gain $K_v = v_o / v_i = i_o R_o / i_i R_i = K_i R_o / R_i = 1.2 \times 10^4 \times 100 / 6.05$ k $= 198$.

We have assumed that the signal source provided forward bias for the Darlington amplifier. If the signal source does not have a dc component which will provide this bias, a resistor must be connected between the base of transistor T_1 and V_{CC} to provide the required bias. Some modern devices provide a Darlington configuration in a single container.

10.2
n-p-n–p-n-p
COMBINA-
TIONS

A dc amplifier can be constructed by alternating n-p-n and p-n-p transistors as shown in Fig. 10.5. This amplifier is diode stabilized and the input voltage is assumed to provide forward bias for the transistors. Observe that the collector current of transistor T_1 is the base current of transistor T_2. Therefore, the input impedance of transistor T_1 is much

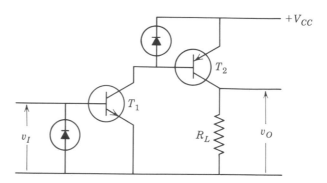

Fig. 10.5. An n-p-n–p-n-p dc amplifier.

lower and the voltage gain much higher for this amplifier as compared with the Darlington amplifier because the input transistor is operating in the common-emitter, instead of the common-collector, configuration. The analysis of this type of amplifier is straightforward, as shown by the following example.

Example 10.3

Let us use the same transistor T_1 as in the Darlington amplifiers of Examples 10.1 and 10.2 and a p-n-p transistor with characteristics similar to T_2 and compare the current gain, input resistance, and voltage gain to the Darlington amplifier values at the same average current values. In Example 10.1 we found that $h_{ie2} = 40 \ \Omega$, $h_{FE2} = 100$, $R_L = 100 \ \Omega$, $h_{ie1} = 1.25 \ k\Omega$, and $h_{FE1} = 120$. Then the current gain is $K_i \simeq \beta_1 \beta_2 = 120 \times 100 = 1.2 \times 10^4$, $R_{in} \simeq h_{ie1} = 1.25 \ k\Omega$, and $K_v = K_i R_L / R_i = 1.2 \times 10^4 \times 100 / 1.25 \ k = 960$. Note that the voltage gain is increased by the same ratio as the reduction of input resistance. Observe also that the dc potential in the output is zero when the dc potential of the input is zero. Sometimes this is a distinct advantage.

The alternating n-p-n–p-n-p arrangement may be extended to include any desired number of transistors. For example, Fig. 10.6 uses three transistors. Notice that the zero signal output and input potential can be the same only if an even number of transistor stages is used in the amplifier.

As mentioned previously, the MOSFET is easy to use in dc amplifiers. Since both the gate and the drain require the same polarity of bias voltage, the circuit would appear as shown in Fig. 10.7. The value of R_{D1} must be chosen so the quiescent value of drain voltage for T_1 is equal to the desired quiescent value of gate voltage for T_2. Resistors could be added between the sources and ground to provide more design flexibility.

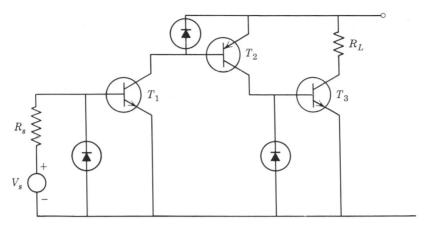

Fig. 10.6. Three-stage dc amplifier.

**10.3
DIFFERENTIAL
AMPLIFIERS**

The next type of dc amplifier to be considered is the differential amplifier. There are two basic types: balanced and unbalanced. The circuit diagram of a balanced differential amplifier is given in Fig. 10.8. In this amplifier the input signal V_i is balanced with respect to ground. With this type of signal, the forward bias of transistor T_1 is increased while the forward bias of transistor T_2 is decreased. If the transistors are matched and linear, the emitter current of one transistor increases by the same amount the emitter current of the other transistor decreases and the current through the common-emitter resistor R_E remains constant. Therefore, the voltage across R_E remains constant and no degeneration is caused by this resistor. If the transistors are not well matched, a small voltage will appear across R_E, but this voltage will tend to degenerate the higher gain transistor and regenerate the lower one, and thus improve the balance. Since the impedance between the base and the emitter of each transistor is h_{ie}, the input impedance from base to base is $2h_{ie}$.

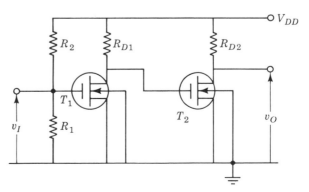

Fig. 10.7. MOSFET dc amplifier.

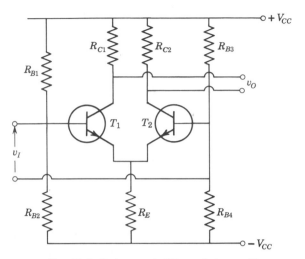

Fig. 10.8. Balanced differential amplifier.

The collector potential of one transistor (Fig. 10.8) increases while the collector potential of the other transistor decreases. The ouptut voltage, which is the difference between the collector potentials, may be fed to a balanced load or to another balanced amplifier. The balanced amplifier which follows may conveniently use the opposite type transistors (p-n-p to follow n-p-n) as shown in Fig. 10.9. However, the transistors may all be of the same type. Note that the two transistors in a balanced amplifier appear to be in series when viewed from the output

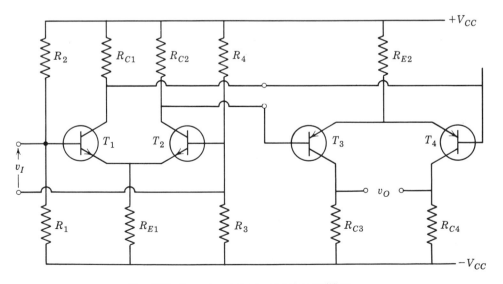

Fig. 10.9. Two-stage balanced amplifier.

terminals as they do when viewed from the input terminals. Therefore, the output resistance of the balanced amplifier is twice that of a single transistor amplifier at the same q-point. However, the current gain and voltage gain of the balanced amplifier are the same as a single transistor similarly biased with the emitter resistor perfectly bypassed.

The main advantage of the balanced amplifier is that in-phase input signals which are applied to the two bases do not produce an output signal, which is the difference between the two collector potentials. These in-phase signals are called *common-mode* signals and include thermally generated currents. Changes in v_{BE}, due to temperature changes and extraneous signals, are not transferred from one stage to the next providing the two transistors are well matched and maintained at the same temperature. Also, the individual transistors may have excellent q-point stability because of the high permissible value of emitter circuit resistance R_E.

The effective value of R_E can be increased greatly, and still allow the desired value of emitter current to flow, if a transistor is used to replace the resistor R_E as shown in Fig. 10.10. Since the collector current of the emitter-circuit transistor T_3 is determined almost entirely by the stabilized bias circuit D_1, R_1, R_2, and R_E', the sum of the emitter currents $2I_E$ of the differential amplifier is held constant and, therefore, the q-point collector currents and voltages of the differential amplifier are held constant. The diode D_1 in the stabilized-bias circuit compensates for the temperature variation of v_{BE} in transistor T_3. In other words, T_3 is a constant current source.

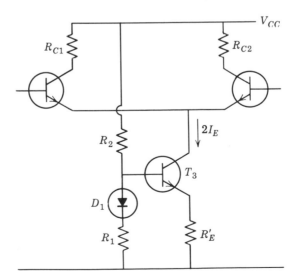

Fig. 10.10. Differential amplifier with improved q-point stabilization.

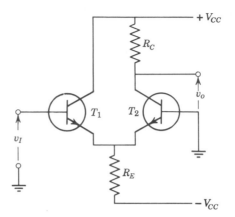

Fig. 10.11. Differential amplifier with unbalanced input and output.

When an input signal is applied to a differential amplifier, each collector voltage varies with respect to V_{CC} or ground, as previously noted. Therefore, an output signal which is referenced to V_{CC} or ground can be obtained from the difference amplifier to drive a single-ended amplifier or load. The amplifier then has a balanced input and an unbalanced output, and the voltage gain is decreased by a factor of two.

The differential amplifier can be used with both the input and the output unbalanced, as shown in Fig. 10.11. In this circuit, the base of one transistor is used as the ground reference. At first glance, it may appear that the unbypassed emitter resistor R_E will cause serious degeneration and low gain in the amplifier. However, the common-base input impedance h_{ib} of transistor T_2 is in parallel with R_E. Since h_{ib} is normally very small in comparison with R_E, the input impedance of transistor T_1 is

$$R_{it} \simeq h_{ie1} + (h_{fe1} + 1) h_{ib2} \qquad (10.3)$$

But if the two transistors are alike, $(h_{fe1} + 1) h_{ib2} = h_{ie1}$. Therefore,

$$R_{it} \simeq 2h_{ie1} \qquad (10.4)$$

When the output of a differential amplifier is unbalanced, the common-mode signals are not canceled in the output because the output voltage is referenced to a fixed point, V_{CC} or ground. However, the amplifier gain for common-mode signals may be very small because h_{ib2} is not in parallel with R_E when in-phase signals are applied to both bases. These in-phase signals could be extraneous signals, in a balanced input, or thermally induced signals ΔI_{CO} and ΔV_{BE}, in either a balanced or unbalanaced input. Thus, an index of the goodness of a differential amplifier is the *common-mode rejection ratio* (*CMRR*) defined as follows:

$$CMRR = \frac{\text{Voltage gain for difference signals}}{\text{Voltage gain for common-mode signals}} \qquad (10.5)$$

The voltage gain for difference signals is

$$K_v = \frac{K_i R_L}{R_i} \simeq \frac{\beta R_L}{2 h_{ie}} \qquad (10.6)$$

The voltage gain for common-mode signals is approxiamtely $R_L/2R_E$, since both emitter currents flow through R_E (which doubles the effectiveness of R_E). Then

$$CMRR \simeq \frac{(\beta R_L/2h_{ie})}{(R_L/2R_E)} = \frac{\beta R_E}{h_{ie}} \qquad (10.7)$$

Therefore, a high common-mode rejection ratio is obtained when R_E is very large in comparison with $h_{ie}/\beta = h_{ib}$. Notice that the unbalanced difference amplifier of Fig. 10.11 is actually a common-collector amplifier directly coupled to a common-base amplifier. As noted previously, the voltage gain of a common-collector amplifier is about one. However, for the unbalanced difference amplifier, the load on T_1 (Fig. 10.11) is the input impedance of a common-base amplifier which is very low. In fact, the input signal is applied across two forward-biased junctions (the base-emitter junctions of T_1 and T_2) in series. As a result, the voltage on the emitters is about one-half of the input voltage. Thus, the voltage gain of the common-collector stage is about 0.5 if two identical transistors are used.

**10.4
THE EMITTER-
COUPLED
AMPLIFIER**

The differential amplifier with unbalanced input and output (Fig. 10.11) is often called an *emitter-coupled amplifier*. This configuration has two distinct advantages over the single common-emitter amplifier in addition to the good thermal stability previously discussed. First, the emitter-coupled amplifier has a much wider bandwidth because the input transistor is in the common-collector configuration, so the Miller effect is essentially eliminated in this transistor. In addition, the common-base amplifier (T_2 of Fig. 10.11) has a very low input impedance and is driven from a low source impedance. Thus, $R_{sh} C_{sh}$ will be small and ω_h will be large. Second, the output signal is well isolated from the input by the grounded base of T_2. Consequently, the emitter-coupled configuration is popular in both *RF* and video amplifiers.

Example 10.4

Two identical transistors are connected as shown in Fig. 10.11. The resistor $R_C = 1$ kΩ and $R_E = 1$ kΩ. The transistor parameters are:

$$h_{ie} = 1 \text{ k}\Omega \qquad\qquad h_{FE} = 100$$

$$h_{re} \simeq 0 \qquad\qquad h_{oe} = 10^{-5} \text{ mhos}$$

Let us determine the characteristics of this amplifier.

First, the input impedance is $2h_{ie} = 2$ kΩ. Since T_2 is a common-base amplifier, the output impedance of transistor T_2 is very high (about $h_{FE}/h_{oe} \simeq 10$ MΩ). Then the output impedance of the amplifier is R_C in parallel with the output impedance of transistor T_2 or $R_o \simeq 1$ kΩ.

The voltage gain of T_1 is $\simeq 0.5$. The input impedance to transistor T_2 is $h_{ib} = h_{ie}/(h_{FE} + 1) = 1000/101 \simeq 10$ Ω. The voltage gain of T_2 is $\simeq R_C/h_{ib} = 1000/10 = 100$. Then the total voltage gain of the amplifier is $0.5 \times 100 = 50$.

The current gain of $T_1 \simeq h_{FE} = 100$, since $1/h_{oe} \gg$ the load on T_1 which is approximately the input impedance to T_2 or h_{ib}. In addition, since $R_C \ll$ the output resistance of transistor T_2, the current gain of $T_2 \simeq \alpha = 0.99 \simeq 1$. Thus, the total current gain of the amplifier is about $100 \times 1 = 100$. As a check, note that $K_v = K_i R_o/R_i = 100 \times 1000/2000 = 50$.

The CMRR is $h_{FE} R_E/h_{ie} = 100 \times 1000/1000 = 100$. Thus, the output signal v_O will be amplified 100 times as much as any temperature-induced voltage at the emitter-base junction of T_1 if T_2 is maintained at the same temperature.

The upper cutoff frequency, or bandwidth, of the emitter-coupled amplifier may be obtained by summing the time constants of the emitter follower and common base amplifiers developed in Chapter 6. Let us begin with Eq. 6.54, repeated below, to obtain the time constant of the emitter-follower input of the emitter-coupled amplifier.

$$V_o = \frac{I_s'}{(G_s'' + y_\mu)\left(1 + \dfrac{G_{ac}}{g_m}\right) + \dfrac{G_{ac}}{g_m} y_\pi} \tag{6.54}$$

where G_s'' is the reciprocal of $(r_x + R_b \| R_s)$ and G_{ac} is the ac load conductance which, for the emitter-coupled amplifier, $G_{ac} = 1/h_{ib} = (h_{fe} + 1)/h_{ie}$. But, since $g_m = h_{fe}/r_\pi$, $G_{ac} \simeq g_m$ for this particular value of load resistance. Making this approximation, Eq. 6.54 becomes

$$V_o = \frac{I_s'}{2(G_s'' + y_\mu) + y_\pi} \tag{10.8}$$

Now, letting $y_\pi = g_\pi + j\omega C_\pi$ and $y_\mu \simeq j\omega C_\mu$, Eq. 10.8 becomes

$$V_o = \frac{I_s'}{2G_s'' + g_\pi + j\omega(C_\pi + 2C_\mu)} \tag{10.9}$$

and the common collector time constant in this application is

$$\tau_{CC} = \frac{C}{G} = \frac{C_\pi + 2C_\mu}{2G_s'' + g_\pi} \tag{10.10}$$

The sum of the time constants for the common base part of the emitter-coupled amplifier may be obtained from Eq. 6.40.

$$\tau_{CB} = \frac{(r_x G_s' + 1) C_\pi}{(r_x G_s' + \beta_0 + 1) g_\pi + G_s'} + \frac{C_\mu}{G_{ac}} \tag{10.11}$$

where G_s' is the output resistance of the emitter-follower driver and G_{ac} is the total ac load conductance on the common base amplifier.

We may now obtain the upper cutoff frequency ω_h of the emitter-coupled amplifier by summing the time constants of Eqs. 10.11 and 10.10 and then taking the reciprocal of this sum.

$$\omega_h - \frac{1}{\dfrac{C_\pi + 2C_\mu}{2G_s'' + g_\pi} + \dfrac{(r_x G_s' + 1) C_\pi}{(r_x G_s' + \beta_0 + 1) g_\pi + G_s'} + \dfrac{C_\mu}{G_{ac}}} \tag{10.12}$$

Equation 10.12 may be more useful if we substitute $1/(R_s + r_x)$ for G_s'' and $(\beta + 1)/(R_s + r_x + r_\pi)$ for G_s'. Then, after some algebraic manipulation, Eq. 10.12 becomes

$$\omega_h = \frac{1}{\dfrac{(C_\pi + 2C_\mu)(R_s + r_x) r_\pi}{R_s + r_x + 2r_\pi} + \dfrac{[R_s + (\beta + 2) r_x + r_\pi] C_\pi}{(\beta + 1)[(R_s + 2r_x + r_\pi) g_\pi + 1]} + R_{ac} C_\mu} \tag{10.13}$$

Let us now compare the emitter-coupled amplifier with the CE, CB, and CC amplifier configurations discussed in Chapter 6 by using the same transistor characteristics.

Example 10.5	Two identical transistors having the following characteristics at the chosen q-point are used in an emitter-coupled amplifier with $R_s = 1$ kΩ $R_{ac} = 1$ kΩ, $r_x = 155$ Ω, $r_\pi = 695$ Ω, $g_m = 0.2$ S, $\beta_0 = 139$, $C_\pi = 10^{-10}$ F and $C_{ob} = 3 \times 10^{-12}$ F. Then,

$$\frac{1}{\omega_h} = \frac{(106 \times 10^{-12})(1155)(695)}{1000 + 155 + 1390}$$

$$+ \frac{(1000 + (141)\,155 + 695)\,10^{-10}}{140\,(1000 + 310 + 695)\,1.44 \times 10^{-3} + 1)} + 3 \times 10^{-9}$$

$$\omega_h = \frac{1}{33.4 \times 10^{-9} + 4.33 \times 10^{-9} + 3 \times 10^{-9}} = 2.455 \times 10^7 \text{ r/s}$$

$$f_h = 2.455 \times 10^7 / 6.28 = 3.91 \text{ MHz}$$

This bandwidth is 3.91 MHz / 0.524 MHz = 7.46 times as wide as the bandwidth of the common-emitter amplifier.

The current gain of the emitter-coupled amplifier, $K_i = (\beta + 1)(\alpha) = \beta_o = 139$ and the voltage gain $K_v = K_i R_L / R_i$. But, $R_i = 2h_{ie} = 1500$ Ω, so $K_v = 139\,(1000/1500) = 92.7$. The power gain $= 139 \times 92.7 = 12885$ and the power-gain bandwidth product is $12.885 \times 3.91 \times 10^9 = 50.4 \times 10^9$, which is $50.4/5.93 = 8.5$ times as great as the CE amplifier with the same characteristics and load.

You may have observed that Eq. 10.13 contains the general form of β, but we used β_o in solving for ω_h in Example 10.5. This approximation was appropriate because we needed to know f_h before we could determine the value of β at f_h from β_o and $f_T = 300$ MHz, given in the data sheet. A more accurate value of f_h may then be obtained by using this more accurate value of β. This update is left as an exercise for the student (Problem 10.8).

In addition to having a wide bandwidth and excellent thermal stability, the emitter-coupled amplifier has very good isolation between its output and input. FETs may be used as source-coupled amplifiers, of course, and their bandwidth may be obtained by letting $r_x = 0$, $r_\pi = \infty$, $G_s' = g_m$, $G_s'' = 1/R_s$, $C_\pi = C_{gs}$ and $C_\mu = C_{gd}$ in Eq. 10.12 to obtain

$$\omega_h = \frac{1}{\dfrac{(C_{gs} + 2C_{gd})}{2} R_s + \dfrac{C_{gs}}{g_m} + C_{gd} R_{ac}} \tag{10.14}$$

10.5
THE CASCODE
AMPLIFIER

An amplifier configuration somewhat related to the unbalanced differential amplifier is shown in Fig. 10.12a. This configuration is known as a *cascode* amplifier. Usually, operation from a single power supply is desirable, so the configuration shown in Fig. 10.12b is used. In this form (Fig. 10.12b), the resistors R_3 and R_4 maintain the base of transistor T_2 at a dc potential above 0 V. The capacitor C is normally connected from the base of transistor T_2 to ground to remove any signals which may be capacitively coupled into this base circuit.

The cascode amplifier is similar to the emitter-coupled amplifier inasmuch as the input amplifier has low gain and the output amplifier is in the common-base configuration. The essential difference is that the

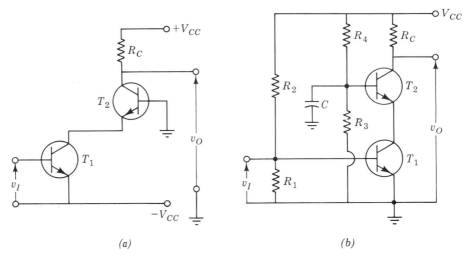

Fig. 10.12. Cascode amplifier: (a) two-power-supply configuration; (b) one-power-supply configuration.

input amplifier of the cascode arrangement is a common-emitter type amplifier having essentially unity voltage gain but high current gain, similar to an emitter follower. Since the load resistance on this amplifier is h_{ib}, the voltage gain equation $K_v = K_i R_L / R_i$ (Eq. 10.6) yields

$$K_{vi} = \frac{h_{fe} h_{ib}}{h_{ie}} \simeq 1 \qquad (10.15)$$

Similarly, if FET amplifiers are used, the load resistance is approximately $1/g_m$, so the voltage gain is essentially $g_m/g_m = 1$. Thus the effective input capacitance is reduced, in comparison with a common-emitter stage, because of the virtual elimination of the Miller effect. This also eliminates the detuning effect in a tuned amplifier.

The voltage gain of the entire cascode bipolar transistor amplifier may be determined from the relationship

$$K_v = \frac{h_{fe} R_\ell}{h_{ie}} \qquad (10.16)$$

When the voltage gain is desired from the driving source voltage v_s to the load, the driving source resistance R_s must be added to h_{ie} in the denominator of Eq. 10.16. This gain expression (Eq. 10.16) is essentially the same as that for a single common-emitter amplifier. However, the output resistance of the cascode configuration is approximately $1/h_{ob}$, since the output transistor is in the common-base configuration, so essentially all of the current $h_{fe} i_b$ passes through R_ℓ.

The following example of a cascode FET amplifier illustrates the principles of biasing all types of cascode amplifiers.

Example 10.6	Two identical FETs are connected as shown in Fig. 10.13. The driving source resistance $R_s = 10^6 \, \Omega$, $R_D = 5 \, k\Omega$, $C_{gs} = 5 \, pF$, and $C_{gd} = 4 \, pF$. The pinch-off voltage of each transistor is two volts and $V_{DD} = 30 \, V$. The g_m of each transistor is 2 mS. Let us determine the characteristics of this amplifier.

The resistors R_1 and R_2 establish the q-point drain voltages V_{DS} across both transistors. In Fig. 10.13, transistor T_1 has zero gate to source bias, so transistor T_2 must also have zero gate to source voltage in order for the drain currents of the two transistors to be identical. Of course, both transistors should have drain voltages V_{DS} higher than the pinch-off voltage V_p. In fact, this requirement can provide the basis for determining R_1 and R_2. With zero gate bias, the voltage drop across R_1 is equal to the q-point drain voltage V_{DS} of transistor T_1. Then,

$$V_{DS1} = \frac{V_{DD} R_1}{R_1 + R_2} \tag{10.17}$$

or

$$R_1 = \frac{V_{DS1} R_2}{V_{DD} - V_{DS1}} \tag{10.18}$$

The pinch-off voltage of our transistors is two volts, so let us choose V_{DS1} at least one or two volts greater than V_p. Let us choose $V_{DS1} = 5 \, V$. We must also choose a value for R_2. Let $R_2 = 220 \, k\Omega$ so it will not load the power supply. Then $R_1 = 5 \times 220 / (30 - 5) = 44 \, k\Omega$. The drain supply voltage for transistor T_2 is $30 - 5 = 25 \, V$. The design of the cascode amplifier can now proceed as though it were a single FET with $V_{DD} = 25 \, V$.

The voltage gain of transistor T_2 is $g_m R_D = 2 \times 10^{-3} \times 5 \times 10^3 = 10$. The voltage gain of the entire amplifier is $1 \times 10 = 10$.

The output impedance of this amplifier is R_D in parallel with the output impedance of transistor T_2. In this example, $R_o \simeq R_D = 5 \, k\Omega$.

Reverse bias will sometimes be desired, depending upon the temperature stability requirements, the magnitude of input signal, or the impedance of the load. A bias resistor and bypass capacitor must then be included in the source circuit of transistor T_1, as shown in Fig. 10.14, to provide the proper bias. Transistor T_2 must have the same drain

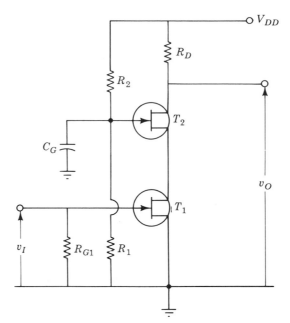

Fig. 10.13. FET cascode amplifier.

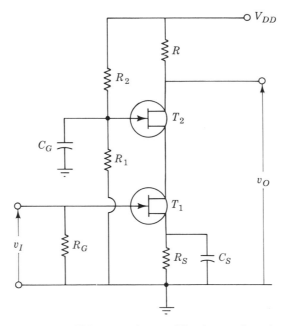

Fig. 10.14. FET cascode amplifier for ac signals.

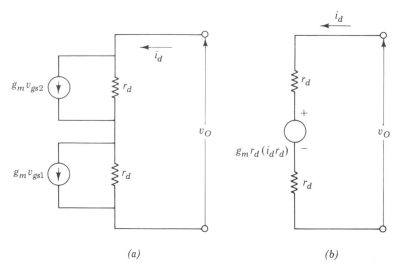

(a)   *(b)*

Fig. 10.15. Equivalent circuit for a cascode FET configuration: (a) current
source model; (b) modified voltage source model.

current and will automatically adjust to the same bias if both transistors
have V_{DS} greater than V_p. However, for the n-channel transistors shown,
the gate potential of transistor T_2 is more negative than the drain potential of transistor T_1 by the amount of the bias voltage V_{GS}. But this same
amount of voltage is dropped across the bias resistor of transistor T_1, so
the voltage drop across R_1 is still equal to the drain-to-source voltage V_{DS}
of transistor T_1. Therefore, the design of the biased cascode amplifier
can proceed in the same manner as for a single FET with source bias, but
the effective drain supply voltage is V_{DD} minus the V_{DS} of transistor T_1.

The absence of coupling between the output and the input circuits of
the cascode amplifier is a great advantage in some applications, particularly in tuned radio-frequency amplifiers. Also, we should note that
stray wiring capacitance should be kept to an absolute minimum since
it adds directly to the FET capacitance, which is very small in the
cascode configuration.

The transistor output impedance of the cascode configuration is very
high, as shown below. The equivalent circuit of the cascode FET amplifier is as shown in Fig. 10.15a. However, the voltage v_o applied to the
output terminals does not cause current to flow in the gate circuit of
transistor T_1 at low and moderate frequencies, so $v_{gs1} \simeq 0$. On the other
hand, the source-to-ground voltage of T_2 is caused by the $i_d r_d$ drop of T_1
in the source circuit of T_2. Therefore, since the gate of T_2 is at ground
potential,

$$v_{gs2} = -v_{sg2} = -i_d r_d \qquad (10.19)$$

The circuit of Fig. 10.15a is simplified to that shown in Fig. 10.15b by transforming the current source $g_m v_{gs2}$ to a voltage source $g_m r_d v_{gs2}$ and then replacing v_{gs2} with $-i_d r_d$. The polarity of the voltage generator is reversed to eliminate the negative sign. Observe from Fig. 10.15b that

$$v_o = i_d (2r_d) + g_m r_d^2 i_d \tag{10.20}$$

and

$$r_{ot} = \frac{v_o}{i_d} = 2r_d + g_m r_d^2 \tag{10.21}$$

Simplifying,

$$r_{ot} = r_d (2 + g_m r_d) \tag{10.22}$$

Since $g_m r_d$ is much greater than one, the output resistance r_{ot} is extremely high. For example, a typical FET with $g_m = 2$ mS and $r_d = 10^5$ Ω would have $r_{ot} = 10^5 (2 + 200) \simeq 20$ $M\Omega$ when placed in the cascode configuration. This is the same order of magnitude as the $1/h_{ob}$ of the bipolar cascode configuration previously discussed.

The bandwidth of a cascode amplifier may be determined by adding the open-circuit time constants, as we did in the emitter-coupled amplifier, and then taking the reciprocal of this sum. We will consider the bipolar cascode configuration first and then adapt our results to the FET configuration as we did for the emitter-coupled amplifier. In fact, we have already developed equations for the open-circuit time constants of the three basic transistor configurations, so we need only adapt them to the cascode arrangement. For example, the input transistor is in the common-emitter configuration and has $g_m R_{ac} = 1$, so inverting Eq. 6.20 to obtain the time constant of the input transistor,

$$\tau_{CE} = \frac{(C_\pi + 2C_\mu)}{G_s'' + g_\pi} \tag{10.23}$$

The output transistor is in the common-base configuration and its driving source conductance G_s' is very small, being approximately h_{oe}. Therefore, we may use Eq. 10.11 to obtain the open-circuit time constant for this transistor. However, with G_s' being very small and $(\beta + 1)g_\pi \simeq g_m$, Eq. 10.11 simplifies to

$$\tau_{CB} = \frac{C_\pi}{g_m} + R_{ac} C_\mu \tag{10.24}$$

Thus, the upper-cutoff frequency of the cascode amplifier is the reciprocal of the sum of these two time constants (Eqs. 10.23 and 10.24),

$$\omega_h = \cfrac{1}{\cfrac{C_\pi + 2C_\mu}{G_s'' + g_\pi} + \cfrac{C_\pi}{g_m} + R_{ac}C_\mu} \tag{10.25}$$

Again, Eq. 10.25 may be adapted to FET cascode amplifiers by letting $r_x = 0$, $r_\pi = \infty$, $C_\pi = C_{gs}$ and $C_\mu = C_{gd}$ as follows:

$$\omega_h = \cfrac{1}{R_s(C_{gs} + 2C_{gd}) + \cfrac{C_{gs}}{g_m} + R_{ac}C_{gd}} \tag{10.26}$$

Example 10.7

Let us compare the bandwidth and the power-gain bandwidth product of the cascode amplifier to that of the emitter-coupled amplifier by using the same transistors and circuit elements for the cascode as we did for the emitter-coupled amplifier of Example 10.5. The needed element values are $r_\pi = 695$ Ω, $r_x = 155$ Ω, $C_\pi = 100$ pF, $C_\mu = 3$ pF, $g_m = .2$ S and $R_s = R_{ac} = 1$ kΩ. Then, using Eq. 10.25, $\omega_h = 1/[106 \times 10^{-12}/(.866 + 1.44) \times 10^{-3}) + 10^{-10}/.2 + 10^3 \times 3 \times 10^{-12}]$ $= 1/(46 + .5 + 3) \times 10^{-9} = 2.02 \times 10^7$ $r/s = 3.22 \times 10^6$ Hz. The current gain is $\beta = 139$ and the voltage gain is $\beta R_{ac}/R_{in} = 139 \times 10^3/750 = 185$, so the power-gain bandwidth product is $139 \times 185 \times 3.22 \times 10^6 = 82.8 \times 10^9$. Thus, in this example, the cascode configuration has slightly less bandwidth, but higher gain-bandwidth product than the emitter-coupled amplifier.

The major advantage of the cascode amplifier is the excellent isolation between its output and input ports, which becomes very important in high-frequency tuned amplifiers. On the other hand, the emitter-coupled amplifier has much better thermal stability. The advantages of both the cascode and emitter-coupled configurations are achieved in the circuit shown in Fig. 10.16. This amplifier is known as an emitter-coupled cascode amplifier.

10.6 COMPLE-MENTARY-SYMMETRY AMPLIFIERS

Another commonly used amplifier configuration is the *complementary-symmetry* amplifier. The most commonly used configuration, the emitter follower, is shown in Fig. 10.17. The name complementary-symmetry is derived from the use of two transistors that have similar (symmetrical) characteristics but opposite (complementary) doping polarities. As seen in Fig. 10.17, the current through the load resistor R_L is the difference between the two transistor currents. Therefore, with zero input voltage and matched transistors in Fig. 10.17, the current through R_L is zero, providing $R_1 = R_4$ and $R_2 = R_3$. (These resistors provide forward bias to the transistors in the zero-signal condition.) However, when the input signal goes positive, the forward bias of transistor T_1

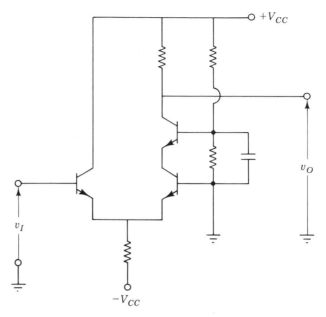

Fig. 10.16. *Emitter-coupled cascode amplifier.*

increases while the forward bias of transistor T_2 decreases. This causes the collector current of T_1 to increase while the collector current of T_2 decreases. The difference between these two currents flows down through R_L and causes its emitter end to go positive. Conversely, when the input signal goes negative, the base bias of transistor T_2 is increased while the base bias of transistor T_1 is decreased. Thus, the direction of

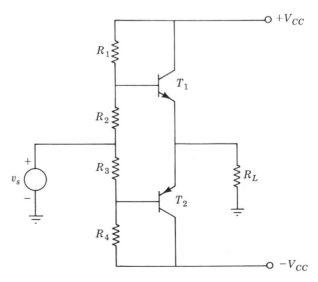

Fig. 10.17. *Complementary-symmetry emitter follower using resistive bias.*

current through R_L is reversed and its emitter terminal goes negative with respect to ground.

Observe that the quiescent collector currents of the complementary-symmetry amplifier do not pass through the load resistance even though there is no blocking capacitor. This is also true for the balanced differential amplifier. However, the significant difference is that the peak load current in the differential amplifier, or any other amplifier that we have previously studied, cannot exceed the quiescent current of the amplifying device. But the load current in the complementary-symmetry amplifier is independent of the quiescent current. Therefore, the quiescent current may be essentially zero while the peak load current is limited only by the power dissipation capability of the transistors and the current available from the power supply. For example, if an alternating signal is applied to the input, the n-p-n transistor will conduct the load current during the positive half cycle while the p-n-p transistor may be cut off, or nonconducting. Conversely, the p-n-p transistor may conduct the entire load current during the negative-going input half cycle while the n-p-n transistor is cut off. Thus, with no input signal, there would be no current through the device. This reduction or elimination of q-point current markedly decreases the power dissipation in the transistors, thus increasing efficiency. Also, the output current and power capability is increased significantly because of the higher efficiency and because the output current is limited only by the load resistance and the power supply voltage and not by the q-point current.

Each transistor in the complementary-symmetry amplifier requires some forward bias voltage in order to eliminate the need for the input signal to reach a few tenths of a volt before the transistor begins to conduct significantly on each half cycle. This forward bias is provided by the voltage drop across resistors R_2 and R_3 in Fig. 10.17. Lack of this forward bias results in a dead zone between half cycles, and the resulting distortion is known as crossover distortion. Figure 10.18a shows how the forward bias voltage may be determined in order to eliminate crossover distortion and yet maintain very small quiescent collector currents. The relatively straight part of the input characteristics is extended to intersect the v_{BE} axis. The voltage at this intersection is the desired projected-cutoff bias. When the p-n-p input characteristic is turned upside down and placed back-to-back with the n-p-n curve so the two projected-cutoff bias points coincide, as shown in Fig. 10.18a, the relatively straight line connecting the two input curves and drawn through the coincident bias point is called the composite input characteristic.

The emitter-follower version of the complementary-symmetry amplifier has essentially unity voltage gain and is usually connected to anoth-

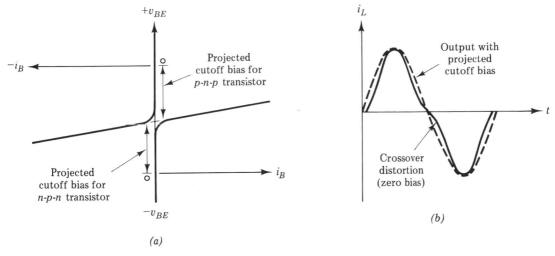

Fig. 10.18. (a) Back-to-back input characteristics showing projected cutoff bias to eliminate crossover distortion. (b) Output current waveform with and without projected-cutoff bias.

er amplifier having voltage gain as shown in Fig. 10.19. In this amplifier, the voltage drop across diodes D_1 and D_2 provides the forward bias to eliminate crossover distortion. The resistor R_C serves as a load resistor for the driving transistor T_3 and also provides forward bias for transistor T_1. At quiescent conditions, this forward bias current is negligible and the driving transistor T_3 may be designed as a regular common-emitter amplifier. The diodes D_1 and D_2 have the effect of subtracting two diode drops from the power-supply voltage. The point between these diodes should be at ground potential under quiescent conditions, so the quiescent drop across R_C should be $+V_{CC} - V_{diode}$. As the driving transistor is driven toward cutoff, the current through R_C decreases and the base of T_1, and hence the output voltage v_o, rise toward $+V_{CC}$. However, when T_3 is cut off, the base of T_1 cannot rise to $+V_{CC}$ because the base current of T_1, which is now maximum, flows through R_C and causes a voltage drop across it. The output voltage v_o is lower than this base voltage by the voltage v_{BE} across T_1. Therefore, if large output voltages are desirable, the circuit designer must choose a value of R_C that will produce only a small voltage drop (perhaps one or two volts) when the maximum base current of T_1 flows through it.

As the driving transistor T_3 is positively biased (toward saturation), the output voltage falls toward $-V_{CC}$. However, the minimum possible base potential of T_2 is $-V_{CC}$, plus the voltage drop across R_E, plus the saturation voltage of T_3. Therefore, for symmetrical clipping in the output, the minimum drop across R_C during the positive half cycle

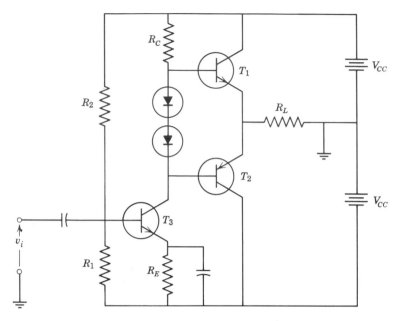

*Fig. 10.19. Voltage amplifier directly coupled to a
complementary-symmetry amplifier.*

should be equal to the drop across R_E plus the saturation voltage of T_3 during the negative half cycle.

The circuit of Fig. 10.19 is suitable for only ac signals because of the blocking and bypass capacitors in the common-emitter driver. A suitable dc amplifier might use an emitter-coupled driving arrangement, as shown in Fig. 10.20. In this circuit, the emitter-coupled driver is inverted because it uses p-n-p transistors, but its operation is essentially identical to that of the common-emitter driver of Fig. 10.19. The voltage drop across R_{E2} should not exceed 2 or 3 volts if large output voltage swings are desirable. The bias resistors R_1 and R_2 should have low enough resistance to provide good thermal stability for the p-n-p stage and to minimize degeneration due to signal currents flowing in these resistors. Power dissipation or power-supply drain are the only limiting factors in choosing low values of resistance for these resistors. Perhaps $S_I = 2$ or 3 would be suitable.

A final complementary-symmetry circuit is shown in Fig. 10.21. This is the common-emitter n-p-n–p-n-p arrangement. The resistor R_E provides necessary q-point stability in the output but causes degeneration, or negative feedback, and thus reduces the gain. This circuit is discussed in greater detail in Chapter 15 after negative feedback has been treated. The efficiency of the various circuits is also discussed in Chapter 15.

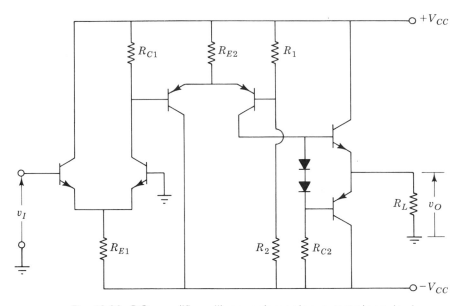

Fig. 10.20. DC amplifier with complementary-symmetry output.

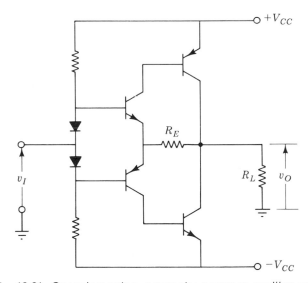

Fig. 10.21. Complementary-symmetry common-emitter amplifier.

Problems   *Section 10.1*

10.1   A given silicon power transistor, which has $h_{ie} = 10 \ \Omega$, $\beta = 120$, and $I_{CO} = -80$ nA at the desired q-point, is driven by a Darlington-connected silicon transistor with $h_{ie} = 1500 \ \Omega$, $\beta = 150$, and $I_{CO} = 5$ nA at its desired q-point. Assume h_{re} and

h_{oe} are negligible. The amplifier is diode stabilized. If the load resistance in the collector circuit of the power transistor is 30 Ω and $V_{CC} = -20$ V, determine the low-frequency input resistance and the voltage gain of the amplifier.

10.2   If the Darlington-connected amplifier of Problem 10.1 is capacitively coupled to its driving source, draw a circuit diagram of the amplifier and determine the value of bias resistance if $I_{C2} = 300$ mA. What should be the values of I_S? Assume that all values are given at $T = 25°C$.

Section 10.2

10.3   The transistors of Problem 10.1 are connected in the n-p-n-p-n-p arrangement of Fig. 10.5 with $V_{CC} = 20$ V and $R_L = 30$ ohms. Determine the voltage gain of the amplifier at the q-point for which the h_{ie} of the input amplifier is 1500 ohms. Compare the voltage gain and input resistance of this amplifier with those of the Darlington amplifier of Problem 10.1.

10.4   Two p-channel MOSFETs are connected as a dc amplifier. Draw the circuit diagram of this amplifier. The characteristics of these MOSFETs are given in Fig. 8.22a. If V_{DD} is -20 V and the desired quiescent values of $v_g = -10$ V, find the values of all resistors in the circuit. What is the voltage gain of each MOSFET?

Section 10.3

10.5   Two 2N3903 transistors (see Appendix I) are used in the emitter-coupled configuration of Fig. 10.11 with $I_C = 5$ mA, $V_{CC} = 15$ V, $-V_{CC} = -15$ V, and $R_s = 1.5$ kΩ. Draw the circuit diagram, determine the circuit component values, and calculate the voltage gain v_o/v_s assuming R_C is the total load resistance.

10.6   Two 2N3903 transistors are connected as shown in Fig. 10.11. $R_C = 2$ kΩ and $R_E = 2$ kΩ. The quiescent collector current of both transistors is 5 mA. Determine R_i, R_o, K_v, K_i, and the CMRR, if the driving source resistance $R_s = 1$ kΩ.

Section 10.4

10.7   Show that Eq. 10.13 may be obtained from Eq. 10.12.

10.8 The transistor used in Example 10.5 has $\beta_o = 139$ and $f_T = 300$ MHz. Determine the value of β for this transistor at $f_h = 3.91$ MHz, the approximate high-frequency cutoff of the amplifier of Example 10.5. Calculate a more precise value of f_h for the amplifier, using your value of β at 3.91 MHz, and determine the error caused by using β_o in Example 10.5.

10.9 Determine the bandwidth for the emitter-coupled amplifier of Problem 10.6.

10.10 A given FET has $g_m = 5$ mS at $I_D = 5$ mA and $V_{GS} = -0.5$ V. $C_{iss} = 7$ pF and $C_{rss} = 3$ pF. Two of these transistors are used in a source-coupled configuration with $R_L = 5$ kohm.

 a. Draw a circuit diagram for the amplifier and determine suitable values for the common source resistance and V_{DD} to provide $I_D = 5$ mA.
 b. Calculate the bandwidth of the amplifier.

Section 10.5

10.11 Two 2N3903 transistors are used in the cascode configuration of Fig. 10.12. If $I_C = 5$ mA, $V_{CC} = 20$ V and $R_s = 1.5$ kohms:

 a. Determine suitable values for the circuit components.
 b. Determine the voltage gain V_O/V_I for the amplifier.
 c. Calculate the bandwidth of the amplifier.
 d. Compare the gain-bandwidth product of this amplifier with that of the emitter-coupled amplifier of Problems 10.6 and 10.9.

10.12 Use the FETs of Problem 10.10 in a cascode configuration with $I_D = 5$ mA and $V_{GS} = -0.5$ V.

 a. Determine suitable values for the circuit components.
 b. Calculate the voltage gain V_O/V_I.
 c. Calculate the bandwidth of the amplifier.
 d. Compare the gain-bandwidth product with that of the amplifier of Problem 10.10.

Section 10.6

10.13 The circuit of Fig. 10.19 has $+V_{CC} = 15$ V and $-V_{CC} = -15$ V with respect to ground. $R_L = 200\ \Omega$ and the h_{fe} of all the silicon transistors $= 100$. Determine the values of all the resistors.

10.14 The circuit of Fig. 10.20 has $R_L = 100$ ohms, $+V_{CC} = 20$ V and $-V_{CC} = -20$ V. All the transistors are silicon and have $\beta_o = 100$.

a. Determine suitable values for all the resistors in the circuit.
b. Determine the peak value of V_I required to provide maximum power output without distortion.

11

Multistage Amplifiers

Most practical amplifiers require more gain than can be obtained from a single stage. Consequently, it is a common practice to feed the output of one amplifier stage into the input of the next stage (Fig. 11.1). When amplifiers are connected in this fashion, they are called *cascaded amplifiers* or *multistage amplifiers*. We have previously considered the gain, bandwidth, and some coupling techniques for two or more stages, but there are a few concepts, unique to cascaded amplifiers, which will be considered in this chapter.

11.1 GAIN AND BANDWIDTH CONSIDERATIONS IN CASCADED AMPLIFIERS

Most of the cascaded amplifier stages are used to obtain either a voltage gain or a current gain. However, in most cascaded amplifiers, it is ultimately the power gain that is important. When the proper level of signal has been obtained, a power amplifier stage is used to produce sufficient power to activate the required load device (loudspeaker, servo motor, antenna, etc.). If a voltage gain is required, we can calculate the total gain by using the equation for voltage gain of one stage. Thus, from Fig. 11.1, the voltage gain for stage 1 is

$$G_1 = \frac{V_2}{V_1} \qquad (11.1)$$

In addition, the voltage gain for stage 2 is

$$G_2 = \frac{V_3}{V_2} \qquad (11.2)$$

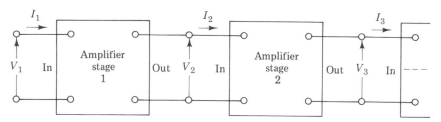

Fig. 11.1. Cascaded amplifier stages.

The gain for additional stages can be written in a similar manner. Then, the total amplifier voltage gain G_A for n cascaded stages is

$$\frac{V_2}{V_1} \times \frac{V_3}{V_2} \times \frac{V_4}{V_3} \times \cdots \frac{V_{n+1}}{V_n} = \frac{V_{n+1}}{V_1} \qquad (11.3)$$

or

$$G_A = G_1 \times G_2 \times G_3 \times \cdots G_n \qquad (11.4)$$

Obviously, a similar derivation could have been achieved for current gains or for power gains. In either of these cases, the total amplifier gain is equal to the product of the individual stage gains as indicated by Eq. 11.4. In general, the individual stage gains are functions of s and, consequently, the amplifier gain G_A is also a function of s. For steady-state sinusoidal signals, s becomes $j\omega$ and G_1, G_2, etc., become magnitudes at given phase angles. Then, G_A will be equal in magnitude to the products of all the magnitudes with a phase angle equal to the sum of the individual stage phase shifts.

A reference gain K has been defined for the individual stages. An amplifier reference gain K_A can be found from the individual stage gains.

$$\pm K_A = (\pm K_1) \times (\pm K_2) \times (\pm K_3) \times \cdots (\pm K_n) \qquad (11.5)$$

The K_A term will be positive if the total number of phase inversions is even and negative if the total number of inversions is odd. Since K is a magnitude only, this relationship does not involve s.

We have previously defined ω_ℓ and ω_h as the lower and upper cutoff frequencies, respectively. These are the frequencies at which the voltage or current gain of one stage has been reduced to 0.707 of its reference value. (Power gain is reduced to 0.5 of its reference value for resistive loads.) Now, if we have an amplifier with two identical stages of amplification, the voltage gain at ω_ℓ wil be reduced by a factor 0.707 in each stage. Thus, thc amplifier gain at ω_ℓ (and also ω_h) will be

$$0.707\,(-K_{v1}) \times 0.707\,(-K_{v2}) = 0.5\,K_A \qquad (11.6)$$

In fact, for n identical cascaded stages of amplification, the gain at ω_ℓ and ω_h will be $(0.707)^n K_A$.

In order to establish a relationship for the bandwidth of the total amplifier, let us define ω_L as the lower-cutoff frequency of the cascaded amplifier and ω_H as the upper-cutoff frequency of the cascaded amplifier. At these frequencies, the gain of the amplifier is $0.707 K_A$. In order to arrive at the relationship between both the single-stage lower-cutoff frequency and ω_L and the single-stage upper-cutoff frequency ω_h and that of the entire amplifier ω_H, consider an RC-coupled amplifier containing n identical cascaded stages. The voltage gain of a single stage for a steady-state sinusoidal input signal in the low-frequency range may be expressed as

$$G = -K \frac{j\omega}{j\omega + \omega_\ell} \tag{11.7}$$

or

$$G = -K \frac{1}{1 - j(\omega_\ell/\omega)} \tag{11.8}$$

where G is a function of frequency and is often expressed as $G(\omega)$. The magnitude of this function may be written as

$$G = -K \frac{1}{[1 + (\omega_\ell/\omega)^2]^{1/2}} \tag{11.9}$$

If there are n cascaded stages, the magnitude of the amplifier gain, from Eq. 11.4, is

$$|G_A| = |G|^n = K^n \left(\frac{1}{[1 + (\omega_\ell/\omega)^2]^{1/2}} \right)^n \tag{11.10}$$

Since $K^n = K_A$, we can write

$$|G_A| = K_A \frac{1}{[1 + (\omega_\ell/\omega)^2]^{n/2}} \tag{11.11}$$

Now, if ω is to be equal to ω_L, the term multiplying K_A must be equal to 0.707 or $1/(2)^{1/2}$. Then

$$2^{1/2} = \left[1 + \left(\frac{\omega_\ell}{\omega_L} \right)^2 \right]^{n/2} \tag{11.12}$$

or

$$2^{1/n} = 1 + \left(\frac{\omega_\ell}{\omega_L} \right)^2 \tag{11.13}$$

This equation is solved for ω_L^2 to yield

$$\omega_L^2 = \frac{\omega_\ell^2}{2^{1/n} - 1} \tag{11.14}$$

or

$$\omega_L = \frac{\omega_\ell}{[2^{1/n} - 1]^{1/2}} \tag{11.15}$$

The sinusoidal steady-state voltage gain of a single stage amplifier in the high-frequency range may be expressed by the following equation.

$$G = -K \frac{\omega_h}{j\omega + \omega_h} = -K \frac{1}{1 + j\omega/\omega_h} \tag{11.16}$$

Starting with Eq. 11.04 and following the procedure we used for the low-frequency range, the following relationship between ω_h and ω_H may be obtained (Problem 11.1).

$$\omega_H = \omega_h [2^{1/n} - 1]^{1/2} \tag{11.17}$$

Table 11.1 gives the relationships between the lower and upper cutoff frequencies of each stage to those of the multi-stage amplifier.

Equation 11.17 and Table 11.1 show that the total bandwidth of two identical stages is equal to 0.644 times the bandwidth of either individual stage. However, the *sum of open-circuit time constants* technique used in Chapter 10 yields 0.5 as the proper ratio because the sum of time constants for two identical stages is double that for the single stage. The *sum of time constant* method is admittedly less accurate but its somewhat pessimistic bandwidth prediction is good for design purposes because it provides a built-in safety factor. Also, the accuracy of the time-constant summation method improves as the stages become more *dissimilar*.

From Eqs. 11.15 and 11.17, or from Table 11.1, ω_L will be greater than ω_ℓ if n is greater than one and ω_H will be less than ω_h if n is greater than one. Thus, the bandwidth of the amplifier decreases as the number of cascaded stages increases. Or, if the amplifier bandwidth is to remain constant, the stage bandwidth must increase as the number of cascaded

Table 11.1.   Ratios ω_ℓ/ω_L and ω_h/ω_H for an n stage amplifier.

n	1	2	3	4	5	6	7	8
ω_ℓ/ω_L	1	0.644	0.510	0.435	0.387	0.348	0.333	0.30
ω_h/ω_H	1	1.55	1.96	2.3	2.58	2.88	3.00	3.33

stages increases. This last statement leads to an interesting dilemma. If a high-gain, very wide-band amplifier is desired, stages must be cascaded to obtain the higher gain. However, as more stages are cascaded, the bandwidth of each stage must be increased. Unfortunately, as noted in Section 6.3, the gain-bandwidth product *may be a constant*. Under these conditions, the increased bandwidth results in reduced gain per stage. Thus, in order to compensate for the reduced gain per stage, more stages of greater bandwidth are required. This process can be carried so far that the total amplifier gain (for a given bandwidth) *may actually decrease as additional cascaded stages are added*.

So far, the work on cascaded amplifiers has been primarily concerned with gain and bandwidth of the amplifier. If a step function is applied to a single-stage amplifier, the rise time was given in Chapter 7 as being approximately equal to $2.2/\omega_h$. If this same concept is to be applied to a cascaded amplifier, the expected rise time will be about $2.2/\omega_H$. In addition, Elmore[1] has shown that the overall rise time of the cascaded amplifier T_{AR} is

$$T_{AR} = (T_{R1}^2 + T_{R2}^2 + T_{R3}^2 + \ldots)^{1/2} \tag{11.18}$$

where

T_{R1} is the rise time of stage one

T_{R2} is the rise time of stage two, etc.

Thus, if *n* identical stages with a rise time of T_{RS} are cascaded, the rise time of the total amplifier T_{AR} will be

$$T_{AR} = T_{RS}(n)^{1/2} \tag{11.19}$$

**11.2
dB GAIN**

Power gain in bel units is defined as

$$B = \log \frac{P_2}{P_1} \tag{11.20}$$

where "log" is the logarithm to the base 10 and B is in *bels*. P_2/P_1 is the power ratio between the points in question.

The bel unit is convenient because it reduces a multiplication problem, in the case of the gain of a cascaded amplifier, to an addition

---

[1] W. C. Elmore, "Transient Response of Damped Linear Network with Particular Regard to Wideband Amplifiers," *Journal of Applied Physics*, Vol. 19, January 1948, pp. 55–62.

problem. Nevertheless, the bel is an inconveniently large unit because a power gain of 10 is only 1 bel. Therefore, the decibel (dB) has been accepted as the practical unit. The dB unit has the additional advantage that a power change of 1 dB in an audio system is barely discernible to the ear, which has a logarithmic response to intensity changes.

$$dB = 10 \log \frac{P_2}{P_1} \tag{11.21}$$

Also

$$dB = 10 \log \frac{V_2^2 / R_2}{V_1^2 / R_1} \tag{11.22}$$

If the resistance is the same at the two points of reference,

$$dB = 10 \log \left(\frac{V_2}{V_1}\right)^2 = 20 \log \frac{V_2}{V_1} \tag{11.23}$$

Similarly,

$$dB = 20 \log \frac{I_2}{I_1} \tag{11.24}$$

The impedance levels are frequently of secondary importance in a voltage or current amplifier. Therefore, the Eqs. 11.23 and 11.24 are sometimes loosely used without regard to the relative resistance levels.

It is often convenient to express a power level in dB with regard to a given reference level. One commonly used reference level is 6 mW. Another commonly used reference level is 1 mW. When this (1 mW) reference level is used, the dB units are usually called *volume units* (vu) or dBm. On the other hand, the open-circuit output voltage of a microphone is usually rated in dB with reference to 1 V when the standard excess acoustical pressure is one microbar, or one-millionth of standard barometric pressure.

11.3
THE
S-DOMAIN
AND
FREQUENCY
DOMAIN

The voltage across an inductor is equal to $L \, di/dt$ and the voltage across a capacitor is equal to $(1/C) \int i \, dt$. Thus, the current or voltage equations of even fairly simple circuits involve differential equations. The Laplace transform provides a simple method for solving differential equations. Table 11.2 provides several Laplace transforms and their corresponding equations in the time domain. Since the Laplace equations contain functions of the Laplace operator s, this region is usually referred to as the *s-domain*. The operator s is equal to $(\sigma + j\omega)$, so the s-domain is also called the *complex frequency domain*.

Table 11.2. Some Laplace transforms.

$f(t)$	Laplace transforms $G(s)$
$Ku(t)$	$\dfrac{K}{s}$
t	$\dfrac{1}{s^2}$
e^{at}	$\dfrac{1}{s-a}$
$\sin \omega t$	$\dfrac{\omega}{s^2+\omega^2}$
$\cos \omega t$	$\dfrac{s}{s^2+\omega^2}$
$e^{-at}\sin \omega t$	$\dfrac{\omega}{(s+a)^2+\omega^2}$
$e^{-at}\cos \omega t$	$\dfrac{(s+a)}{(s+a)^2+\omega^2}$
$\sinh at$	$\dfrac{a}{s^2-a^2}$
$\cosh at$	$\dfrac{s}{s^2-a^2}$

The solution of a differential equation comprises two components. One of these components is the *steady-state component* of the current or voltage waveform. The usual steady-state solution of an ac circuit will produce this steady-state component of the waveform. The second component in a differential equation is the *transient component*. This transient component serves to connect the initial voltage or current waveform to the final steady-state waveform.

In this chapter, we will be mainly concerned with steady-state ac waveforms. Then, the complex-frequency operator s can be replaced by $j\omega$ and we will be operating in the frequency domain. Of course, the transient component of the solution will be lost when this simplification is used.

11.4 STRAIGHT-LINE APPROXIMATIONS OF GAIN AND PHASE CHARACTERISTICS (BODE PLOTS)

The analysis given in Section 11.1 is sufficient for the analysis and design of identical cascaded stages. This section will present a method of analysis which can be used on any amplifier. However, as an introduction, we will apply this method to a single RC-coupled stage. The transfer function of a single RC stage amplifier can be written by combining the equations derived for each frequency range of the single stage amplifier (Eqs. 11.7 and 11.16). A single equation containing both of the poles and zeros in these equations is written in the s-domain that follows.

$$G = -K \frac{s\,\omega_h}{(s + \omega_\ell)(s + \omega_h)} \tag{11.25}$$

Now, consider the high-frequency range with $s = j\omega$ for steady-state sinusoidal signals and $\omega \gg \omega_\ell$. Then, the term $(s + \omega_\ell) \rightarrow j\omega$ and Eq. 11.25 becomes

$$G = -K \frac{j\omega\omega_h}{j\omega(j\omega + \omega_h)} = -K \frac{\omega_h}{j\omega + \omega_h} \tag{11.26}$$

Note that Eq. 11.26 is the same as Eq. 11.16, which was derived for the high-frequency range.

The mid-frequency range occurs where $s = j\omega$ and $\omega \ll \omega_h$, but $\omega \gg \omega_\ell$. Then, the term $(s + \omega_\ell) \rightarrow j\omega$ and the term $(s + \omega_h) \rightarrow \omega_h$. With these substitutions, Eq. 11.25 becomes

$$G = -K \frac{s\,\omega_h}{s\,(\omega_h)} = -K \tag{11.27}$$

Note that G reduces to the mid-frequency, or reference, gain as it should. The negative sign results from the assumed CE configuration.

Finally, in the low-frequency range, $s = j\omega$ and $\omega \ll \omega_h$. Then, the term $(s + \omega_h) \rightarrow \omega_h$ and Eq. 11.25 reduces to

$$G = -K \frac{s\,\omega_h}{(s + \omega_\ell)\,\omega_h} = -K \frac{j\omega}{j\omega + \omega_\ell} \tag{11.28}$$

Since this equation is identical to Eq. 11.7, the general equation 11.25 is valid for all frequency ranges.

Now, let us examine the very-low frequency range where $s = j\omega$ and ω is small compared to ω_ℓ. Then, $(s + \omega_\ell) \rightarrow \omega_\ell$ and $(s + \omega_h) \rightarrow \omega_h$. Eq. 11.24 then becomes

$$G = -K \frac{s\,\omega_h}{\omega_\ell\omega_h} = -K \frac{j\omega}{\omega_\ell} \tag{11.29}$$

Thus, the gain of the amplifier is proportional to frequency and the phase angle of G is nearly 270°. An interesting method of expressing the magnitude relationship exists. In musical terms, the frequency doubles every octave. Thus, from Eq. 11.29 the gain doubles for every octave increase in frequency. When Eq. 11.23 or 11.24 is used, 20 log 2 is approximately 6. Thus, we can also state the *voltage gain increases 6 dB per octave frequency increase or 20 dB per decade.* A plot of dB gain versus frequency over the range where Eq. 11.29 applies will be a straight line if frequency is plotted on a logarithmic scale. As noted, Eq. 11.29 is valid for $\omega \ll \omega_\ell$. However, as ω approaches ω_ℓ, the accuracy of this approximation decreases. Nevertheless, as an approximation, let us

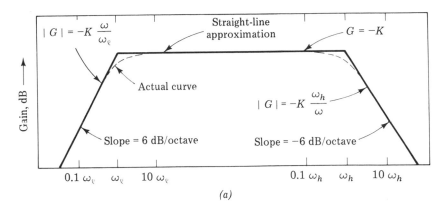

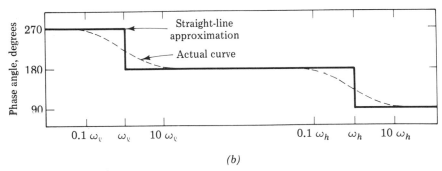

Fig. 11.2. Straight line approximations for $G = -Ks\omega_h / (s + \omega_\ell)(S + \omega_h)$; (a) gain; (b) phase shift.

assume that Eq. 11.28 is valid for $\omega \leq \omega_\ell$ in Fig. 11.2. Then, the plot of G versus ω and phase angle versus ω will be as shown for $\omega \leq \omega_\ell$ in Fig. 11.2.

Now, if $s = j\omega$ and $\omega_\ell \ll \omega \ll \omega_h$, then $(s + \omega_\ell) \approx j\omega$ and $(s + \omega_h) \approx \omega_h$. Then, Eq. 11.27 applies.

$$G = -K \tag{11.27}$$

In this range, the gain is a constant (the midband gain) and is independent of frequency while the phase angle remains constant at about 180°. If this condition is assumed to exist (again, this is a rough approximation) for $\omega_\ell \geq \omega \geq \omega_h$, the plots will have the form given in the center segment of Fig. 11.2.

Finally, if $s = j\omega$ and $\omega \gg \omega_h$, then $(s + \omega_\ell) \approx j\omega$ and $(s + \omega_h) \approx j\omega$. Under these conditions, Eq. 11.25 becomes

$$G = -K \frac{\omega_h}{j\omega} \tag{11.30}$$

Table 11.3. Correction factors near pole frequencies ω_p (either ω_ℓ or ω_h).

ω or f	$0.3\ \omega_p$	$0.5\ \omega_p$	$0.7\ \omega_p$	ω_p	$1.4\ \omega_p$	$2\ \omega_p$	$3\ \omega_p$
Ratio	0.96	0.9	0.82	0.7	0.82	0.9	0.96
Departure	0.5 dB	1 dB	2 dB	3 dB	2 dB	1 dB	0.5 dB

In this range, G is reduced by one-half for each octave frequency increase or G decreases by 6 dB per octave or 20 dB per decade frequency increase. In addition, the phase angle becomes nearly 90°. If these approximations are assumed to be valid for $\omega \geq \omega_h$, the plots will be as shown (for $\omega \geq \omega_h$) in Fig. 11.2.

The actual plots of gain versus frequency are shown as dashed lines in Fig. 11.2. The straight-line approximations are known as *Bode Plots*. They are seen to be fairly good approximations. In fact, with the correction factors given in Table 11.3 near the singularities, or corners, accurate plots may be easily obtained.

The straight-line approximation for the phase angle (Fig. 11.2b) is a poor approximation to the actual phase plot. Fortunately, a very good straight-line approximation can be drawn, as shown in Fig. 11.3. In this figure, the first straight-line approximation is drawn as a dotted curve with a value of 0 for $\omega \leq \omega_p$ and a value of $\pi/2$ for $\omega \geq \omega_p$. The actual phase angle departures from this first straight-line approximation are given in Table 11.4. The second straight-line approximation is given in Fig. 11.3 as a dashed line. This second approximation has a value of zero for $\omega \leq 0.1\ \omega_p$. Then, from $\omega = 0.1\ \omega_p$ and phase 0, a straight line is drawn to the point where $\omega = 10\ \omega_p$ and the phase angle $= \pi/2$. (Of course, the ω axis must be plotted on a logarithmic scale for this straight-line approximation or for any other Bode plot.) This second straight line approximation is very close to the actual phase plot (the solid line in Fig. 11.3). The largest error in this approximation is 5.7° and occurs at $\omega = 0.1\ \omega_p$ and $10\ \omega_p$. A slight fillet at these locations produces a very good approximation to the actual curves.

Bode plots are especially useful when several stages are connected in cascade. The half-power frequencies must first be determined for each stage, using the techniques described in Chapters 7, 9, and 10, and then a Bode plot may be drawn to obtain the gain and phase characteristics of

Table 11.4. Departures of the actual phase plot from the first straight-line approximation.

Frequency	$0.1\ \omega_p$	$0.3\ \omega_p$	$0.5\ \omega_p$	$0.7\ \omega_p$	ω_p	$1.4\ \omega_p$	$2\ \omega_p$	$3.3\ \omega_p$	$10\ \omega_p$
Departure	5.7°	17.5°	26.6°	35.3°	$+45°$	$-35.3°$	$-26.6°$	$-17.5°$	$-5.7°$

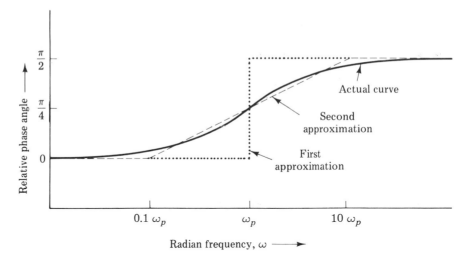

Fig. 11.3. A plot of $\log(j\omega + \omega_p)$ showing the first and second straight-line approximations.

the entire amplifier. Figure 11.4 shows the gain-magnitude and phase characteristics of a typical 3-stage amplifier in which direct coupling is used to eliminate one of the low-frequency poles. The frequency scale must be logarithmic, but may be given in either radians/s or Hz. The gain-magnitude scale must also be logarithmic, but may be given in either dB increments or gain ratio, as indicated in Fig. 11.4. The relative phase may be given in either radians or degrees, as shown. The midband phase is used as the 0° reference. The actual midband phase depends upon the number of common-emitter stages used, but the 0° reference is commonly used for convenience. The following example illustrates the construction techniques for both the straight-line approximations and the actual curves.

Example 11.1	Figure 11.4 contains the Bode plot for an amplifier having the transfer function or gain equation $G_v = 1.131 \times 10^{24}$ $s^2/(s + 2\pi \times 10)$ $(s + 2\pi \times 30)(s + 2\pi \times 10^6)(s + 2\pi \times 2 \times 10^6)(s + 2\pi \times 10^7)$. From Eq. 11.25, the constant in the numerator is equal to the midband gain multiplied by the three different values of ω_h. Consequently, the midband gain is 3000 or 69.5 dB. Therefore, the horizontal segment of the straight-line plot is drawn at this level from the highest of the low-frequency poles (30 Hz) to the lowest of the high-frequency poles (10^6 Hz). The low-frequency part of the Bode plot is constructed by first drawing a straight line to the left and downward at a slope of 20 dB/decade from the highest of the low-frequency poles (30 Hz) until

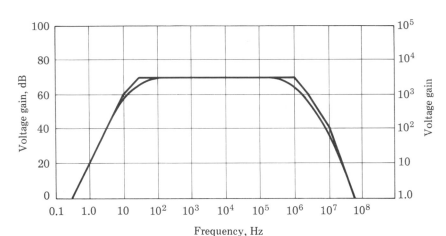

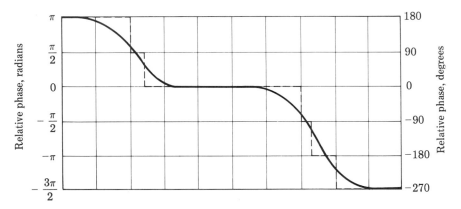

Fig. 11.4. Bode plots and actual curves of gain-magnitude and phase for a 3-stage amplifier including a directly-coupled stage.

the next-lower pole frequency (10 Hz) is reached. This identifies the location of that pole, which is the lowest frequency pole in this example. The final low-frequency segment continues to the left of this pole at a slope of 40 dB/decade until it runs off the graph. The slope of the 20 dB/decade segment is 1 and the slope of the 40 dB/decade segment is 2, providing the graph spacing is the same for both the frequency decades and the gain decades. Sometimes this terminology is used, although incorrectly, when the spacings are different.

The high-frequency segments of the gain curve are drawn in a similar manner to the low-frequency segments. The segment from the lowest of the high-frequency poles (10^6 Hz) is drawn to the right and downward at a -20 dB/decade slope until the next higher pole at

2×10^6 Hz is reached. Then the slope is changed to -40 dB/decade until the highest frequency pole at 10^7 Hz is reached. Beyond this pole, the slope is -60 dB/decade until the edge of the plot is reached. These slopes may be given as -1, -2, and -3, respectively.

The actual gain curve is drawn by adding the corrections given in Table 11.3 for each of the poles in the vicinity. For example, the departure at the 10^6 Hz pole is 3 dB plus 1 dB for the pole at 2×10^6 Hz $= 4$ dB. Similarly, the correction at 2×10^6 Hz is 4 dB because of the 1 dB correction for the pole at 10^6 Hz plus the 3 dB correction due to the 2×10^6 pole. Halfway between these poles at 1.4 MHz, the frequency ratio to the 1.0 MHz is 1.4 and to the 2 MHz pole is .7. The correction for each of these poles is therefore 2 dB, so the total is 4 dB. Correction in this area for the 10^7 Hz pole is negligible. After making similar corrections at, or near, the low-frequency poles, the actual gain was plotted in Fig. 11.4.

The phase corrections can be made in a manner similar to the gain corrections, using Table 11.4. For example, at $f = 10$ Hz (Fig. 11.4), the phase correction is $45°$ for the pole at 10 Hz plus $17.5°$ for the pole at 30 Hz $= 62\frac{1}{2}°$ total below the π or $180°$ phase line. Similarly, the phase correction at 30 Hz is $45° - 17.5° = 27.5°$ below the $90°$ phase line. At 10^6 Hz, the phase correction is $45° + 26.6° + 5.7° = 77°$ below the $0°$ line due to the three high-frequency poles. At 2 MHz, the correction is $45° - 26.6° + (\sin^{-1} 0.2 = 11.5°) = 29.9°$ below the $-90°$ phase line, and at 10^7 Hz, the correction is $45° - 11.5° - 5.7° = 27.8°$ below the $180°$ phase line. Observe that the phase corrections for poles having lower frequency than the one under consideration subtract from the correction for the one under consideration, providing the correction is made from the straight-line approximation *above* the actual curve. Under the same circumstances corrections for higher-frequency poles add to the correction for the one under consideration.

It is usually easiest to begin the Bode plot in the constant gain mid-frequency region. Observe that at each frequency where a pole in the gain equation exists, there is a 20 dB/decade change in the slope of the straight-line curve. If the direction of increasing frequency is used, the slope decreases at 20 dB/decade at each pole frequency. If two poles of the gain equation exist at a given frequency, the change in slope will be 40 dB/decade.

11.5
AMPLIFIER
NOISE

The maximum usable gain of a cascade amplifier is usually determined by the noise generated in the amplifier. High gain is required when the available input signal is very weak. However, if the input signal is not stronger than the noise generated in the first stage of a cascaded amplifier, the noise may make the signal unusable. The signal source always generates some noise and therefore has a finite ratio of signal to noise. Ideally, the amplifier would not generate noise, and therefore, the signal-to-noise ratio in the output of the amplifier would be the same as the signal-to-noise ratio out of the source. However, the amplifier always adds noise of its own and thus the signal to noise ratio in the output of the amplifier is always lower than the signal-to-noise ratio of the source.

Three fundamental sources of noise in a transistor amplifier are:

1. Diode noise which results from the random injection of charge carriers across the depletion region.

2. Resistor noise which results from the random motion of electrons in a resistance at temperatures above 0°K.

3. A third type of noise is known as $1/f$ noise because its magnitude is approximately inversely proportional to frequency. This type of noise is effective only at lower audio frequencies, usually 1 kHz or less. This $1/f$ noise results primarily from surface leakage in transistors and semiconductor diodes. This type of noise varies quite widely among units of the same type device and has not been theoretically characterized. Improvements in transistor surface treatment have greatly reduced the $1/f$ noise in comparison with earlier models. In applications which require a very low-noise amplifier, the transistor should be hand picked for low $1/f$ noise.

In resistors and diodes, the average currents are very predictable when a known constant voltage is applied to the device, but small, random fluctuations about the average values result from the random motion of electrons or charge carriers. This random motion is due to the kinetic energy of the carriers which is proportional to the absolute temperature of the material. These random variations are called noise because they are amplified along with the signal and cause background noise radiation from a loud-speaker or headset in an audio system. These noise signals also cause fuzzy oscillograph tracings or "snow" on a TV picture tube. The noise will not be audible or the snow noticeable if the signals are very large in comparison with the noise. Therefore, the *signal-to-noise ratio* determines whether or not the noise will be disturbing. Of course, the tolerance level of the user will depend upon the program material. For example, a noisy but intelligible long-distance voice communication may be satisfactory, but a noticeable background noise accompanying a musical concert may be annoying.

Since the first stage of an amplifier amplifies its own noise as well as the input signal, the combined noise and signal power available as an input signal to the second stage is usually large in comparison with the noise power generated by the second stage. Therefore, the first stage is the crucial one in considering noise contribution in an amplifier. If we let N_s be the noise power delivered by the source to the first, or input, stage, and N_{a1} be the noise power generated in the first amplifier referenced to its input, then the noise power in the output of the first stage is

$$N_{o1} = (N_s + N_{a1}) G_{p1} \tag{11.31}$$

where G_{p1} is the power gain of the first stage. This output noise of the first stage becomes the input source power for the second stage. Therefore, if we let N_{a2} be the noise power generated in the second amplifier, referred to its input, the noise power output of the second stage is

$$N_{o2} = [(N_s + N_{a1}) G_{p1} + N_{a2}] G_{p2} \tag{11.32}$$

where G_{p2} is the power gain of the second stage. Thus it may be seen from Eq. 11.32 that the noise-power contribution N_{a2} of the second stage is small in comparison with the noise power output $(N_s + N_{a1}) G_{p1}$ of the first stage, providing that the two amplifiers generate comparable amounts of noise and the power gain G_{p1} of the first stage is reasonably high. The noise contributions of succeeding stages are even less significant.

**11.6
RESISTOR
NOISE**

First, we will determine the noise voltage of a resistor. This noise voltage fluctuates randomly and, therefore, does not have discrete frequency components such as one might obtain from the Fourier analysis of a periodic fluctuation. Instead, the noise voltage or current has a continuous noise spectrum over a very broad band of frequencies extending well beyond the upper frequency limits of transistor amplifiers. Therefore, the noise power in the output of an amplifier is proportional to the bandwidth of the amplifier in Hz. Since Boltzmann's constant k relates temperature to energy, a reasonable realtionship between noise power p_n, temperature T, and bandwidth Δf is

$$p_n = kT\Delta f \tag{11.33}$$

The Δf is used instead of B for bandwidth because the effective noise power bandwidth is different from the half-power bandwidth we have previously defined. The difference between the noise bandwidth and the half-power bandwidth is illustrated in Fig. 11.5. The half-power bandwidth, as used heretofore, is the frequency at which the power is one-half the low-frequency value in a low-pass amplifier. However, Δf

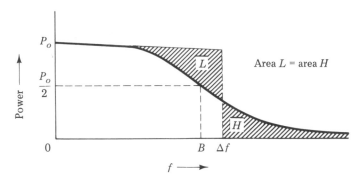

Fig. 11.5. Power-versus-frequency curve showing the difference between half power bandwidth B and noise bandwidth Δf.

is the bandwidth which makes the rectangular area $p_o \Delta f$ (Fig. 11.5) equal to the total area under the total power response curve. In mathematical terms,

$$p_o \Delta f = \int_o^\infty p \; df \qquad (11.34)$$

Thus the area labeled L in Fig. 11.5 is equal to the area H. For a single RC-coupled amplifier, Δf is approximately 1.5 B. However, Δf approaches B as the number of stages, or frequency-discriminating circuits, are added because the slopes of the response curves become steeper. Therefore, little error is introduced if Δf is assumed equal to B in most applications.

Nyquist[2] postulated that the noise power given by Eq. 11.33 is the maximum power that can be transferred from a noisy (normal) resistor to a noiseless resistor. Since maximum power is transferred when the load resistance is equal to the source resistance, a Thévenin's equivalent circuit can be drawn for the noisy resistor, as shown in Fig. 11.6. A simple calculation will show that this equivalent circuit will transfer the maximum power $p_n = kT\Delta f$. Of course, if two normal resistors are connected together, they both generate noise power so there is no net transfer of energy if they are at the same temperature. The voltage $\overline{v}_n$ is not an rms voltage in the usual sense because it cannot represent a frequency component. Therefore, the bar is placed over v_n to indicate that it is a fictitious voltage which will yield the proper noise power transfer. Therefore, $\overline{v}_n$ can be treated as an rms voltage.

The Thévenin's equivalent noise source can be transformed to a Norton's or current source, as shown in Fig. 11.7. A simple calculation

[2] H. Nyquist, "Thermal Agitation of Electric Charge in Conductors," *Physical Review*, Vol. 32, 1928, p. 110.

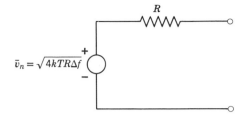

Fig. 11.6. A Thévenin's equivalent circuit for the noisy resistor.

will show that the short-circuit current of the Thévenin's circuit (Fig. 11.6) is

$$\bar{i}_n = \frac{\sqrt{4kTR\Delta f}}{R} = \sqrt{4kTG\Delta f} \qquad (11.35)$$

**11.7
DIODE NOISE**

The noise generated by a semiconductor diode can be deduced from resistor noise and a basic law of thermodynamics. If the diode and a resistance, at the same temperature, are connected together with no external voltage applied, no net noise power will flow between them. Otherwise, one of them must get hotter and the other cooler. Therefore, the noise current flow resulting from noise generated in the diode must be equal to the noise current flow resulting from noise generated in the resistor. But the noise in a diode, exclusive of the noise generated in the ohmic resistance of the doped semiconductor, is due to the random variations in current across the junction. However, in a diode with no bias voltage, the saturation current flows in one direction and an injection current equal to the saturation current flows in the opposite direction, as discussed in Chapter 3. Although the average current is zero, the noise components of the two oppositely directed currents are additive because the random variations depend only on the magnitude of current and not on the direction. Thus the magnitude of noise-producing current in the unbiased diode is $2I_s$.

As discussed in Chapter 3, the dynamic conductance of a diode with zero bias is

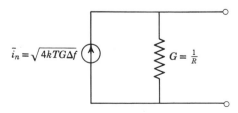

Fig. 11.7. A Norton's equivalent noise source for a conductor.

$$G_o = \frac{q}{kT} I_s \tag{11.36}$$

If this value of conductance is substituted into Eq. 11.35, the noise component of an unbiased diode becomes

$$\overline{i_{nd}} = \sqrt{4qI_s\Delta f} \tag{11.37}$$

But the noise-producing current in an unbiased diode is $I = 2I_s$, as discussed above. Therefore, the noise component of diode current can be expressed in terms of the noise-producing diode current by substituting $I_s = I/2$ in Eq. 11.33 to obtain

$$\overline{i_{nd}} = \sqrt{2qI\Delta f} \tag{11.38}$$

Schottky and others have shown that this (Eq. 11.38) relationship holds for any value of noise-producing diode current I. When reverse bias is applied, $I = I_s$. When forward bias is applied, $I = I_s + I_I$, where I_I is the total injection current. But I_I is normally very large compared with I_s, so I is essentially the q-point or average value of diode current when the diode is forward biased.

The noise bandwidth of a diode usually exceeds the useful frequency range of the diode, so the noise bandwidth Δf is usually the noise bandwidth of the amplifier or measuring instrument which follows the diode.

11.8
TRANSISTOR
NOISE

We are now prepared to identify the noise sources in a transistor and to make some noise voltage (or equivalent resistance) calculations. Thermal currents (I_{CO}, I_{EO}) have noise components and, therefore, increase the noise of a transistor. Thus, silicon transistors have less noise than germanium transistors if other characteristics are similar. Consequently, we will assume that the transistor which is chosen for low noise is silicon and the thermal currents may be neglected. The main noise sources in a silicon transistor are then the forward-biased emitter junction (diode noise) and the ohmic resistance in the base r_x. A hybrid-π equivalent circuit showing these noise sources is given in Fig. 11.8. The emitter junction diode noise has two effects. First, it produces a noise voltage across the junction. The equivalent noise current $\overline{i_{nb}}$ is placed in parallel with r_π to produce the proper contribution to the noise voltage $\overline{v}$ across the junction. The second effect of the emitter junction noise current is the direct transmittal of this current, reduced by the ratio, α, to the collector circuit. This component, which is not amplified by the transistor, is represented by the noise current source $\overline{i_{nc}}$ in the collector circuit. The resistors r_o and r_π are noiseless resistors because they are fictitious resistors which account for the transistor characteristics but do not represent ohmic resistance.

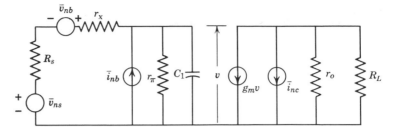

Fig. 11.8. A noise equivalent circuit for the transistor.

The following conclusions can be drawn from the foregoing discussions.

1. The collector circuit noise current $\bar{i}_{nc}$ can be small if the q-point collector current is small.

2. The noise current $\bar{i}_{bn}$ can be minimized for a given q-point collector current if h_{FE} is large.

3. The contribution of the base resistance noise is small if r_x is small in comparison with the source resistance R_s.

4. The noise in the output of the amplifier can be minimized by restricting the bandwidth of the amplifier to that required by the signal.

Thus, a silicon transistor with high h_{FE} at low values of I_C will have a low noise figure when driven by a source with $R_s \gg r_x$. The transistor should be selected for low $1/f$ noise if very low noise is required.

A noise figure of merit for an amplifier is the *spot noise figure, F.* This figure is defined as

$$F = \frac{\text{Noise power delivered by an amplifier to a load}}{\text{Noise power delivered if } R_s \text{ (at 290°K) is the only noise source}}$$

This ratio is expressed for a narrow-frequency band at a specific frequency because the amplifier noise is a function of frequency. The noise figure is usually given in dB at 1 kHz.

The noise contributed by both the amplifier and the driving source can be represented by a noise voltage $\bar{v}_n = (\bar{v}_{ns}^2 + \bar{v}_{na}^2)^{1/2}$ at the input of the amplifier, as shown in Fig. 11.9, where $\bar{v}_{na}$ is the equivalent noise voltage of the amplifier referred to the amplifier input. The equivalent noise voltage $\bar{v}_n$ can be represented as the noise voltage which will be produced by a resistance R_n at 290°K (the standard reference temperature). Then the noise figure F is equal to R_n / R_s, as shown by the

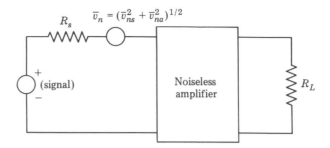

Fig. 11.9. Representation of amplifier and driving-source noise by an equivalent noise voltage $\bar{v}_n$.

following relationships. The equivalent input noise current to the amplifier is

$$\bar{i}_n = \frac{\bar{v}_n}{R_s + Z_i} = \frac{(4kTR_n \Delta f)^{1/2}}{R_s + Z_i} \tag{11.39}$$

where Z_i is the input impedance of the transistor at the specified frequency. The noise output voltage is then

$$\bar{v}_{no} = \bar{i}_n G_i R_L = \frac{(4kTR_n \Delta f)^{1/2} \, G_i R_L}{R_s + Z_i} \tag{11.40}$$

Similarly, the noise voltage in the output which results only from the noise voltage of the source is

$$\bar{v}_{nos} = \frac{(4kTR_s \Delta f)^{1/2} \, G_i R_L}{R_s + Z_i} \tag{11.41}$$

The noise power in the load is $\bar{v}_{no}^2 / R_L$, whereas if the amplifier were noiseless, the noise power in the load would be $\bar{v}_{nos}^2 / R_L$. By definition, the spot noise figure is the ratio of these two powers. Then, using Eqs. 11.40 and 11.41, the spot noise figure is

$$F = \frac{4kTR_n \Delta f G_i^2 R_L^2}{4kTR_s \Delta f G_i^2 R_L^2} = \frac{R_n}{R_s} \tag{11.42}$$

Observe that the noise figure F is independent of the load resistance. Also, the input impedance Z_i does not appear explicitly in Eq. 11.42. However, the equivalent noise resistance R_n is a function of the transistor input impedance.

We will now find the equivalent noise resistance R_n in terms of the transistor parameters given in Fig. 11.8. Since R_n represents the total noise resistance including the source resistance, let us define a noise

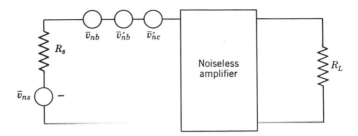

Fig. 11.10. An equivalent noise circuit for the transistor.

resistance R_a which, at 290°K, will produce noise equivalent to the noise generated by the amplifier. Then $R_n = R_s + R_a$, and in terms of R_a

$$F = \frac{R_s + R_a}{R_s} \tag{11.43}$$

We will first find the noise voltage $\bar{v}_{na}$ which, if placed in the input of the transistor in series with the driving source, will produce the same output noise current as the transistor. Figure 11.8 shows that there are three noise sources in the transistor—one voltage source $\bar{v}_{nb}$ and two current sources $\bar{i}_{nb}$ and $\bar{i}_{nc}$. The voltage $\bar{v}_{nb}$ is already in series with the driving source, and we need to convert the two current sources to equivalent voltage sources in series with $\bar{v}_{nb}$ as shown in Fig. 11.10.

The equivalent voltage $\bar{v}'_{nb}$ results from the current source $\bar{i}_{nb}$. This current source can be easily transformed to the equivalent voltage source $\bar{v}'_{nb}$ with the aid of the equivalent circuit given in Fig. 11.11, wherein all noise sources except $\bar{i}_{nb}$ have been turned off. The voltage $\bar{v}'_{nb}$ in the Thévenin's circuit (Fig. 11.11b) is the open-circuit voltage at terminals $b' - E$ of the Norton's circuit (Fig. 11.11a). This voltage is

$$\bar{v}'_{nb} = \bar{i}_{nb}(R_s + r_x) \tag{11.44}$$

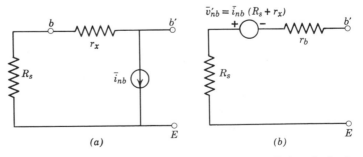

Fig. 11.11. Equivalent circuits (a) Norton and (b) Thévenin for the base current noise component i_{nb}.

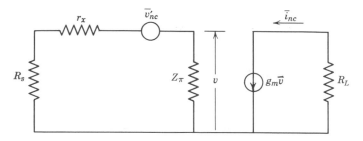

Fig. 11.12. An equivalent circuit used to determine $\overline{v}'_{nc}$ terms of $\overline{i}_{nc}$.

The equivalent noise voltage $\overline{v}'_{nc}$ can be found from the equivalent circuit given in Fig. 11.12. Since $\overline{i}_{nc}=g_m\overline{v}$, $\overline{i}_{nc}$ can be expressed in terms of $\overline{v}'_{nc}$ since $\overline{v} = \overline{v}'_{nc} Z_\pi / (R_s+r_x+Z_\pi)$.

$$\overline{i}_{nc}=\frac{g_m\overline{v}'_{nc}Z_\pi}{R_s+r_x+Z_\pi} \tag{11.45}$$

Note that Z_π is the impedance of r_π and $j\omega C_1$ in parallel. Using Eq. 11.45 to obtain $\overline{v}'_{nc}$ explicitly,

$$\overline{v}'_{nc} = \overline{i}_{nc}\frac{(R_s+r_x+Z_\pi)}{g_mZ_\pi} \tag{11.46}$$

Now the effective noise voltage $\overline{v}_{na}$ which we are seeking is the effective sum of the three noise components we have found. But these component noise voltages are uncorrelated; that is, their phase or frequency components are unrelated. However, the noise power of each component is proportional to the square of the component voltage, and these noise powers can be added directly. Therefore,

$$\overline{v}_{na}^2=\overline{v}_{nb}^2 + (\overline{v}'_{nb})^2 + (\overline{v}'_{nc})^2 \tag{11.47}$$

or, using Eqs. 11.44 and 11.46,

$$\overline{v}_{na}^2=\overline{v}_{nb}^2 + \overline{i}_{nb}^2\,(R_s+r_x)^2 + \frac{\overline{i}_{cn}^2\,|R_s+r_x+Z_\pi|^2}{g_m^2\,|Z_\pi|^2} \tag{11.48}$$

We now need to express these voltage components in terms of the base resistance r_x and the q-point diode current components I_B and I_C which cause them. Letting $\overline{v}_{na}^2=4kTR_a\Delta f$, as previously suggested,

$$4kTR_a\Delta f = 4kTr_x\Delta f + 2qI_B\Delta f\,(R_s+r_x)^2$$
$$+ \frac{2qI_C\Delta f\,|R_s+r_x+Z_\pi|^2}{g_m^2\,|Z_\pi|^2} \tag{11.49}$$

Dividing Eq. 11.49 through by $4kT\Delta f$,

$$R_a = r_x + \frac{q}{2kT}\, I_B\,(R_s + r_x)^2 + \frac{qI_C}{2kT}\,\frac{|R_s + r_x + Z_\pi|^2}{g_m^2\,|Z_\pi|^2} \qquad (11.50)$$

This expression for R_a can be simplified if the relationships $I_B = I_C / h_{FE}$ and $qI_C / kT = g_m$ are used. Then,

$$R_a = r_x + \frac{g_m\,(R_s + r_x)^2}{2h_{FE}} + \frac{g_m\,|R_s + r_x + Z_\pi|^2}{2(g_m Z_\pi)^2} \qquad (11.51)$$

Equation 11.50 shows that R_a, and therefore the noise figure F is a function of frequency because Z_π decreases as the frequency increases. We have found that $\beta_o = g_m r_\pi$, and in general $\beta = g_m Z_\pi$; therefore, Z_π is equal to β / g_m and $h_{ie} = r_x + Z_\pi$. So Eq. 11.51 can be written

$$R_a = r_x + \frac{g_m\,(R_s + r_x)^2}{2h_{FE}} + \frac{g_m\,(R_s + h_{ie})^2}{2\beta^2} \qquad (11.52)$$

This expression points up the importance of small I_C and, hence, small g_m with large β and large h_{FE} (dc beta) for low noise at low frequencies.

Sometimes the amplifier noise is expressed as a *noise temperature* T_a instead of as a noise figure. The noise temperature is the temperature which the source resistance R_s must have in order to produce the same noise as the amplifier. Then

$$(4kT_a R_s \Delta f)^{1/2} = [4k\,(290°K)\,R_a \Delta f]^{1/2} \qquad (11.53)$$

and

$$T_a = 290\,\frac{R_a}{R_s}\,°K \qquad (11.54)$$

The noise figure increases with frequency above f_β because the signal gain decreases while the transistor noise remains constant. This effect appears in the last term of Eq. 11.52 because β and h_{ie} are both inversely proportional to frequency for frequencies above f_β. However, as Eq. 11.52 shows, R_a does not increase appreciably with frequency until the magnitude of h_{ie} becomes smaller than R_s.

11.9 NOISE OPTIMIZATION FOR LOW-FREQUENCY TRANSISTOR AMPLIFIERS

We have considered some general principles which can be used to design a low-noise amplifier, but do not yet know how to determine an optimum driving-source resistance for a given q-point (I_C) or vice versa. Let us investigate this optimization by writing an expression for the noise temperature, using Eqs. 11.54 and 11.52.

$$T_a = 290\left[\frac{r_x}{R_s} + \frac{g_m\,(R_s + r_x)^2}{2R_s h_{FE}} + \frac{g_m\,(R_s + r_x + r_\pi)^2}{2R_s \beta_o^2}\right] \qquad (11.55)$$

The minimum value of T_a can be obtained by differentiating T_a with respect to R_s and equating this derivative to zero to find the optimum R_s, assuming I_C and hence g_m and β_o to be constant. However, the derivative can be taken more readily if the squared terms in Eq. 11.55 are expanded and the approximation $r_x \ll (R_s + r_\pi)$ is made, as follows.

$$T_a = 290 \left[\frac{r_x}{R_s} + \frac{g_m}{h_{FE}} \left(\frac{R_s}{2} + r_x + \frac{r_x^2}{2R_s} \right) + \frac{g_m}{\beta_o^2} \left(\frac{R_s}{2} + r_\pi + \frac{r_\pi^2}{2R_s} \right) \right] \quad (11.56)$$

Collecting terms,

$$T_a = 290 \left[\frac{1}{R_s} \left(r_x + \frac{g_m r_x^2}{2h_{FE}} + \frac{g_m r_\pi^2}{2\beta_o^2} \right) + R_s \left(\frac{g_m}{2h_{FE}} + \frac{g_m}{2\beta_o^2} \right) + \frac{g_m r_x}{h_{FE}} + \frac{g_m r_\pi}{\beta_o^2} \right] \quad (11.57)$$

Differentiating,

$$\frac{dT_a}{dR_s} = 290 \left[-\frac{1}{R_s^2} \left(r_x + \frac{g_m r_x^2}{2h_{FE}} + \frac{g_m r_\pi^2}{2\beta_o^2} \right) + \left(\frac{g_m}{2h_{FE}} + \frac{g_m}{2\beta_o^2} \right) \right] = 0 \quad (11.58)$$

Using the relationships $r_\pi = \beta_o / g_m$ and $\beta_o^2 \gg h_{FE}$,

$$-\frac{1}{R_s^2} \left(r_x + \frac{g_m r_x^2}{2h_{FE}} + \frac{1}{2g_m} \right) + \frac{g_m}{2h_{FE}} = 0 \quad (11.59)$$

Solving for R_s^2,

$$R_s^2 = \frac{2h_{FE}}{g_m} r_x + r_x^2 + \frac{h_{FE}}{g_m^2} \quad (11.60)$$

and

$$R_{s(opt)} = \left[\frac{h_{FE}}{g_m} \left(2r_x + \frac{1}{g_m} \right) + r_x^2 \right]^{1/2} \quad (11.61)$$

In low-noise amplifiers with small g_m, $r_x \ll h_{FE} / g_m$, so the r_x^2 term in Eq. 11.61 can usually be neglected. Then

$$R_{s(opt)} \simeq \frac{\sqrt{h_{FE} (2r_x g_m + 1)}}{g_m} \quad (11.62)$$

Very frequently, the driving-source resistance is given and the value of q-point collector current is desired. The optimum g_m and hence I_C can be determined in terms of R_s from Eq. 11.60. Thus

$$R_s^2 g_m^2 = 2h_{FE} r_x g_m + h_{FE} \quad (11.63)$$

and

$$g_m^2 - \frac{2h_{FE} r_x g_m}{R_s^2} - \frac{h_{FE}}{R_s^2} = 0 \quad (11.64)$$

Using the quadratic equation,

$$g_m = \frac{h_{FE}r_x}{R_s^2} \pm \left[\left(\frac{h_{FE}r_x}{R_s^2} \right)^2 + \frac{h_{FE}}{R_s^2} \right]^{1/2} \tag{11.65}$$

The positive sign is needed in Eq. 11.65 to give positive values of g_m. Then

$$g_{m\,(opt)} = \frac{h_{FE}r_x + \sqrt{(h_{FE}r_x)^2 + h_{FE}R_s^2}}{R_s^2} \tag{11.66}$$

Manufacturers frequently give noise-figure contours similar to the one shown in Fig. 11.13 for their low-noise transistors. These contours greatly simplify the problem of selecting the best value of I_C for a fixed source resistance. For example, if $R_s = 2$ kΩ, a good value of I_C is 50 μA, which will yield a noise figure less than 3 dB at $f = 1$ kHz.

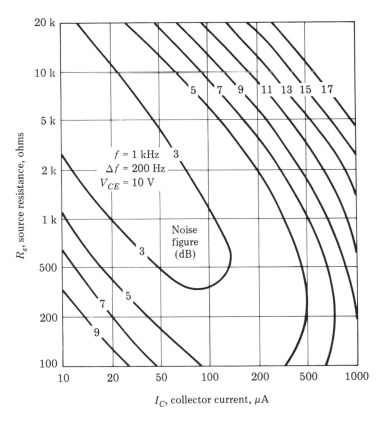

Fig. 11.13. Constant noise-figure contours for a typical low-noise transistor (2N2443).

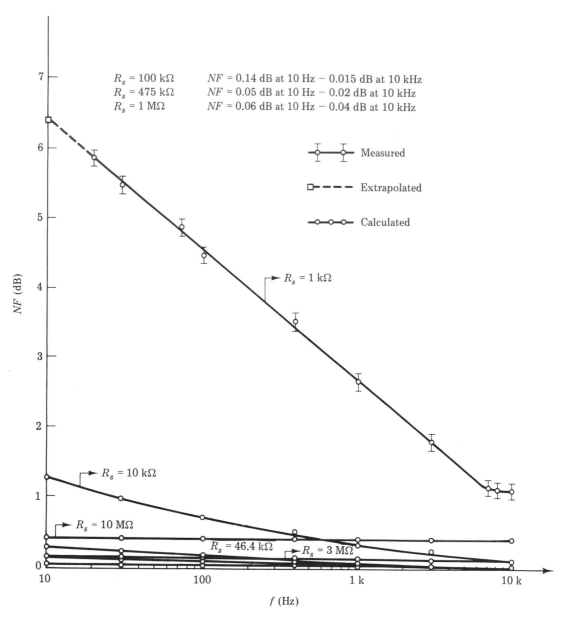

Fig. 11.14. Noise figure of a typical FET amplifier as a
function of R_s and frequency.

**11.10
NOISE IN FET
AMPLIFIERS**

The field-effect transistor has only the channel resistance and surface leakage as principal sources of noise, since the only diode current is the thermal saturation current of the reverse-biased gate diode. Therefore, the noise figure of a silicon FET may be very low. Since the diode current is very small, the medium-frequency (1 kHz) noise figure is determined primarily by the ratio r_{ch}/R_s where r_{ch} is the channel resistance. A typical value for F is 1.1 for high values of source resistance (of the order of 1 MΩ). The $1/f$ noise is present, however, and the noise figure rises at low frequencies. Also, there is some diode current, so the noise figure increases as the power gain decreases at high frequencies. The noise figure for a typical FET is given as a function of frequency and driving-source resistance in Fig. 11.14.

The noise generated by an electronic circuit may be much higher than the theoretical values obtained by the foregoing equations because of defects in the amplifying device or circuit. For example, faulty electrode contacts in transistors may generate noise. Faulty resistors may generate noise far above the predicted level. These unusually noisy components need to be located and replaced.

11.11 SIGNAL-TO-NOISE RATIO CALCULATIONS

An example will illustrate the calculation of signal-to-noise ratio (SNR) at both the input and output of an amplifier. The SNR is a power ratio.

Example 11.2

A microphone is rated as having −60 dB (1 mV) open-circuit output voltage, 25 kΩ impedance (resistance), and frequency response from 50 to 15,000 Hz (assumed noise bandwidth). Determine its signal-to-noise ratio (SNR). (Assume $T = 290°K$.) $SNR = \bar{v}_o^2/4kTR\,\Delta F = 10^{-6}/4 \times 1.38 \times 10^{-23}\ (290)(2.5 \times 10^4)(1.5 \times 10^4) = 1.65 \times 10^5$. This ratio is usually expressed in dB. Thus $SNR = 10 \log (1.65 \times 10^5) = 52$ dB.

The SNR at the output of the amplifier is reduced by the noise figure F of the amplifier since the output noise, referred to the input, is represented by $(R_s + R_a)$. Then, if $1/f$ noise is neglected, the signal-to-noise ratio of an amplifier with $F = 2$, or 3 dB, when used with this microphone, is 8.3×10^4 (or 49 dB). Note that the noise figure in dB is subtracted from the input SNR in dB to obtain the output SNR in dB.

Problems   *Section 11.1*

11.1   Derive Eq. 11.17 from Eq. 11.16.

11.2   A three-stage amplifier is constructed of identical stages. Each individual stage has a pass band from 30 Hz to 20,000 Hz. Determine low cutoff and high cutoff frequency of the three-stage amplifier.

11.3   We desire to make a four-stage amplifier. Determine the low and high cutoff frequencies for each stage if the amplifier must have a bandwidth of 60 Hz to 15,000 Hz.

11.4   If gain × bandwidth = 409×10^6 rad/s, and $f_H = 12$ MHz, calculate K_A for 3, 7, and 8 cascaded stages. Determine the bandwidth per stage (f_h) and gain per stage for each amplifier.

Section 11.2

11.5   An amplifier consists of four stages, each of which has a voltage gain of 20. What is the dB gain of each stage? What is the total gain in dB? Assume that the resistance levels of each stage are the same.

11.6   What is the dB level at the half-power frequencies, f_ℓ and f_h, compared with the midfrequency or reference level?

11.7   What is the power level of 30 dB? 40 vu?

11.8   What is the open-circuit output voltage of a microphone which has -56 dB level? Assume the excess acoustical pressure to be 1 microbar.

Section 11.4

11.9   Make Bode plots of the gain and phase for the amplifier whose gain equation is

$$G = \frac{10^{15}}{(s + 200)(s + 400)(s + 10^5)(s + 10^6)}$$

What are the low and high cutoff frequencies for this amplifier?

11.10   Make Bode plots of the gain and phase for the amplifier whose gain equation is

$$G = \frac{10^{14}}{(s + 400)^2(s + 10^5)^2}$$

What are the low and high cutoff frequencies for this amplifier? How well do these values correspond to those given by Eq. 11.15 and Eq. 11.17?

Section 11.6

11.11   Verify that the noise generator of Fig. 11.6 transfers maximum power $kT\Delta f$.

11.12 Determine the effective noise voltage of a 100 kΩ resistor at 300°K, as measured with a noiseless voltmeter with a 1 MHz bandpass. $k = 1.38 \times 10^{-23}$ j /°K.

Section 11.7

11.13 Determine the noise current of a diode at 300°K with 1 mA average current, as measured with a meter with 1 MHz bandwidth.

Section 11.8

11.14 A transistor with $r_x = 200$ Ω, $\beta = 100$, and $h_{FE} = 80$ at $I_C = 0.1$ mA is used in an amplifier with $R_s = 10$ kΩ. Determine the equivalent noise resistance R_a and the noise figure F.

11.15 Determine the noise temperature of the amplifier in Problem 11.14.

11.16 Determine the noise figure of the amplifier in Problems 11.14 and 11.15 at $f = 20\, f_\beta$.

Section 11.9

11.17 A transistor with $h_{FE} = 80$ and $r_x = 400$ Ω at $I_C = 100$ μA is used as a low-noise amplifier at the given I_C. Determine the optimum driving-source resistance.

11.18 A low-noise amplifier is needed for a microphone with 10 kΩ internal resistance. Use the transistor of Problem 11.17 to determine optimum I_C, assuming h_{FE} remains 80.

Section 11.11

11.19 A given strain gage has 10 kΩ internal resistance and produces a 100 microvolt rms open-circuit signal. This signal is amplified by a bipolar silicon transistor which has $r_x = 200$ Ω, $\beta_o = 100$, $h_{FE} = 80$, and $f_T = 5$ MHz at $I_C = 100$ μA $V_{CE} = 10$ V. Determine the theoretical signal-to-noise ratio both into the amplifier and out of the amplifier, neglecting $1/f$ noise, if the noise bandwidth of both the source and the amplifier is 10 kHz. Assume $T = 290$°K.

11.20 A given dynamic microphone is rated −57 dB output (open circuit), 20 kΩ resistance, and 40 Hz to 15 kHz frequency response. Choose a suitable transistor, or transistors, and design a low-noise amplifier which will provide approximately 1 V rms output. Determine the approximate SNR in the output of your amplifier.

11.21 A given capacitor microphone is rated -74 dB and has $C = 10^{-9}$ F as an internal capacitance. This microphone will provide uniform frequency response from 20 to 20,000 Hz if it feeds into an amplifier having a high input resistance. Design an amplifier which will provide approximately 1 V rms output over the 20–20,000 Hz range when used with this microphone. You choose the transistors. Determine the approximate SNR in the output of your amplifier.

12

Negative Feedback

The performance of an amplifier can be altered by the use of feedback; that is, by adding part or all of the output signal to the input signal. If there is an even number of polarity reversals (or no polarity reversals) between the input and the output of the amplifier, the feedback is said to be positive. This type of feedback is used in oscillator circuits, bootstrapping circuits, and some active filters. On the other hand, if there is an odd number of polarity reversals in the amplifier so the feedback signal tends to cancel the input signal in the mid-frequency range, the feedback is said to be negative. This negative feedback, which is the main subject of this chapter, can reduce distortion, increase the bandwidth, change the output impedance and the input impedance, and stabilize the gain of an amplifier. All of these improvements are obtained at the expense of reduced mid-frequency gain.

12.1
THE EFFECT OF
FEEDBACK
ON GAIN,
DISTORTION,
AND
BANDWIDTH

The block diagram of an amplifier with feedback is given in Fig. 12.1. An expression for the gain of the amplifier with feedback will be developed without regard to whether the feedback is negative or positive. The voltage v_a which actually drives the amplifier is the algebraic sum of the input voltage v_i and the feedback voltage v_f, or

$$v_a = v_i + v_f \tag{12.1}$$

The feedback factor $\mathcal{F}_v$ is the ratio of the feedback voltage v_f to the output voltage v_o. This ratio is usually obtained by a resistive voltage divider, as shown later. Thus, $\mathcal{F}_v$ is usually a real number, but in some

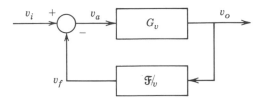

Fig. 12.1. Block diagram of an amplifier with negative feedback.

applications, such as in active filters, $\mathcal{F}_v$ will be a function of frequency. Then,

$$v_f = \mathcal{F}_v\, v_o \tag{12.2}$$

We will let G_v be the voltage gain or amplification of the amplifier without feedback. Thus,

$$v_o = G_v\, v_a \tag{12.3}$$

The voltage gain G_{vf} of the amplifier with feedback is

$$G_{vf} = \frac{v_o}{v_i} \tag{12.4}$$

But, using Eq. 12.1, $v_i = v_a - v_f$ and substituting the value of v_o from Eq. 12.3,

$$G_{vf} = \frac{G_v\, v_a}{v_a - v_f} \tag{12.5}$$

Substituting $\mathcal{F}_v\, v_o$ for v_f (Eq. 12.2) and then dividing both the numerator and denominator of Eq. 12.5 by v_a, we have

$$G_{vf} = \frac{G_v\, v_a}{v_a - \mathcal{F}_v\, v_o} = \frac{G_v}{1 - \mathcal{F}_v\, (v_o/v_a)} \tag{12.6}$$

Finally, substituting G_v for v_o/v_a in Eq. 12.6,

$$G_{vf} = \frac{G_v}{1 - G_v\, \mathcal{F}_v} \tag{12.7}$$

In the preceding work, the symbol G_v was used to represent the voltage gain of the amplifier. Thus, G_v must be a function of frequency $j\omega$, or s, in agreement with the gain expressions in earlier chapters. Also, G_v must include a negative sign if the amplifier has an odd number of common-emitter stages (the negative feedback configuration) so a polarity reversal occurs in the amplifier. Then $G_v = -K_v\, g\,(s)$, where K_v is the reference or mid-frequency gain and $g\,(s)$ is the part of the gain expression which varies with frequency. Using this expression, Eq. 12.7 becomes

$$G_{vf} = \frac{-K_v g(s)}{1 + K_v \mathcal{F}_v g(s)} \qquad (12.8)$$

This is the basic equation for *negative* feedback. Because of the polarity reversal, the input signal to the amplifier is the arithmetic *difference* between the input signal and the feedback signal at mid-frequencies. The positive sign in the denominator identifies the negative feedback. In most applications we are interested in the characteristics of the amplifier in the mid-frequency range. Thus the mid-frequency voltage gain with feedback is

$$K_{vf} = \frac{K_v}{1 + K_v \mathcal{F}_v} \qquad (12.9)$$

As in Chapter 7, K_{vf} is the magnitude of the mid-frequency gain with feedback. This gain is accompanied by a negative sign (as in Chapter 7) when we wish to consider the signal inversion due to an odd number of polarity inversions between the input and output signals. Equation 12.9 is used for most applications because it yields a simple number for the voltage gain. However, when considering frequency response (or similar problems), the general relationship of Eq. 12.8 must be used.

Equation 12.9 could be written in terms of current gains K_{if}, K_i and a current feedback ratio $\mathcal{F}_i$ instead of their voltage counterparts. Making these substitutions into Eq. 12.9,

$$K_{if} = \frac{K_i}{1 + K_i \mathcal{F}_i} \qquad (12.10)$$

In a given amplifier, the feedback reduces both the current gain and the voltage gain by the same factor, since the relationship $K_{vf} = K_{if} Z_L / Z_i$ holds for any amplifier, with or without feedback. Therefore, $1 + K_v \mathcal{F}_v = 1 + K_i \mathcal{F}_i$ and $K_v \mathcal{F}_v = K_i \mathcal{F}_i$. Thus the term $K\mathcal{F}$ can mean either $k_v \mathcal{F}_v$ or $K_i \mathcal{F}_i$. This reduction in gain is the disadvantage of negative feedback, and additional gain must be provided in the amplifier to compensate for the feedback. Several advantages accrue, however, which make negative feedback very attractive.

One desirable characteristic of negative feedback is improved gain stability. You may observe from Eq. 12.9 or Eq. 12.10 that the product $K\mathcal{F}$ (meaning either $K_v \mathcal{F}_v$ or $K_i \mathcal{F}_i$) may be large in comparison with one. Then

$$K_f = \frac{K}{1 + K\mathcal{F}} \simeq \frac{K}{K\mathcal{F}} = \frac{1}{\mathcal{F}} \bigg|_{K\mathcal{F} \gg 1} \qquad (12.11)$$

Thus the gain becomes almost independent of the amplifier character-istics and depends primarily on the resistance ratio of a voltage divider.

Another desirable characteristic of negative feedback is the reduction of harmonic or nonlinear distortion. The reason for this reduction may be seen from Fig. 12.2, where the amlifier is assumed to distort the sinusoidal input voltage by flattening the peaks. The feedback voltage v_f has the same waveform as the output voltage. Therefore, the flattened peaks of the feedback voltage subtract less from the input voltage, thus accentuating the peaks of the amplifier input voltage v_a and predis-torting v_a in a manner which will partially compensate for the flattening caused by the amplifier.

The amount of distortion reduction caused by negative feedback can be determined with the aid of Fig. 12.3. The amplifier with gain K has distortion D_o appearing in its output without feedback. After feedback

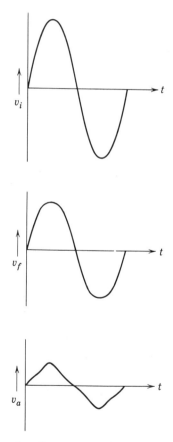

Fig. 12.2. How the feedback voltage v_f predistorts the amplifier input voltage v_a to partially compensate for the amplifier distortion.

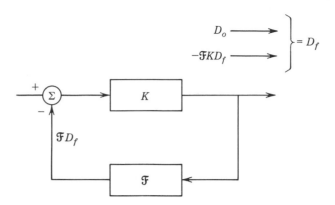

Fig. 12.3. Block diagram illustrating the reduction of distortion in the output of an amplifier.

is applied, distortion D_f appears in the output. But distortion D_f is first reduced by the factor $\mathcal{F}$ and then applied to the input of the amplifier. The amplifier, in turn, amplifies $\mathcal{F}D_f$ by the gain and adds it, in opposite polarity, to the original distortion D_o to obtain the total distortion D_f. Thus,

$$D_f = D_o - K\mathcal{F}D_f \tag{12.12}$$

Solving explicitly for D_f in terms of D_o,

$$D_f = \frac{D_o}{1 + K\mathcal{F}} \tag{12.13}$$

Note that the distortion is reduced by the same factor as the gain. Frequently, the feedback factor $\mathcal{F}$ is chosen to reduce the distortion a given amount.

The bandwidth of an amplifier is also increased by the use of negative feedback. We will use the expression of voltage gain for an RC-coupled stage to show this. You may recall that the expression for gain (voltage or current) for the low and middle-frequency range is

$$G = \frac{-K}{1 - jf_\ell/f} \tag{12.14}$$

where K is the reference, or mid-frequency, gain and f_ℓ is the low half-power frequency. Using this expression for G in the general feedback formula, Eq. 12.8,

$$G_f = \frac{-K/(1 - jf_\ell/f)}{1 + [K\mathcal{F}/(1 - jf_\ell/f)]}$$

$$= \frac{-K}{1 - j\,(f_\ell/f) + K\mathcal{F}}$$

$$= \frac{-K}{1 + K\mathcal{F} - jf_\ell/f} \tag{12.15}$$

Then dividing both numerator and denominator of Eq. 12.15 by $1 + K\mathcal{F}$,

$$G_f = \frac{-K/(1 + K\mathcal{F})}{1 - (jf_\ell/[(1 + K\mathcal{F})f\,])} \tag{12.16}$$

You may observe from Eq. 12.16 that the low-frequency cutoff for the amplifier with feedback is $f_{\ell f} = f_\ell/(1 + K\mathcal{F})$.

Similarly, we can determine the effect of feedback on the upper cutoff frequency of an RC-coupled amplifier, since we know the gain expression for middle and high frequencies is

$$G = \frac{-K}{1 + jf/f_h} \tag{12.17}$$

Using this value of G in the feedback Eq. 12.8,

$$G_f = \frac{[-K/(1 + jf/f_h)]}{[1 + K\mathcal{F}/(1 + jf/f_h)]} \tag{12.18}$$

Equation 12.18 can be rearranged in the same manner as Eq. 12.13 to give

$$G_f = \frac{K/(1 + K\mathcal{F})}{1 + jf/(1 + K\mathcal{F})f_h} \tag{12.19}$$

Observe that the upper half-power frequency of the amplifier with feedback is $f_{hf} = (1 + K\mathcal{F})f_h$. Sometimes negative feedback is used for the specific purpose of increasing the bandwidth of an amplifier. However, when several stages are included in the feedback loop, or transformer coupling is used, the improvement in bandwidth is not so simply related to the feedback factor as in the preceding example. These cases will be considered later.

An example of a feedback amplifier may help clarify some of the foregoing concepts.

Example 12.1

An amplifier which is assumed to consist of a single RC-coupled stage is connected as shown in Fig. 12.1. The characteristics of the amplifier without feedback are: voltage gain $K_v = 1000$, distortion $= 6\%$, $f_\ell = 20$ Hz, and $f_h = 20$ kHz. Determine the characteristics of the amplifier with negative feedback when the voltage feedback ratio $\mathcal{F}_v = 0.01$.

The value of $1 + K_v \mathscr{F}_v$ or $(1 + K\mathscr{F})$ appears in each equation, so let us evaluate this term. Then $1 + K\mathscr{F} = 1 + 1000 \times 0.01 = 1 + 10 = 11$. From Eq. 12.9, $K_{vf} = K_v / (1 + K\mathscr{F}) = 1000/11 = 90.9$. (If we use the approximation given by Eq. 12.11, $K_{vf} \approx 1/.01 = 100$.) The distortion with feedback (Eq. 12.13) is $D_f = D_o / (1 + K\mathscr{F}) = 6\%/11 = 0.54\%$. The lower half-power frequency with feedback $f_{\ell f}$ is $f_\ell / (1 + K\mathscr{F}) = 20/11 = 1.8$ Hz. The upper half-power frequency is $f_{hf} = (1 + K\mathscr{F})f_h = 11 \times 20$ kHz $= 220$ kHz with negative feedback.

12.2
THE EFFECT OF NEGATIVE FEEDBACK ON INPUT IMPEDANCE AND OUTPUT IMPEDANCE

Negative feedback will either increase or decrease the input impedance of an amplifier depending upon whether the feedback signal is added in series or parallel with the input signal. Let us first consider the series connection shown in Fig. 12.4. The input impedance Z_{if} is the ratio of input voltage v_i to the input current i_i. Thus

$$Z_{if} = \frac{v_i}{i_i} = \frac{v_a + v_f}{i_i} \tag{12.20}$$

But $v_f = K\mathscr{F}v_a$, so

$$Z_{if} = \frac{v_a + K\mathscr{F}v_a}{i_i} = \frac{v_a}{i_i}(1 + K\mathscr{F}) \tag{12.21}$$

However, v_a / i_i is the impedance Z_i of the amplifier without feedback. Therefore, in terms of Z_i,

$$Z_{if} = Z_i (1 + K\mathscr{F}) \tag{12.22}$$

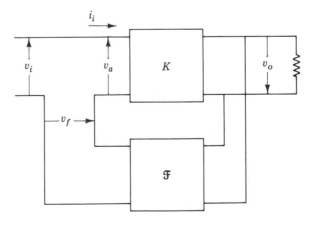

Fig. 12.4. Series input connection for a feedback circuit.

Example 12.2

An amplifier with $Z_i = 1$ kΩ has a voltage gain $K_v = 1000$ and $\mathcal{F}_v = .01$. With the feedback connection shown in Fig. 12.4, the input impedance with feedback is $Z_{if} = 1$ k$(11) = 11$ kΩ.

We will now consider the parallel arrangement of input and feedback circuits shown in Fig. 12.5. In the parallel circuit, the input current i_i is the sum of amplifier input current i_a and the feedback current i_f. Thus, the input admittance with *feedback* is

$$Y_{if} = \frac{i_i}{v_i} = \frac{i_a + i_f}{v_i} \tag{12.23}$$

But, using the amplifier current gain K_i and the current feedback factor $\mathcal{F}_i = i_f / i_o$, the feedback current $i_f = i_a K_i \mathcal{F}_i$. Making this substitution for i_f in Eq. 12.23,

$$Y_{if} = \frac{i_a + i_a K_i \mathcal{F}_i}{v_i} = \frac{i_a}{v_i}(1 + K_i \mathcal{F}_i) \tag{12.24}$$

Since the input admittance of the amplifier without feedback Y_i is i_a / v_i, Y_{if} can be written in terms of Y_i.

$$Y_{if} = Y_i(1 + K\mathcal{F}) \tag{12.25}$$

As mentioned previously, $K_i \mathcal{F}_i = K_v \mathcal{F}_v$. We can also show this by recalling that $K_v = K_i R_L / R_i$. Also, $\mathcal{F}_v = v_f / v_o = i_f R_i / i_o R_L = \mathcal{F}_i R_i / R_L$. Thus, the product $K_v \mathcal{F}_v = K_i \mathcal{F}_i$, providing the input resistance R_i as seen from the feedback network is the same as the input resistance as seen by the driving source. This equality can exist only if the input

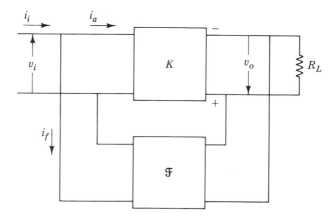

Fig. 12.5. Parallel input connection for a feedback circuit.

current is the driving-source current and the source resistance R_s is included as part of the input resistance. This subject will be discussed in greater detail and examples will be given in Section 12.3. At the moment let us accept the fact that $K_i \mathscr{F}_i$ may equal $K_v \mathscr{F}_v$.

Example 12.3

An amplifier has $K_v = 1000$, $\mathscr{F}_v = .01$ and $Z_i = 1000\ \Omega$ and is connected with parallel feedback as shown in Fig. 12.5. Then the input admittance without feedback, $Y_i = 1/Z_i = 10^{-3}$ S and the input admittance with feedback $Y_{if} = 10^{-3}\ (11) = 1.1 \times 10^{-2}$ S. The input impedance with feedback is $Z_{if} = 1/Y_{if} = 90.9\ \Omega$.

Up to this point, the feedback network has been shown as connected in parallel with the output terminals. This parallel connection is commonly known as *voltage* feedback because the feedback quantity (either current or voltage) is proportional to the output voltage. This voltage feedback is illustrated in Fig. 12.6. The output impedance with feedback is desired and can be determined by turning the input current source off and applying a voltage source v_2 to the output terminals. The current i_2 which flows is the sum of the current i_o through the output resistance, which may include the load resistance, and the current i_a which flows into the amplifier as a result of the feedback. The current flow into the feedback network should be, and usually is, negligible. Then

$$Y_{of} = \frac{i_2}{v_2} = \frac{i_o + i_a}{v_2} \qquad (12.26)$$

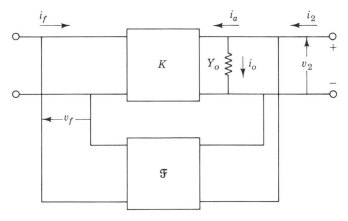

Fig. 12.6. Parallel output connection, or voltage feedback.

But with the driving-current source off, the feedback current i_f is also the amplifier input current, so $i_a = i_f K_i = i_o \mathcal{F}_i K_i$. Making this substitution in Eq. 12.26, Y_{of} is

$$Y_{of} = \frac{i_o + i_o \mathcal{F}_i K_i}{v_2} = \frac{i_o}{v_2}(1 + K_i \mathcal{F}_i) \qquad (12.27)$$

But i_o/v_2 is the output admittance Y_o without feedback. Therefore, Y_{of} can be written in terms of Y_o

$$Y_{of} = Y_o(1 + K\mathcal{F}) \qquad (12.28)$$

Example 12.4

An amplifier is connected as shown in Fig. 12.6 and has $K_v = 1000$ and $\mathcal{F} = .01$. The output resistance $R_o = 500$ Ω before feedback is applied. Then, with feedback, $Y_{of} = 2 \times 10^{-3}(11) = 2.2 \times 10^{-2}$ S and $R_{of} = 45$ Ω.

You may recall that we assumed that the loading of the feedback circuit on the output of the amplifier was neglected. If this loading is not negligible it can be included as part of Z_o.

You may have observed the regularity with which the factor $(1 + K\mathcal{F})$ appears as a modifier in negative feedback circuits. Let us now investigate the final feedback connection—the series arrangement of feedback and load in the amplifier output. This connection, which is known as current feedback because the feedback quantity (either current or voltage) is proportional to the output current, as shown in Fig. 12.7. Again, voltage v_2 is applied to the output terminals and current i_2 flows as a result. But i_2 is the sum of the current i_o through the output re-

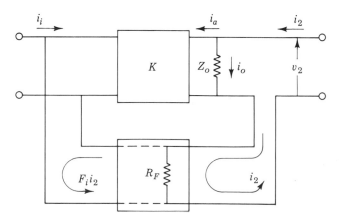

Fig. 12.7. Series output connection, or current feedback.

sistance and the current i_a which flows into the amplifier as a result of the feedback. Note that the feedback curent $\mathscr{F}_i i_2$ is opposite in direction to the assumed amplifier input current i_i. But with the driving-source current turned off, the feedback current is the amplifier input current, so $i_i = -\mathscr{F}_i i_2$ and $i_a = -K_i \mathscr{F}_i i_2$. Then

$$i_2 = i_o + i_a = i_o - K_i \mathscr{F}_i i_2 \qquad (12.29)$$

or

$$i_2 (1 + K\mathscr{F}) = i_o \qquad (12.30)$$

If the voltage drop $i_2 R_F$ across the feedback resistor is small in comparison with v_2, then

$$i_o \simeq \frac{v_2}{Z_o} \qquad (12.31)$$

and

$$i_2 (1 + K\mathscr{F}) \simeq \frac{v_2}{Z_o} \qquad (12.32)$$

Thus, since the output resistance with feedback Z_{of} is v_2 / i_2,

$$Z_{of} = \frac{v_2}{i_2} = Z_o (1 + K\mathscr{F}) \qquad (12.33)$$

Example 12.5 | The amplifier of the preceding example with $K_v = 1000$, $\mathscr{F}_v = .01$ and $R_o = 500\ \Omega$ has the series output connection, or current feedback. Then, the output resistance with feedback, $R_{of} = 500\,(11) = 5500\ \Omega$.

12.3 FEEDBACK CIRCUITS   Many types of feedback circuits can be devised but only a few typical circuits will be discussed here. We will first consider the circuit of Fig. 12.8 which has the parallel output connection, or voltage feedback and the series input connection. The feedback is negative because one polarity reversal occurs betwen the emitter of transistor T_1, where the feedback is applied, and the collector of transistor T_2, where the output voltage is obtained. Since $i_f = v_o / (R_F + R_1)$ and $v_f = i_f R_1$, the voltage feedback factor is approximately

$$\mathscr{F}_v = \frac{v_f}{v_o} = \frac{R_1}{R_F + R_1} \qquad (12.34)$$

The impedance looking into the emitter circuit of T_1 is in parallel with R_1 so far as the feedback signal is concerned, and this may cast

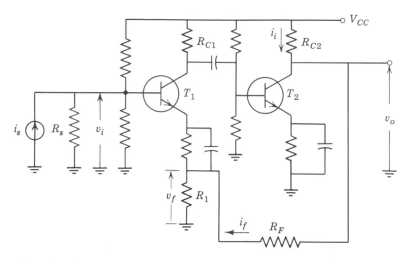

Fig. 12.8. Feedback circuit with parallel output connection, or voltage feedback, and series input connection.

serious doubts upon the accuracy of Eq. 12.34. However, the feedback voltage v_f is nearly equal to the input voltage v_i when normal amounts of feedback are used, so the signal voltage between the emitter and base is small in comparison with v_f. Therefore, most of the feedback current i_f flows through R_1 and Eq. 12.34 is fairly accurate. The resistor R_1 should be small enough to permit good voltage gain in the first stage, and R_F should be large in comparison with the load impedance Z_L so R_F will not seriously load the output.

Example 12.6

Let us use the foregoing assumptions and analyze the circuit shown in Fig. 12.8. Both transistors have $h_{fe} = 150$, $h_{ie} = 2$ kΩ, $h_{re} \approx 0$, $h_{oe} \approx 10^{-5}$ S. The value of R_{C2} is 1 kΩ and is essentially the total load impedance of transistor T_2. The value of $R_s = 10$ kΩ and $R_{C1} = 2$ kΩ. The bias circuit resistance $R_b = 12$ kΩ.

We will choose $R_1 = 100$ Ω, so the input impedance of transistor T_1 is $Z_{i1} = h_{ie} + (h_{fe} + 1) R_1 = 17$ kΩ. The total shunt resistance at the input of stage 1 is the parallel combination of R_s, R_b, and Z_{i1}, so $R_{sh} = 1/(1/10\text{ k} + 1/12\text{ k} + 1/17\text{ k}) = 4.14$ kΩ. Then the current gain for stage 1 is $K_{i1} = i_{c1}/i_s = h_{fe} R_{sh}/Z_{i1} = 150 \times 4.14\text{ k}/17\text{ k} = 36.4$. The input impedance to transistor T_2 is $Z_{i2} = h_{ie2} = 2$ kΩ. The total shunt load on transistor T_1 is the parallel combination of R_{C1}, R_b and h_{ie2} (the value of $1/h_{oe}$ is much greater than this parallel combination so it can safely be ignored). This shunt resistance is about 925 Ω. The current gain for stage 2 is $K_{i2} = h_{fe} R_{sh}/R_i = 150 \times 925/2{,}000 \simeq$

69. Then the total current gain of the amplifier $= K_i = K_{i1} K_{i2} = 36.4 \times 69 = 2510$, and the total voltage gain is $K_v = v_o/v_i = K_i R_{L2}/R_{sh} = 2510 \times 10^3/4.14 \times 10^3 = 606$.

We will assume that the feedback is to be used to reduce the distortion of this amplifier by a factor of 6. Then the value of $1 + K_v \mathcal{F}_v = 6$, so $K_v \mathcal{F}_v = 5$. Thus, $\mathcal{F}_v = 5/K_v = 5/606 = 0.00825$. Then from Eq. 12.34, we have $R_F = (R_1/\mathcal{F}_v) - R_1 = (100/.00825) - 100 \simeq 12$ kΩ. This value of R_F is more than ten times the load resistance R_{C2} so we will assume it is large enough to have negligible loading effect on R_{C2}. Then Z_o without feedback is $\simeq R_{C2}$ (since $1/h_{oe} \gg R_{C2}) = 1{,}000$ Ω. The output resistance with feedback is $Z_{of} \simeq 1$ k$/(1 + K_v \mathcal{F}_v) = 1$ k$\Omega/6 = 167$ Ω. Also, the input impedance to transistor T_1 is $Z_{if} = Z_i (1 + K\mathcal{F}) = 17$ k$\Omega \times 6 = 102$ kΩ. The current gain K_{if} is $2510/6 = 420$ and the voltage gain K_{vf} is $606/6 = 101$.

A typical circuit which uses both parallel input connection and parallel output connection is shown in Fig. 12.9. Observe that an odd number of common-emitter stages must be included in the feedback loop to obtain negative feedback. We will analyze this circuit in the following example.

Example 12.7

A circuit is connected as shown in Fig. 12.9. All three transistors have $h_{fe} = 100$, $h_{ie} = 1$ kΩ, $h_{re} \approx 0$, $h_{oe} = 10^{-5}$ S. The bias circuit resistance of each stage is $R_b = 10$ kΩ. The source resistance $R_s = 2$ kΩ, $R_{C1} = R_{C2} = 2$ kΩ, and $R_{C3} = 1{,}000$ Ω. We wish to reduce the distortion of this circuit to one-tenth of its open-circuit value and then determine the characteristics of the circuit with feedback.

First, we must determine the amplifier characteristics with R_F

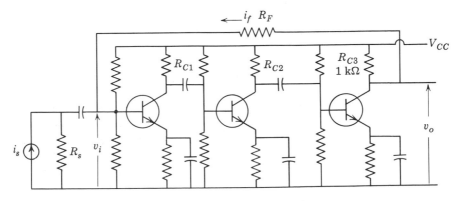

Fig. 12.9. Feedback circuit with parallel input and output connections.

open. We note that the values of R_C for each stage are much less than $1/h_{oe}$ so we can neglect h_{oe}. The input impedance to each transistor is h_{ie} which is equal to 1 kΩ. The shunt resistance on the input of the first stage is the parallel combination of R_s, R_b, and h_{ie}. Thus $R_{sh} = 1/(1/2\ k + 1/10\ k + 1/1\ k) = 625\ \Omega$. The current gain of the first stage is $K_{i1} = i_{c1}/i_s = h_{fe}R_{sh}/h_{ie} = 100 \times 625/1,000 = 62.5$. Note that if the first transistor is regarded as a current source and R_{C1} is used for R_s, the circuit for the second stage is identical to the circuit for the first stage. Therefore, $K_{i2} = K_{i1} = 62.5$. A similar observation for the circuit driving the third stage indicates $K_{i3} = K_{i1}$ also. Thus, the total open-loop current gain is $K_i = K_{i1} \times K_{i2} \times K_{i3} = 62.5^3 = 2.42 \times 10^5$. Since $R_{C3} \ll 1/h_{oe}$, the output impedance, $Z_o = 1,000\ \Omega$. Then the total open-loop voltage gain is $K_v = K_i Z_o/R_{sh} = 2.42 \times 10^5 \times 10^3/625 = 3.87 \times 10^5$.

Since we desire a reduction of distortion by a factor of 10, the term $(1 + K\mathcal{F}) = 10$. Then $K\mathcal{F} = 9$. Now, since the output voltage v_o is very large (3.87×10^5 times as large) in comparison with the input voltage v_i, the feedback current i_f is very nearly equal to v_o/R_F. The output signal current $i_o = v_o/R_{C3}$. Therefore, the feedback current factor, or ratio, is

$$\mathcal{F}_i = \frac{i_f}{i_o} \simeq \frac{v_o/R_F}{v_o/R_C} = \frac{R_C}{R_F} \qquad (12.35)$$

Now, $K_i = 2.42 \times 10^5$ and $K_i\mathcal{F}_i = 9$. Consequently, $\mathcal{F}_i = 9/2.42 \times 10^5 = 3.71 \times 10^{-5}$. From Eq. 12.35, we note that $R_F = R_C/\mathcal{F}_i = 10^3/(3.71 \times 10^{-5}) = 2.69 \times 10^7\ \Omega$. The input impedance of the amplifier, including R_s, can be found from Eq. 12.25 as $Z_{if} = Z_i/(1 + K\mathcal{F}) = Z_i/10 = 625/10 = 62.5\ \Omega$. The output impedance (from Eq. 12.28) is 100 Ω.

Let us now find out if $K_i\mathcal{F}_i$ does equal $K_v\mathcal{F}_v$ for this circuit. For this amplifier, $\mathcal{F}_v = Z_i/(R_F + Z_i) = 625/2.69 \times 10^7 = 2.32 \times 10^{-5}$. Then, $K_v\mathcal{F}_v = 2.32 \times 10^{-5} \times 3.87 \times 10^5 = 9$.

You may have observed in the preceding examples that the voltage gains, with and without feedback, were considered from driving source to load, or $K_v = v_o/v_s$. Similarly, the current gains were the ratios of load current to source current, or $K_i = i_o/i_s$. Only these ratios reveal the total effect of the feedback on the current and voltage gains of the amplifier and verify that they are both reduced by the same factor, namely $(1 + K\mathcal{F})$ where $K\mathcal{F} = K_v\mathcal{F}_v = K_i\mathcal{F}_i$. Therefore, the decision to use either voltage ratios or current ratios in the design of a feedback circuit should depend upon the ease of obtaining the desired ratios.

Figure 12.10 shows a feedback system with series output connection, or current feedback, and parallel input connection. Since the out-

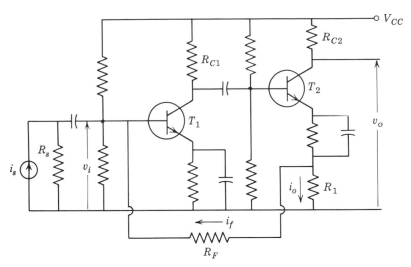

Fig. 12.10. Feedback circuit with series output connection (current feedback) and parallel input connection.

put current flows through the resistor R_1 in the emitter circuit of the output stage, the voltage across this resistance is proportional to the output current. Therefore, the feedback current is proportional to the output current. If the signal voltage across R_1, $i_o R_1$, is large in comparison with v_i, the current feedback factor is approximately

$$\mathscr{F}_i = \frac{i_f}{i_o} = \frac{R_1}{R_F} \tag{12.36}$$

Negative feedback may be used to improve the q-point stability or thermal stability of a *direct-coupled* amplifier. The need for this kind of improvement in the case of a common-emitter type complementary-symmetry amplifier was mentioned in Chapter 10. The circuit diagram of this amplifier which includes a simple feedback circuit to stabilize the dc output voltage is given in Fig. 12.11. The output voltage stabilization improves by the same ratio as the reduction of dc voltage gain. The capacitor C (Fig. 12.11) may be used to eliminate ac feedback and thus allow high signal-frequency gain but low dc gain, providing the input signal v_i contains no essential dc components. In the event equal gain is desired for both dc and ac signal components, the capacitor C should be removed.

The feedback resistor R_F (Fig. 12.11) should be large in comparison with R_L to avoid altering the load on the amplifier. The voltage feedback factor $\mathscr{F}_v$ is equal to $R_E / (R_F + R_E)$ at very low frequency but decreases because of C as the frequency increases. The magnitude of $(1 + K_v \mathscr{F}_v)$ is equal to $\sqrt{2}$ at the lower half-power frequency f_1.

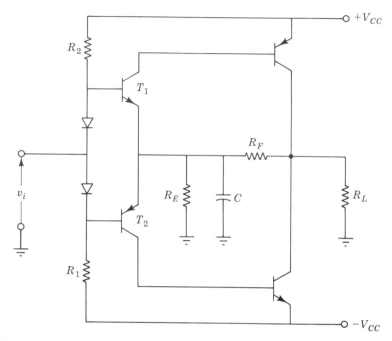

Fig. 12.11. Common-emitter type complementary-symmetry amplifier with negative feedback applied to provide output q-point stabilization.

**12.4
STABILITY OF
FEEDBACK
CIRCUITS**

The general equation for gain with negative feedback was developed in Section 12.1 and is repeated here for convenience.

$$G_f = \frac{-Kg\,(s)}{1 + K\mathcal{F}g\,(s)} \qquad (12.37)$$

where G_f is the gain with feedback, or closed-loop gain; $-Kg\,(s)$ is the gain without feedback, or forward open-loop gain; and $\mathcal{F}$ is the feedback factor or ratio. Observe that the denominator of Eq. 12.37 will be equal to zero if the term $K\mathcal{F}g\,(s) = -1$. Then the gain G_f will be infinite, or output will occur with no input and the amplifier will oscillate or be unstable. This situation can arise at either a high frequency or a low frequency where the relative phase shift of the amplifier is 180° with respect to the mid-frequency phase, or the feedback voltage is actually in phase with the signal voltage. Thus, if the magnitude of total loop gain $K\mathcal{F}g\,(s)$ is equal to at least *one* at a frequency for which the relative phase shift is 180°, oscillation will occur. Therefore, as previously stated, the condition for oscillation is

$$K\mathcal{F}g\,(s) = -1 = 1 \quad \pm\underline{/180} \qquad (12.38)$$

Root locus is one technique used to determine the stability of an amplifier, or system. As the name implies, root locus is the locus of all possible roots, in the s domain, of the denominator of Eq. 12.37. Therefore, the root locus is an s-plane plot of all points where $Kg(s)\mathcal{F}(s) = -1 = 1 \pm\underline{/180°}$. The locus has as many branches as $Kg(s)\mathcal{F}(s)$ has poles. Each branch begins on a pole and terminates on a zero, either finite or at infinity, of $Kg(s)\mathcal{F}(s)$. The root locus progresses along its path toward the zero as the reference K increases from zero toward infinity. The amplifier with feedback is unstable for values of $K\mathcal{F}$ which cause the locus to cross the $j\omega$ axis. This is because sustained oscillation will then occur since poles will occur in the feedback amplifier at real frequencies. The development of the rules for plotting a root locus are beyond the scope of this book. However, the root-locus plots will be given as visual aids and for the benefit of those who have studied the root-locus technique. The Bode plots, discussed below, are used as the basic analysis and design tool.

The gain-phase plots, or Bode plots, discussed in Chapter 11, can be used to predict the stability of an amplifier with feedback. In fact, with the aid of a Bode plot, the feedback network can be designed so that the amplifier will not only be stable, but will provide near-optimum frequency and transient response. This design will be the subject of this section.

Example 12.8

We will now consider the stability of the three-stage RC-coupled amplifier of Fig. 12.12. Both the low and high half-power frequencies are determined for each stage, either by calculation or by measurement, as discussed in Chapters 7 and 11. Let us assume that all three stages have a low half-power frequency at 100 Hz and have three different upper half-power frequencies at 10^5, 3×10^5 and 10^6 Hz. The gain magnitude and phase plots are sketched in Fig. 12.13 by using the asymptote technique discussed in Chapter 11. The mid-frequency current gain of the amplifier is assumed to be 10^5 and the total load resistance in the output of the amplifier is assumed to be 1 kΩ. The current-gain axis is labeled in dB as well as numerical gain. The relative phase is plotted on the same frequency scale as the gain magnitude. Observe that the relative phase shift is 180° at two frequencies, one low and the other high. These frequencies are approximately 50 Hz and 6×10^5 Hz. At the lower frequency, the forward current gain $|G_i|$ of the amplifier is 7×10^3 or 77 dB and at the higher frequency $|G_i| = 5 \times 10^3$ or 74 dB at 180° relative phase. But the total loop gain $G\mathcal{F}$ must be less than one at these 180° relative phase frequencies or the amplifier will oscillate when feedback is applied. Therefore, the magnitude of the feedback factor $\mathcal{F}_i$ which will cause marginal oscillation, considering each frequency individually, is

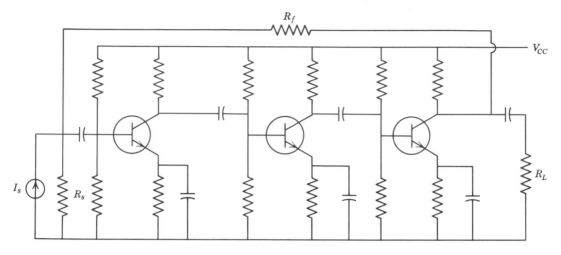

Fig. 12.12. Three-stage amplifier with feedback.

$\mathscr{F}_i = 1/7 \times 10^3 = 1.4 \times 10^{-4}$, or -77 dB at 50 Hz, and $\mathscr{F}_i = 2 \times 10^{-4}$, or -74 dB, at 6×10^5 Hz. Note that the smaller amount of feedback is unstable at the lower frequency because the half-power frequencies of all the stages are the same. Therefore, if this smaller feedback is used, the amplifier will be stable at the higher frequency, but will marginally oscillate at the lower frequency. Of course, oscillation is intolerable at any frequency, so the feedback factor must be reduced not only to provide stability but also to provide satisfactory frequency response and transient response. The root-locus sketch of Fig. 12.14 provides some insight into the reduction of $\mathscr{F}$, or rather $K\mathscr{F}$, required to produce satisfactory response. Those familiar with root locus will recognize that the damping ratio $\zeta = \cos\theta$ where θ is the angle between a line drawn from the origin to a given point on the locus and the negative real axis. But $Q = 1/2\zeta$, as will be shown later, and Q should not exceed about 1.0 for good transient response. Therefore, θ should not exceed about 60°, and typical values of loop gain $K\mathscr{F}$ found by a spirule, or algebraic calculation, at the intersection of the 60° radial and the root locus are about 20 percent of the value found at the intersection with the $j\omega$ axis. This gain reduction depends upon the amount of separation between the amplifier poles, but the factor of 5 is a commonly accepted value. Thus the total loop gain magnitude should not exceed about 0.2 or -14 dB at the frequencies which give 180° phase shift. Then, there are no peaks in the frequency response and no more than 5 percent overshoot in the transient response. This reduction of gain (a factor of 5, or 14 dB) is known as *gain margin*. Smaller values of gain magin may be used if peaks in the frequency response or ringing in the transient response can be tolerated. A sketch of the time and frequency response of a

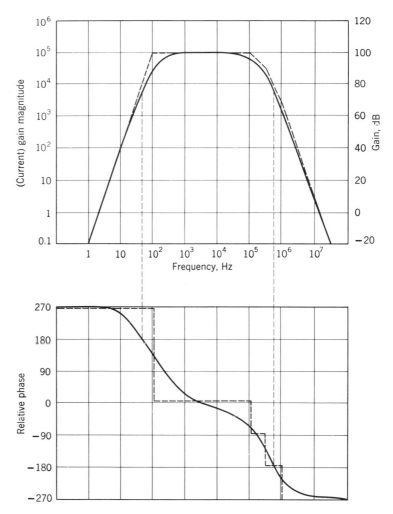

Fig. 12.13. Bode plot for the amplifier of Fig. 12.12.

feedback amplifier with 3 dB gain margin in contrast with the same amplifier with 14 dB gain margin is given in Fig. 12.15. A step current input is used to excite the transient and only the low-frequency transient is shown. Gain margin values between 6 dB and 14 dB are freuqently used.

We will complete the feedback design for the amplifier of Fig. 12.12 using a 14 dB or factor of 5 gain margin. The feedback factor required to reduce the loop gain magnitude to 0.2 or -14 dB at 40 Hz is $\mathscr{F}_i = 0.2 / 7 \times 10^3 = 2.8 \times 10^{-5}$ or -77 dB $- 14$ dB $= -91$ dB. The total mid-frequency loop gain with this magnitude of feedback is $K_i \mathscr{F}_i = 10^5 \times 2.8 \times 10^{-5} = 2.8$, or 100 dB $- 91$ dB $= 9$ dB.

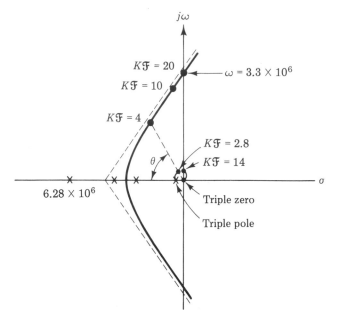

Fig. 12.14. *Root-locus sketch for the amplifier of Fig. 12.12 (not to scale).*

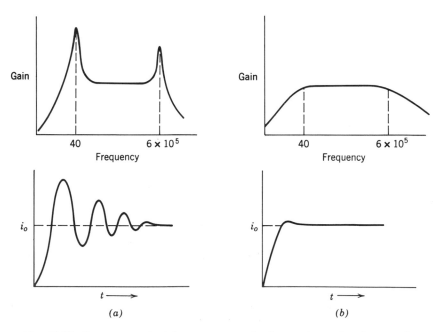

Fig. 12.15. *Frequency and time response for two values of gain margin:*
(a) 3dB gain margin; (b) 14 dB gain margin.

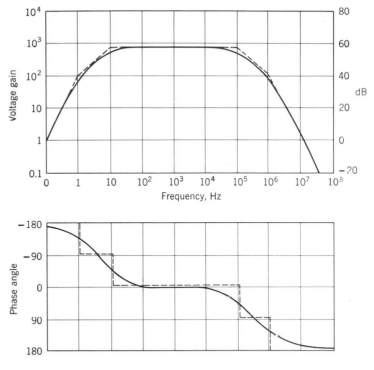

Fig. 12.16. Bode plot for the amplifier of Fig. 12.8.

In Example 12.8 the value of permissible loop gain is disappointingly small because of the modest available improvement in the amplifier characteristics. A smaller gain margin would permit somewhat higher loop gain at the expense of peaks in the frequency response or ringing in the transient or time response. A capacitor can be used in the feedback circuit to improve the characteristics of this amplifier. This technique is known as phase-lead compensation and will be discussed after we have considered a two-stage amplifier.

Example 12.9

The stability of the two-stage RC-coupled amplifier shown in Fig. 12.8 will now be considered. We will assume that there are two low half-power frequencies at 1 and 10 Hz and two high half-power frequencies at 10^5 and 10^6 Hz. The mid-frequency voltage gain was previously calculated to be 833. A Bode plot for this amplifier is given in Fig. 12.16 and a root-locus sketch is given in Fig. 12.17. Observe that the phase plot does not cross the $\pm 180°$ values and the root-locus branches do not cross the $j\omega$ axis. Therefore, the amplifier will be stable for any finite value of loop gain. Thus, the amplifier is said to

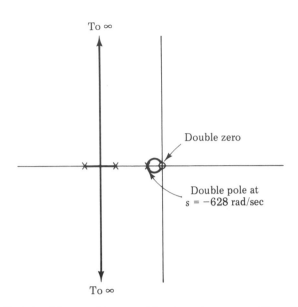

Fig. 12.17. Sketch of the root locus of the two-stage amplifier of Example 12.9
(not to scale).

be *unconditionally stable*. However, the transient response may be unsatisfactory or severe peaks may occur in the frequency response if the loop gain is not limited. The *gain margin* specification given for the three-stage amplifier is not appropriate because there is no frequency at which the phase is 180° except zero or infinity, where the gain is zero. Therefore, another margin known as *phase margin* is used to obtain the desired transient characteristics. For example, a 45° phase margin means that the amplitude of the total loop gain is one when the phase shift is 45° less than 180°, or ±135°. For the Bode plot of Fig. 12.16, the forward voltage gain of the amplifier is approximately 60 or 36 dB at both the low- and high-frequency 135° phase points. This phase margin compares roughly with the 14 dB gain margin as can be seen in Fig. 12.13. Thus, the value of feedback factor $\mathscr{F}_v$ required to reduce the loop gain to one at this phase margin is $\mathscr{F}_v = 1/60 = .0167$ or -36 dB. The mid-frequency value of loop gain is therefore $K_v \mathscr{F}_v = 833 \times .0167 = 13.9$ or 59 dB $- 36$ dB $= 23$ dB for this 45° phase margin. This is a satisfactory value of loop gain for many applications. Of course, higher loop gain can be realized with a smaller phase margin.

**12.5
PHASE-LEAD
COMPEN-
SATION**
The relatively large amount of feedback that can be applied to the two-stage amplifier in comparison with the three-stage amplifier may lead us to believe that no more than two stages should be included in the feedback loop. However, the technique known as phase-lead com-

pensation, mentioned previously, can make the phase characteristics of the three-stage amplifier comparable with the uncompensated two-stage amplifier, and thus permit a similar closed-loop gain. You may have observed that the relative phase shift in the amplifier causes the instability and that each *pole*, or term of the form $K / (s + a)$, contributes a relative phase shift up to 90°. The two-stage amplifier was unconditionally stable because only two such terms appear at high frequencies, due to the shunt capacitance, and two appear at low frequencies, due to the coupling and bypass capacitance. The phase shift at low frequencies due to the coupling capacitors can be eliminated by using direct coupling. Therefore, the low-frequency phase characteristics of the three-stage amplifier will be similar to the characteristics of the two-stage amplifier if direct coupling is used to eliminate the coupling and emitter bypass capacitors in one stage of the three-stage amplifier. Phase-lead compensation will be used to effectively eliminate one of the high-frequency poles in the amplifier gain expression.

The phase-lead network consists simply of a small capacitance C_F in parallel with the feedback resistor R_F as shown in Fig. 12.18. Since the impedance (Z_F) of R_F and C_F in parallel is usually very much larger than either R_L or R_i, particularly in a three-stage amplifier, the output current is very nearly $I_L = V_o / R_L$ and the feedback current I_f is very nearly V_o / Z_F. Therefore, the current feedback factor is

$$\mathscr{F}_i = \frac{I_f}{I_L} = \frac{R_L}{Z_F} \tag{12.39}$$

But, in the s domain

$$Z_F = \frac{R_F (1 / sC_F)}{R_F + 1 / sC_F} = \frac{R_F}{1 + sR_F C_F} \tag{12.40}$$

Since the time constant $R_F C_F = 1 / \omega_f$, where ω_f is the corner-frequency or crossover frequency of the network, $R_F = 1 / \omega_f C_F$, then

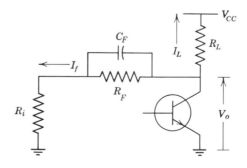

Fig. 12.18. Phase-lead compensated feedback network.

$$Z_f = \frac{R_F}{1 + s / \omega_f} = \frac{1}{C_F} \frac{1}{(s + \omega_f)} \qquad (12.41)$$

and, using Eq. 12.39,

$$\mathcal{F}_i' = R_L C_F (s + \omega_f) = \frac{R_L}{R_F} R_F C_F (s + \omega_f) = \mathcal{F}_i \frac{s + \omega_f}{\omega_f} \qquad (12.42)$$

where $\mathcal{F}_i$ is the mid-frequency current feedback factor. Observe that the capacitor C_F creates a zero in the feedback network.

We will now write the complete equation for the high-frequency gain of the three-stage RC-coupled amplifier with phase-compensated feedback.

$$G_{if} =$$
$$\frac{K_1 K_2 K_3 \, \omega_{h1} \, \omega_{h2} \, \omega_{h3} / (s + \omega_{h1}) (s + \omega_{h2}) (s + \omega_{h3})}{1 + K_1 K_2 K_3 \omega_{h1} \, \omega_{h2} \, \omega_{h3} \, \mathcal{F}_i (s + \omega_f) / \omega_f (s + \omega_{h1}) (s + \omega_{h2}) (s + \omega_{h3})} \qquad (12.43)$$

Observe that the term $(s + \omega_f)$ in the numerator, or zero, of the term $G_i \mathcal{F}_i'$ will cancel one of the poles of the form $(s + \omega_h)$ in the denominator if ω_f is equal to one of the values of ω_h. Let us make $\omega_f = \omega_{h2}$, then the $G_i \mathcal{F}_i'$ term is

$$G_i \mathcal{F}_i' = \frac{K_1 K_2 K_3 \mathcal{F}_i \, \omega_{h1} \, \omega_{h3}}{(s + \omega_{h1}) (s + \omega_{h3})} \qquad (12.44)$$

This term has the gain magnitude characteristics of a three-stage amplifier but the relative high-frequency phase characteristics (for the total loop gain) of a two-stage amplifier. However, the forward open-loop gain magnitude can be plotted as though the $(s + \omega_f)$ term is in the amplifier gain G_i instead of the feedback path $\mathcal{F}_i'$. Then the gain magnitude can also be plotted as though the three-stage amplifier has only two high-frequency break points or poles. This plot will not give the correct high-frequency forward gain but will give the correct feedback design. The zero may cause the proper phase correction even though it is not located precisely at a pole frequency but is only in the same vicinity.

Example 12.10

We will now apply phase-lead compensation to the three-stage amplifier with feedback (Example 12.8) shown in Fig. 12.12, letting $f_f = f_{h2} = 3 \times 10^5$ Hz. Also, direct coupling will be used to couple two of the stages. The modified forward gain magnitude is sketched in Fig. 12.19. The relative phase is sketched as though the phase compensation is in the amplifier. We will allow 45° phase margin which

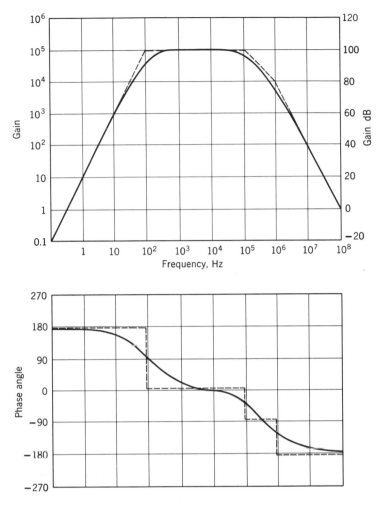

Fig. 12.19. Bode plot for the three-stage compensated-feedback amplifier.

occurs at approximately $f = 30$ Hz and $f = 10^6$ Hz. Note that the low-frequency 45° phase margin point has higher forward gain than the comparable high-frequency point. As previously mentioned, this situation results from the two break points or poles occurring at the same frequency (100 Hz). Therefore, the gain-phase characteristics would be improved and a higher closed-loop gain permitted if one of the coupling circuits had a considerably different value of f_1; for example, 10 Hz. Of course, additional direct coupling could eliminate the pole completely. Therefore, we will be primarily concerned with the high-frequency gain at the 45° margin. This current gain magnitude is 6×10^3 or 77 dB. Therefore, the value of $\mathscr{F}_i$ which will give 45° phase margin is $\mathscr{F}_i = 1/6 \times 10^3 = 1.67 \times 10^{-4}$ or -77 dB. The mid-frequency closed-loop gain is, therefore, $K_i \mathscr{F}_i = (10^5)(1.67 \times 10^{-4}) = 16.7$, or 100 dB $-$ 77 dB $= 23$ dB, and the amplifier distortion is reduced by the factor $(1 + K\mathscr{F}) = 17.7$. This is a very satisfactory value of loop gain for most applications. The value of R_F (Fig. 12.12) required to give this value of feedback is (with $R_L = 1$ kΩ) $R_F = R_L / \mathscr{F}_i = 6 \times 10^5$ Ω. The value of C_F required to cancel the pole or break-frequency at 3×10^5 Hz is $C_F = 1 / R_F \omega_f = 0.8$ pF. This value of capacitance is near the minimum value obtainable from a parts store. The value of R_F could be decreased and the value of C_F increased proportionally if the feedback is taken from a tap in the load circuit as shown in Fig. 12.20. In this circuit R_F is decreased by a factor of 10 and C_F is increased by a factor of 10 compared with the value calculated when the feedback is obtained directly across the output.

The value of $K\mathscr{F}$ determined at a specified phase margin may be larger than the designer desires to use. Smaller values of $K\mathscr{F}$ can be used, of course, with improved stability (larger gain or phase margins).

The capacitor C_F in the compensating circuit actually creates a pole as well as a zero in the feedback factor $\mathscr{F}_i'$. The pole did not appear because we neglected R_i. when $R_F \gg R_i$, the pole can usually be neglected because the frequency of the pole is high compared with the frequency of the zero. Then, the pole has little influence on the Bode plot at frequencies where $|K\mathscr{F}| > 1$. This pole frequency is discussed in Chapter 13.

The gain-phase plots of an amplifier which has been constructed can be readily obtained by the use of a good signal generator and a dual trace oscilloscope, so the output and input voltages can be compared as to

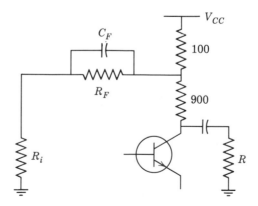

Fig. 12.20 Illustration of a tapped output to reduce R_F and increase C_F.

magnitude and phase. The gain and phase plots for an amplifier design can be easily obtained by using a computer program known as ECAP (Electronic Circuit Analysis Program) if a computer is available.

**12.6
LOG GAIN
PLOTS**

The Bode plots discussed in the preceding section permit the design of stable feedback circuits which provide satisfactory frequency and transient response, but they do not show the frequency characteristics or bandwidth of the amplifier after the feedback has been applied. Therefore, we will develop in this section a technique which will show the closed-loop and open-loop gain characteristics and, as an added bonus, make the feedback design process easier. This proposed technique is just a slight extension and simplification of the Bode plot method previously discussed.

Let us consider the amplifier of Figure 12.12 without the compensating capacitor C_F. The Bode plot for this amplifier is given in Fig. 12.21. Observe that the amplifier gain without feedback is about 2000 at a phase margin of 45°, as shown by the dashed lines. Then the feedback factor which will provide this 45° phase margin is $1/2000 = 5 \times 10^{-4}$, as previously discussed. Also, as previously discussed, the gain of an amplifier *with feedback* is approximately $1/\mathscr{F}$ so long as $K\mathscr{F}$ is large in comparison with 1. Therefore, within the limits of this approximation, the gain of the amplifier *with* feedback is 2,000 which is the *same* as the gain without feedback at the frequency which produces $-135°$ relative phase shift. In other words, the straight-line approximation of the *closed-loop gain* curve (with feedback) intersects the open-loop forward gain curve at the frequency at which $|G\mathscr{F}| = 1$. This relationship should be expected if one reviews the rules for making straight-line approximations. At frequencies above the point of intersection $|G\mathscr{F}|$ is less than one, so the closed-loop and open-loop gains are equal, within the limits of the straight-line approximation. Therefore, the *phase margin* is the difference between the relative phase of the amplifier at the frequency of the intersection of the straight-line approximations of the open-loop and closed-loop gains and 180°. The upper cutoff frequency, or bandwidth, of the amplifier with feedback is approximately at the first break-frequency in the straight-line approximation of the *closed-loop* gain curve. This break-frequency is at the intersection of the closed- and open-loop gain curves, as shown in Fig. 12.21 at $f = 5 \times 10^5$ Hz, when the feedback does not include compensation.

A close relationship exists between the relative phase of the open-loop amplifier and the break-frequencies of the straight-line gain plot of the amplifier. For example, the $-135°$ relative phase occurs at approximately the second break-frequency of the open-loop amplifier, as shown in Fig. 12.21. Therefore, the two gain curves must intersect at a

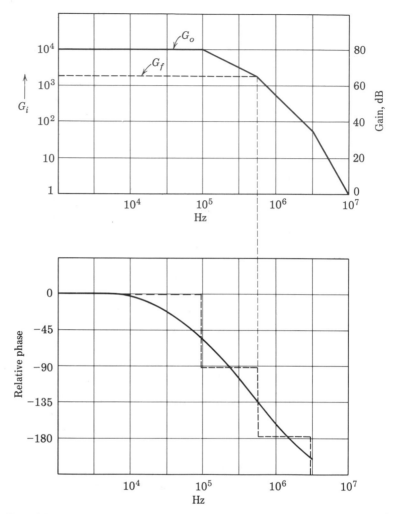

Fig. 12.21. Bode plot for the uncompensated amplifier of Problem 12.9.

frequency no greater than the second break-frequency of the open-loop gain curve if no compensation is used and the desired phase margin is at least 45°. Thus, the phase plot is not required for the feedback design, provided the break-frequencies are reasonably well separated.

The effect which compensation in the feedback circuit has on the bandwidth of the closed-loop amplifier can now be shown. Since the gain with feedback $G_f \simeq 1 / \mathscr{F}(s)$, and $\mathscr{F}(s) = \mathscr{F}(s + \omega_f) / \omega_f$ as given in Eq. 12.42, $G_f = \omega_f / \mathscr{F}(s + \omega_f)$ and therefore has a break-frequency or pole at f_f, as shown in Fig. 12.22. The value of f_f was chosen to be 5×10^5 Hz, the second pole frequency, the same as in Example 12.10. Note that the bandwidth of the amplifier with feedback is the same as f_f because of the

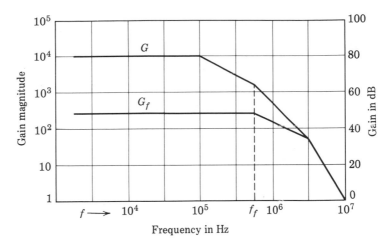

Fig. 12.22. Straight-line approximations for G and G$_f$ for the amplifier of Fig. 12.21 with compensated feedback.

pole at f_f. The zero in the feedback circuit corrects the phase by approximately 90° at frequencies well above f_f. Thus, the intersection of the G and G$_f$ curves may occur at the third break-frequency of G instead of the second break-frequency. Observe that the maximum difference in the slopes of the two curves at their intersection is the same for both the compensated and uncompensated feedback systems, Figs. 12.21 and 12.22. In other words, the difference in the slopes of G and G$_f$ (straight-line approximations) at their intersection should not exceed 20 dB per decade, or 6 dB per octave, if the phase margin is to be at least 45°. Note that G$_f$ = 200 and $\mathscr{F}$ is about ten times as high as the uncompensated case with similar phase margin. Also observe that the bandwidth of the feedback amplifier could be increased if the frequency f_f were increased to the neighborhood of 10^6 Hz, or perhaps 2×10^6 Hz. However, the phase margin is maximum when f_f is the same as the center break-frequency.

The actual closed-loop gain G$_f$ is within a few dB of the straight-line approximation if the phase margin is at least 45°. However, small gain or phase margins cause severe peaking of the actual frequency response where $G\mathscr{F}$ approaches −1, as previously discussed. This peaking is illustrated in Fig. 12.23 for several values of phase margin and is discussed in detail in the following section.

**12.7
PEAKING
OF THE
FREQUENCY
RESPONSE**   Let us investigate the peaking phenomenon of the frequency response of a two stage amplifier with feedback. We will assume that the amplifier is direct-coupled and we therefore need to consider only the high-frequency poles at ω_{h1} and ω_{h2} of the two stages. Then the s-domain gain equation with feedback is

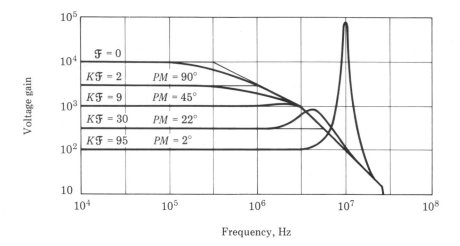

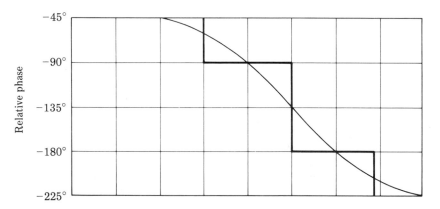

Fig. 12.23. Frequency response for several values of phase margin.

$$G_f(s) = \frac{\dfrac{-K\omega_{h1}\,\omega_{h2}}{(s + \omega_{h1})\,(s + \omega_{h2})}}{1 + \dfrac{K\mathcal{F}\omega_{h1}\,\omega_{h2}}{(s + \omega_{h1})\,(s + \omega_{h2})}}$$

$$= \frac{-K\omega_{h1}\,\omega_{h2}}{(s + \omega_{h1})\,(s + \omega_{h2}) + K\mathcal{F}\omega_{h1}\,\omega_{h2}} \qquad (12.45)$$

where K is the reference gain of the amplifier. Expanding the denominator and collecting like powers of s,

$$G_f(s) = \frac{-K\omega_{h1}\,\omega_{h2}}{s^2 + (\omega_{h1} + \omega_{h2})\,s + (1 + K\mathcal{F})\,\omega_{h1}\,\omega_{h2}} \qquad (12.46)$$

The standard form of a second order equation is

$$G_f(s) = \frac{-K\omega_{h1}\,\omega_{h2}}{s^2 + 2\zeta\omega_n\,s + \omega_n^2} \tag{12.47}$$

where ζ is the damping ratio and ω_n is the natural, or undamped resonant frequency. We can see by comparison of Eq. 12.46 with Eq. 12.47 that

$$\omega_n = \sqrt{(1 + K\mathscr{F})\,\omega_{h1}\,\omega_{h2}} \tag{12.48}$$

and

$$\zeta = \frac{\omega_{h1} + \omega_{h2}}{2\omega_n} = \frac{\omega_{h1} + \omega_{h2}}{2\sqrt{(1 + K\mathscr{F})\,\omega_{h1}\,\omega_{h2}}} \tag{12.49}$$

Observe from Eq. 12.48 and Eq. 12.49 that the resonant frequency ω_n increases and the damping ratio ζ decreases as the feedback factor $\mathscr{F}$ is increased. We will factor the denominator of Eq. 12.47 in order to locate the closed-loop poles. Using the quadradic equation,

$$s_1, s_2 = \zeta\omega_n \pm \sqrt{\zeta^2\omega_n^2 - \omega_n^2} \tag{12.50}$$

Equation 12.50 shows that the amplifier with feedback is underdamped and the poles are complex conjugates when the damping ratio ζ is less than 1. Let us express Eq. 12.47 in factored form for the underdamped case.

$$G_f(s) = \frac{-K\omega_{h1}\,\omega_{h2}}{(s - \zeta\omega_n - j\omega_n\sqrt{1 - \zeta^2})\,(s - \zeta\omega_n + j\omega_n\sqrt{1 - \zeta^2})} \tag{12.51}$$

The s-plane plot of Eq. 12.51 is given in Fig. 12.24. As the frequency response of the amplifier is plotted, $s = j\omega$ and therefore s moves along

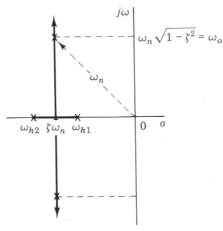

Fig. 12.24. Underdamped poles of a two-stage amplifier with feedback.

the $j\omega$ axis from zero toward ∞. Fig. 12.24 shows that the maximum response is expected when $\omega = \omega_n\sqrt{1 - \zeta^2}$ because the $j\omega$ axis is nearest to the pole at this frequency which is sometimes designated ω_o. The response is easier to find analytically at ω_n than at ω_o, however, so let us substitute $j\omega_n$ for s in Eq. 12.51 to obtain

$$G_f(s)\Big|_{s = j\omega_n} =$$

$$\frac{-K\omega_{h1}\,\omega_{h2}}{(j\omega_n - \zeta\omega_n - j\omega_n\sqrt{1 - \zeta^2})\,(j\omega_n - \zeta\omega_n + j\omega_n\sqrt{1 - \zeta^2})} \qquad (12.52)$$

Simplifying,

$$G_f(s)\Big|_{s = j\omega_n} =$$

$$\frac{-K\omega_{h1}\,\omega_{h2}}{\omega_n^2\,[-\zeta + j\,(1 - \sqrt{1 - \zeta^2})][-\zeta + j\,(1 + \sqrt{1 - \zeta^2})]} \qquad (12.53)$$

which reduces to

$$G_f(s)\Big|_{s = j\omega_n} = \frac{-K\omega_{h1}\,\omega_{h2}}{\omega_n^2\,(\zeta^2 + 2j\zeta - \zeta^2)} = \frac{-K\omega_{h1}\,\omega_{h2}}{2j\zeta\omega_n^2} \qquad (12.54)$$

substituting the expression for ω_n given in Eq. 12.48 into Eq. 12.54

$$G_f(s)\Big|_{s = j\omega} = \frac{-K}{j2\zeta\,(1 + K\mathscr{F})} \qquad (12.55)$$

But $-K/(1 + K\mathscr{F})$ is the midband, or reference, gain K_f of the amplifier. Therefore the ratio of the gain at ω_n to the midband gain is

$$\frac{G_f(s)\Big|_{s = j\omega_n}}{K_f} = \frac{1}{j2\zeta} = \frac{1}{2\zeta}\,\angle{-90°} \qquad (12.56)$$

Also, as shown in Chapter 17, $1/2\zeta = Q$, so the peak response is Q times the reference response. Equation 12.56 shows that the relative gain magnitude is always $1/2\zeta$ and the relative phase shift is always $-90°$ with respect to midband values at $\omega = \omega_n$. In Fig. 12.25, the relative gain magnitude and phase are plotted as functions of ω, normalized to ω_n, for several values of ζ. Observe that values of ζ in the area of 0.5 to 0.6 give superior frequency response characteristics. For a given amplifier, the

Fig. 12.25. Magnitude and phase of $\omega_n{}^2/(s^2 + 2\zeta\omega_n s + \omega_n{}^2)$: (a) magnitude; (b) phase.

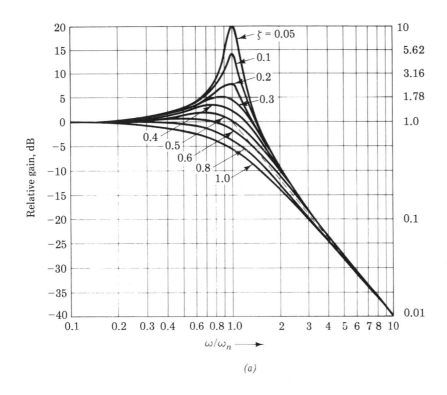

(a)

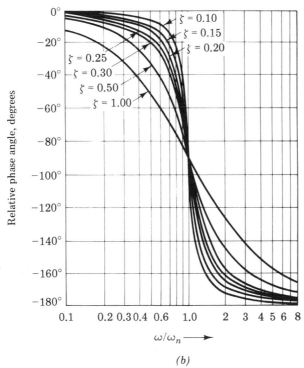

(b)

value of feedback factor $\mathcal{F}$ that would provide the desired value of ζ may be determined from Eq. 12.49. A phase margin of 45° will provide a value of ζ in this general area. Optimum time response is obtained when the damping ratio ζ is approximately 0.7. Observe that the amplifier bandwidth is somewhat greater than ω_n for damping ratios less than 0.8.

Amplifiers having more than two poles behave like the two-pole amplifier because the complex poles always appear in conjugate pairs. The pair with the smallest damping ratio primarily determines the peaking in the frequency response and the overshoot in the time response. This peaking may be used to advantage in the design of active bandpass filters treated in Chapter 14.

**12.8
PHASE-LAG
COMPEN-
SATION**

The preceding feedback examples have shown that the stability problems decrease and more feedback can be applied to the amplifier when the half-power frequencies of the amplifier are widely separated, in contrast with their occurrence in a narrow frequency range. Therefore, stabilization can be obtained by adding shunt capacitance in one of the stages, so its upper cutoff frequency is low in comparison with the other stages. This type of compensation is known as *phase lag* compensation. The benefit of this technique can be seen from the log gain plot of Fig. 12.26. As illustrated, the open-loop gain of the amplifier may decrease greatly before the next higher stage cutoff frequency is reached. As noted, the requirement for stability is that the closed-loop gain G_f intersect the open-loop gain curve G at a higher gain, or lower frequency, than the second break-frequency when no compensation is employed in

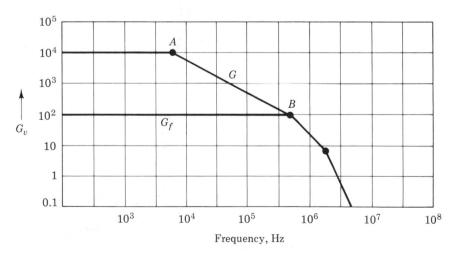

Fig. 12.26. *Illustration of stabilization by lowering the upper cutoff frequency of one stage.*

the feedback network. Then, the closed-loop gain G_f can be small in comparison with the open-loop gain G over the mid-frequency range or, in other words, $K\mathscr{F}$ can be large. In the amplifier of Fig. 12.26, $K\mathscr{F} \approx$ 100. One convenient design technique is to determine the desired value of feedback factor $\mathscr{F}$ from some criterion such as distortion reduction or the fixing of the closed-loop gain. The closed-loop gain line is then drawn at the value $1/\mathscr{F}$. The second break point B is then placed slightly below the closed-loop gain line G_f to allow adequate phase margin. This second break-frequency is determined from either theoretical calculations or lab measurements, as previously discussed. It was given as 5×10^5 Hz for the amplifier of Fig. 12.26. The desired open-loop gain curve G is then constructed by drawing a straight line from this desired second break point B, with a slope -20 dB per decade, until it intersects the horizontal part of the open-loop gain curve. This intersection is shown as point A in Fig. 12.26, and the horizontal part of the open-loop gain curve represents the mid-frequency gain of the open-loop amplifier. The upper cutoff frequency of that stage which had the lowest corner frequency or break point must then be lowered by adding capacitance to coincide with the frequency indicated at point A (5 kHz in Fig. 12.26).

The upper cutoff frequency of an amplifier can be easily lowered by increasing the shunt capacitance in the amplifier. One method is to add a capacitor in parallel with the load resistor as shown in Fig. 12.27a. You may recall that the upper cutoff frequency of a stage is determined by the total shunt resistance and the effective shunt capacitance. The total shunt resistance R_{sh} in parallel with the capacitor C is the load

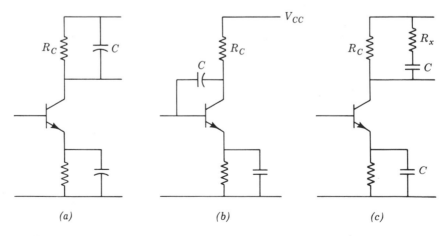

Fig. 12.27. Typical methods of reducing the upper cutoff frequency of an amplifier: (a) capacitor shunt of collector; (b) capacitor shunt from collector to base; (c) RC shunt of collector.

resistor R_C in parallel with both the output admittance (approximately h_{oe}) of the preceding transistor and the input resistance (approximately h_{ie}) of the following transistor. The capacitor C is usually much larger than the other capacitances in the amplifier. Therefore, since $\omega_h = 1/R_{sh} C$,

$$C = \frac{1}{R_{sh}\,\omega_h} \tag{12.57}$$

where ω_h is the desired upper cutoff frequency of the amplifier, determined at point A by the compensation technique illustrated in Fig. 12.27a. For example, the lowest break-frequency, point A, in Fig. 12.26 is 5×10^3 Hz or 3.14×10^4 rad/sec. Thus, if $R_C = 10$ kΩ, $h_{oe1} = 20$ μS, and $h_{ie2} = 4$ kΩ, $R_{sh} \simeq 2.7$ kΩ, and $C = 1/2.7 \times 3.14 \times 10^7 = 0.12$ μF.

Another appropriate location for the phase-lag compensating capacitor is between the collector and the base of the amplifier, as shown in Fig. 12.27b. The advantage of this location is that the value of the capacitance may be much smaller than the capacitor in Fig. 12.27a because it is subject to the Miller effect and therefore its effective capacitance is $(1 + K_v)$ times its actual capacitance. The resistance facing this Miller capacitance is the shunt resistance between the base and ground, as discussed in Chapter 6.

If a small resistance R_X is placed in series with the added capacitance C, as shown in Fig. 12.27c, a zero is produced in the amplifier and the relative phase shift can be decreased at frequencies in the neighborhood of the zero and higher, with a consequent increase of phase margin. The zero occurs at the radian frequency $\omega_x = 1/R_X C$. Therefore, ω_x should be approximately equal to $\omega_B = 2\pi f_B$, where f_B is the second break, or pole, frequency at point B, as in Fig. 12.26. If this relationship is used in the example of Fig. 12.26 where $\omega_B = 100\,\omega_A$, then $R_X = R_{sh}/100 = 27$ Ω. The zero created by R_X would then tend to cancel the effect of the pole at ω_B.

Problems   *Section 12.1*

12.1 Repeat Example 12.1 if $\mathscr{F}_v = .02$.

Answer: $(1 + K\mathscr{F}) = 21$, $K_{vf} = 47.6$, $D_f = .286\%$, $f_{\ell f} = 0.95$ Hz, $f_{hf} = 420$ kHz.

12.2 Assume that the acceptable distortion level in the amplifier of Example 12.1 is 1 percent. Determine the feedback factor $\mathscr{F}_v$ which will reduce the distortion from 6 percent to 1 percent. What is the voltage gain of the amplifier with this amount of feedback?

Answer: $\mathscr{F}_v = 0.005$, $K_{vf} = 167$.

Section 12.2

12.3 An emitter follower can be viewed as an amplifier with feedback, where $\mathscr{F} = 1$. Using this approach, derive expressions for the reference voltage gain and input impedance of an emitter follower.

12.4 An amplifier with self bias as shown in Fig. 12.28 can be viewed as an amplifier with feedback. Determine the current gain and the input impedance (for small signals) if $\beta_o = 100$, $h_{ie} = 1.2$ kΩ, $R_L = 5$ kΩ, $R_B = 250$ kΩ, $R_s = 2$ kΩ and $r_x = 200$ Ω, using the feedback approach.

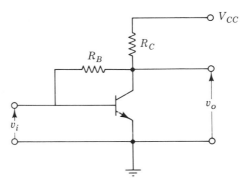

Fig. 12.28. Circuit configuration for Problem 12.4.

12.5 A given amplifier has $K_v = 10^4$, $R_i = 3$ kΩ, and $R_o = 3$ kΩ. This amplifier is to be inserted in a transmission line with characteristic impedance $Z_o = 150$ Ω. If feedback is used to make the impedance of the amplifer match the line in both directions, determine the feedback factor, the type connections (series or parallel), and the voltage gain with feedback.

Answer: $\mathscr{F}_v = 1.9 \times 10^{-3}$, $K_{vf} = 500$.

Section 12.3

12.6 Repeat Example 12.6 if the h_{fe} of both transistors is 200. Assume that we wish to reduce the distortion to 10 percent of the value without feedback.

Answer: $K_i \approx 367$, $K_v \approx 84$, $\mathscr{F}_v = 0.0119$, $R_F \approx 8.3$ kΩ, $Z_{of} \approx 100$ Ω, $Z_{if} = 220$ kΩ.

12.7   Find a value of R_F which will reduce the distortion of the amplifier in Example 12.7 by a factor of 50. Find the value of K_{if}, K_{vf}, Z_{if} and Z_{of} with this value of R_F connected in the circuit.

Answer: $R_F = 4.95$ MΩ, $K_{if} = 4.84 \times 10^3$, $Z_o = 20$ Ω.

12.8   Assume that the amplifier of Fig. 12.10 uses the same components as the amplifier of Example 12.6. The unbypassed resistor $R_1 = 100$ Ω has been changed from the input stage to the output stage, however, so the input impedance as seen by the current source i_s is the parallel combination of R_s, R_b and h_{ie} or 1.33 kΩ for the values given previously. We will assume that the total current gain (i_o/i_s) is the same as before, or 2,590. Then, if the desired value of $K_i \mathcal{F}_i = 9$, determine the required value of R_F. Since the output impedance without feedback was assumed to be approximately 1 kΩ, determine the output impedance and the input impedance, as seen by the current-driving source I_s, with the feedback specified.

Answer: $R_F = 2.88 \times 10^4$ Ω, $R_{of} = 10$ kΩ, $R_{if} = 133$ Ω.

12.9   The amplifier of Fig. 12.11 has $R_L = 100$ Ω, $R_1 = R_2 = 200$ kΩ, $R_E = 100$ Ω, h_{fe} of all transistors $= 100$, $+V_{cc} = 20$ V, $-V_{cc} = -20$ V and driving source resistance $R_s = 10$ kΩ. Assume that $h_{ie} = 2.5$ kΩ for T_1 and T_2.

   a.   Determine the values for R_F and C that will provide dc voltage gain $= 10$ and $f_1 = 30$ Hz.
   b.   Will the circuit operate if R_E is removed from the circuit and/or if C is removed from the circuit? What will be the approximate dc and ac voltage gain in each case?

Section 12.4

12.10 The phase margin of the amplifier in Example 12.9 is reduced to 30°. Determine the mid-frequency loop gain $K_v \mathcal{F}_v$.

Answer: 27.8 or 29 dB.

Section 12.5

12.11 A given three-stage transistor amplifier has stage upper cutoff frequencies of 10^5, 5×10^5, and 3×10^6 Hz. The amplifier input resistance $R_i = 10$ kΩ, the load resistance $R_L = 1$ kΩ, and the driving-source resistance $R_s = 5$ kΩ. The current gain $I_L/I_s = 10^4$. Direct coupling is used so stability at low frequencies is not a problem. A feedback circuit similar to the one given in Fig. 12.12 is used except a capacitor C_F is placed in parallel with R_F and C_F will be chosen so the zero in the feedback network will cancel the amplifier pole at 5×10^5 Hz. Make a Bode plot similar to Fig. 12.19 and determine the values of R_F, C_F, and $K\mathcal{F}$ for a 45° phase margin.

Answer: $R_F = 200$ kΩ, $C_F = 1.6$ pF, $K\mathcal{F} = 50$.

12.12 Derive an expression for $\mathcal{F}_i$ that includes R_i (Fig. 12.18) instead of neglecting it. What is the ratio of the frequency of the pole to the frequency of the zero?

Answer: Ratio $= R_F/(R_i \| R_F)$.

Section 12.6

12.13 A given dc amplifier has an open-loop voltage gain $K_v = 5,000$ in the midfrequency range. Its three stages have upper cutoff frequencies of 10^5, 3×10^5 and 10^6 Hz.

    a.   What maximum value of feedback factor $\mathcal{F}_v$ can be used, assuming a satisfactory phase margin is maintained and no compensation is used? What is the approximate upper cutoff frequency of the amplifier with this feedback?

    b.   What maximum value of feedback factor $\mathcal{F}_v$ can be used (with satisfactory phase margin) if phase-lead compensation is used in the feedback network? What is a good, or optimum, value of the $R_F C_F$ time constant? What voltage gain and bandwidth will the amplifier have with this feedback?

    c.   It is desired that the voltage gain of the amplifier with feedback be 100. Assume that you will decrease the bandwidth of one stage sufficiently to obtain good stability with an uncompensated feedback network. What should be the upper cutoff frequency of the modified stage? What value of shunt capacitance is needed if $h_{oe1} = 40$ μS, $R_C = 5.0$ kΩ, and $h_{ie2} = 2.0$ kΩ (R_b, if any, can be neglected) and C is placed in parallel with R_c as shown in Fig. 12.27a?

    d.   What will be the upper cutoff frequency of the amplifier of part c? What value of resistance R_X (Fig. 12.27c) will nullify the effect of the pole at $f = 3 \times 10^5$ Hz?

13

Linear Integrated Circuits

One advantage a transistor has over a vacuum tube is the great size reduction possible for a given power output. Thus, it is quite surprising to open a conventional transistor container and discover that the actual portion of germanium or silicon is so much smaller than the total container. Most of the volume is used for mounting the leads and protecting the silicon or germanium chip from an unfavorable environment. To utilize some of this waste space, some semiconductor manufacturers began mounting several diodes in one container. Then, two transistors in one case began to appear on the market. By making two internal connections, two transistors can be mounted in a Darlington configuration as shown in Fig. 13.1. As noted in Chapter 10, the current gain of a Darlington configuration is approximately the product of the current gains of the individual transistors. Thus, very high current gains from a single, three-lead "transistor" case are possible.[1] While the manufacturing process varies depending on which type of active devices are used (MOSFET, Bipolar, or FET), essentially all of these devices are constructed as *monolithic circuits*. In monolithic circuits the entire circuit, including transistors, is constructed in a single chip of semiconductor material.

13.1
MONOLITHIC
CIRCUITS

Most of the integrated circuits on the market at the present time use a silicon chip for the substrate material. However, galium-arsenide chips and other substrate materials are being used where high temperature or

---

[1] For example, Motorola advertises Uniblock Darlingtons with h_{FE} from 5,000 to 75,000.

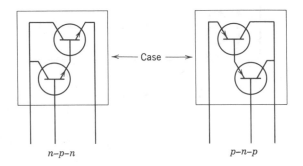

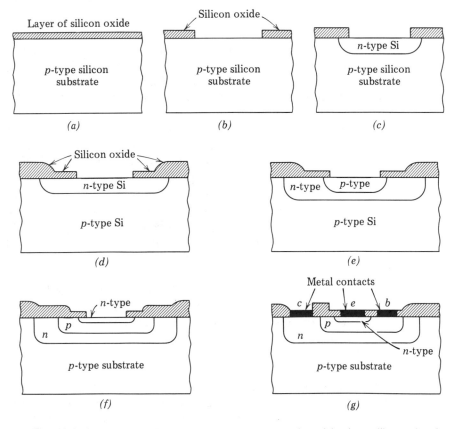

Fig. 13.1. Two transistors internally connected to form a Darlington configuration.

Fig. 13.2. Process used in constructing an n-p-n transistor in a silicon structure.

very high-frequency operation is desired. The n and p regions are diffused into this substrate chip to produce the desired circuit elements. The procedure used in constructing a transistor in a chip will be used to help clarify this process. As shown in Fig. 13.2a, a substrate of p-doped silicon is subjected to an oxidizing atmosphere (a small amount of water vapor speeds up the process considerably) at an elevated temperature. As a result of this step, a thin layer of silicon oxide (or glass) is formed over the substrate (Fig. 13.2a). By the use of a photoengraving process, the layer of silicon oxide is removed over the required collector area, as shown in Fig. 13.2b. The silicon chip is then placed in an n-type atmosphere at an elevated temperature. The n-type atoms diffuse into the silicon (but do not diffuse into the silicon oxide) creating a layer of n-type silicon, as shown in Fig. 13.2c. The density of n-atoms is much greater than the original p-atoms, so the p-atoms are neutralized and then overwhelmed by the n-atoms. Again, a layer of silicon oxide is formed over the entire chip. (The old silicon-oxide layer becomes a little thicker.) The photoengraving process is repeated to remove the silicon oxide over the required base region, as shown in Fig. 13.2d. The chip is then placed in a p-type atmosphere at an elevated temperature and a layer of p-type silicon is formed in the n-type collector region, as shown in Fig. 13.2e. Again, a layer of silicon oxide is formed over the entire chip. The photoengraving process is repeated to remove the silicon oxide over the required emitter area. Then the chip is subjected to an n-type atmosphere until a layer of n-silicon is formed in the base region p-silicon (Fig. 13.2f). A final silicon-oxide layer is formed over the chip and the silicon oxide is again removed over the points where ohmic contact is to be made with the silicon (emitter, base, and collector lead junctions). By vacuum evaporation or sputtering, a layer of metal is deposited onto the chip as shown in Fig. 13.2g. This metal is also deposited over the insulating silicon oxide to form connecting leads to the other circuit elements. If the junction between the substrate and the collector is maintained in a reverse-bias condition, the leakage current will be small and the transistor is effectively insulated from the substrate. In Fig. 13.2, the substrate should be connected directly to the negative lead of the power supply to ensure proper bias of the substrate. The configuration shown in Fig. 13.2e could be used (with proper ohmic contacts) for a diode, and the configuration shown in Fig. 13.2c could be used for a resistor. In the latter case, the n material is the resistor, and an increase of length or decrease of width of this region would increase the resistance.

Of course, the techniques described above could be used to produce field-effect transistors (FET) or MOSFET transistors. Actually, fewer steps are required to construct MOSFET circuits than for bipolar cir-

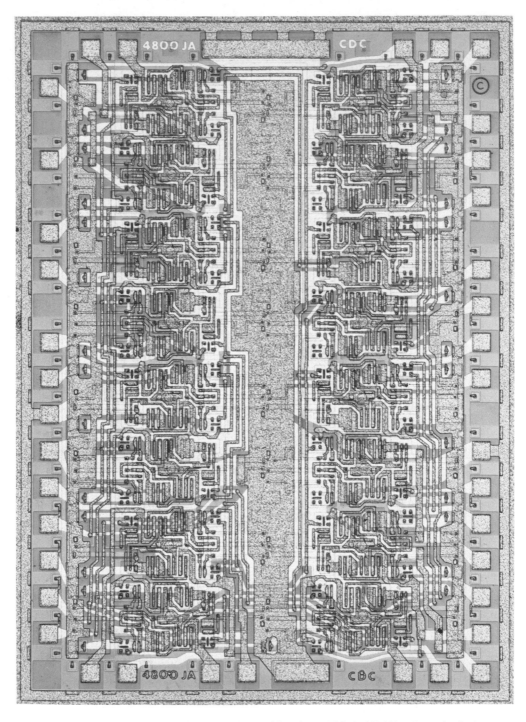

Fig. 13.3. Monolithic integrated circuit. (Courtesy of Fairchild Semiconductor)

cuits. Both types of circuits are presently being manufactured on a commercial basis.

A typical monolithic integrated circuit is shown in Fig. 13.3.

The design of integrated circuits is beyond the scope of this book. However, some interesting considerations which are used in designing these integrated circuits will be noted. In the past, the active circuit elements (tubes, transistors, etc.) have been relatively expensive when compared to the passive circuit elements (resistors, capacitors, and inductors). Consequently, conventional designs usually incorporate as few active elements as possible. In contrast, a whole new concept is used in designing monolithic circuits with deposited active elements. All of the transistor collectors in the circuit can be formed at the same time and with the same process. Similarly, all the bases are formed simultaneously and all the emitters are formed in a given time period. Consequently, an integrated circuit with ten or fifteen transistors can be produced at a cost essentially the same as an integrated circuit with four or five transistors. In addition, the cost of the chip with the circuit in place is very small (10 percent or so) in comparison to the cost of testing and packaging the circuit. Thus, when designing monolithic circuits, *the cost of additional active elements is almost negligible.* In fact, since a transistor uses less space on the silicon chip than a resistor, it is desirable to replace resistors with transistors where possible! Also, it is possible to use the junction capacitance of a reverse-biased diode instead of constructing a capacitor in the circuit.

As shown in Fig. 13.2, a p-type substrate is used for n-p-n type monolithic circuits. Conversely, an n-type substrate is required for p-n-p type monolithic transistors. Therefore, it seems impossible to produce both n-p-n and p-n-p transistors on the same monolithic substrate, and thus take advantage of complementary symmetry and other convenient n-p-n–p-n-p arrangements. However, a *lateral* p-n-p transistor which uses a p-type substrate has been developed, as shown in Fig. 13.4, thus making possible the use of both n-p-n and p-n-p transistors on the same monolithic chip. The lateral p-n-p transistor has very low

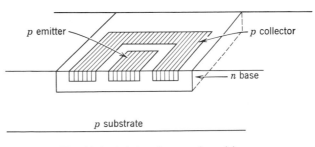

Fig. 13.4. A lateral p-n-p transistor.

β (of the order of one), however, so a high gain n-p-n driver is always used with it to provide the needed current gain.

Thermally-stable, high-resistance resistors were, until recently, difficult to produce and space consuming. Therefore, a transistor, acting as a current source, is often used as a high load resistance for another transistor. However, high-resistance, low-volume, thermally stable resistors can now be created by diffusing chromium into silicon.

13.2 BASIC LINEAR INTEGRATED CIRCUITS

The integrated-circuit manufacturer attempts to fill many of the needs of the user with a single type of integrated circuit, thereby creating large sales volume for each type. Since the differential amplifier, discussed in Chapter 10, can have excellent thermal stability and is useful for either dc or ac amplifiers in either balanced or single-ended operation, the differential amplifier is used in almost all linear integrated circuits.

A versatile amplifier should have high input impedance, so it will not load the commonly used driving sources; low output impedance, so it can drive a wide variety of load resistance; high gain (which can be controlled over wide limits with feedback); and broad bandwidth which can also be controlled by feedback and compensation. The techniques used for accomplishing or controlling these characteristics will be discussed in the remainder of this chapter.

A simple, basic, linear integrated circuit (IC) is shown in Fig. 13.5. This circuit consists of a differential amplifier (T_1 and T_2) with a current

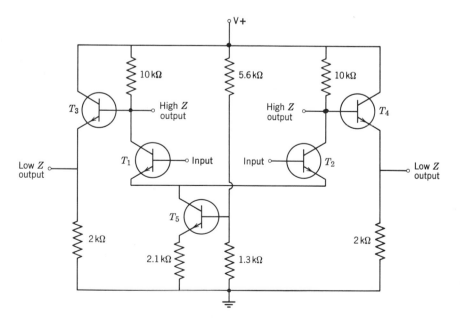

Fig. 13.5. Schematic diagram of Fairchild μA 730 IC.

source T_5 acting as a high emitter impedance, as discussed in Chapter 10, and a pair of emitter followers (T_3 and T_4) to provide low output impedance. This circuit may be ideally suited for some applications, but has the following disadvantages for other applications:

1. Insufficient voltage or current gain.

2. Only moderate input impedance.

3. Large (volts) offset voltage between both the input and output terminals and ground, probably requiring the use of blocking capacitors.

4. Bias currents for transistors T_1 and T_2 must be provided from the external circuit.

The integrated circuit shown in Fig. 13.6 eliminates most of the shortcomings listed above. In this circuit, the input differential ampli-

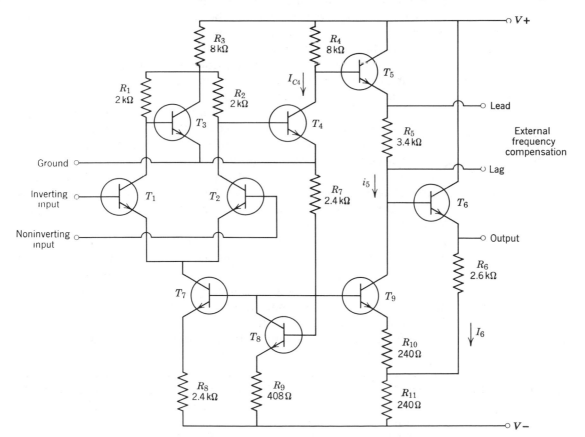

Fig. 13.6. Schematic diagram of Fairchild μA702IC.

fier (T_1 and T_2) is followed by a second differential amplifier (T_3 and T_4) to provide additional gain, and two emitter followers (T_5 and T_6) are connected in cascade to provide low-impedance output. The resistor R_5 is placed in the emitter circuit of transistor T_5 to provide a dc voltage drop so that the output terminal will be at approximately dc ground potential. The input terminals are at approximately dc ground potential because of the balanced supply voltage $V+$ and $V-$. The transistors T_7, T_8, and T_9 are used to control the bias current in the other transistors, as discussed later. Transistor T_9 also provides a *bootstrapping* function which increases the voltage swing available at the output terminal and actually provides voltage gain between the base of transistor T_4 and the output of the amplifier. We will illustrate this bootstrapping technique by assuming that the base of transistor T_5 is driven positive, therefore causing the emitter of T_5 and the base of T_6 to become more positive. If the bootstrapping circuit were missing, the potential at the base of T_6, and hence the output potential, would not rise as much as the emitter potential of T_5 because of the increased IR drop across R_5 as the current through it would increase. However, the increased emitter current of the output transistor T_6 flows through resistor R_{11} in the emitter circuit of T_9 and thus tends to reverse bias T_9. But the reduced forward bias of T_9 reduces its collector current which flows through R_5. Therefore, the *total* current through R_5 actually *decreases* and the base potential (and hence the emitter potential) of T_6 experiences a *greater* rise than either the base or emitter potential of T_5. Another way to look at the output circuit is to recognize that positive feedback is applied from the output through T_9 (as a common-base amplifier) to the input of T_6. The loop gain must be less than one, so oscillation will not occur.

Transistors T_7, T_8, and T_9 provide stabilized bias to all the amplifiers in the IC, as illustrated in Fig. 13.7. This bias circuit is identical to its counterpart in Fig. 13.6, but is rearranged slightly for convenience. Transistor T_8 is connected as a diode, so the voltage drop across $R_7 + R_9$ is $V-$ minus V_{EB}. Thus, the collector and emitter currents of T_8 are fixed almost completely by the supply voltage $V-$ if the transistor current gains are high. Therefore, the base currents of T_9 and T_7 can be neglected. Also, since the transistors are essentially identical, their values of V_{EB} are very nearly equal, so their emitter potentials are almost identical. The emitter, and hence collector, currents of the biasing transistors are controlled almost completely by their emitter circuit resistors.

| Example   13.1 | The bias currents for the circuit in Figs. 13.6 and 13.7 will be determined as an example of the bias circuit design and operation. We will assume that $V- = -8$ V and $V_{BE} = 0.5$ V for all transistors. Then for transistor T_8, $I_C \simeq |I_E| = (8 - 0.5)$ V$/(2.4$ k $+ 480)$ $\Omega = 2.6$ mA. The voltage drop across the emitter resistor of T_8 is therefore approxi- |

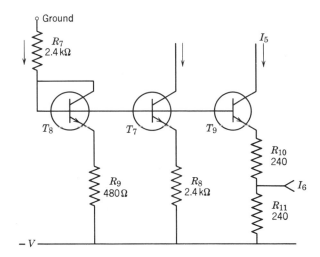

Fig. 13.7. Stabilized-bias circuit of Fig. 13.6.

mately (2.6 mA) (480 Ω) = 1.25 V. The voltage across the 2.4 kΩ must also be 1.25 V, so the emitter current of transistor T_7 is 1.25 V / 2.4 kΩ = 0.52 mA. The collector current of T_7 is essentially 0.52 mA, and this current divides equally between T_1 and T_2 whose collector currents must therefore each be about 0.26 mA. The voltage drop across R_1 and R_2 is (2 kΩ) (0.26 mA) = 0.52 V. The voltage drop across R_3 must therefore be V_+ (8 V) − (0.52 V + V_{BE}) = 6.98 V. The current through R_3 is then 6.98 V / 8 kΩ = 0.87 mA and the collector current of transistor T_3 = (0.87 − 0.52) mA = 0.35 mA. The collector current of transistor T_4 must also be 0.35 mA since T_3 and T_4 form a balanced differential amplifier. The emitter current of transistor T_9 is not so easy to calculate because only current I_5 (approximately) flows through R_{10}, but both I_5 and I_6 flow through R_{11}. We do know that the sum of the voltage drops in the emitter circuit is 1.25 V. Therefore,

$$I_5 R_{10} + (I_5 + I_6) R_{11} = 1.25 \qquad (13.1)$$

We can find the value of I_5 required to provide zero dc output voltage, which is desired. If we neglect the base currents, Fig. 13.6 shows that

$$I_5 R_5 + I_{C4} R_4 = V_+ - 2V_{BE} \qquad (13.2)$$

We have already found I_{C4} to be approximately 0.35 mA.

**13.3
OPERATIONAL
AMPLIFIERS**

The term *operational amplifier (op amp)* was coined by people in the analog computer field and was used to designate an amplifier which has very high open-loop gain, high input impedance, and low output impedance. This type of amplifier is very versatile because feedback can

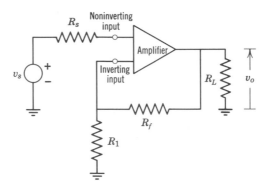

Fig. 13.8. Operation amplifier used as a linear amplifier with stable gain and very high input impedance.

be used to control its characteristics, thus making it useful for a wide variety of applications. In Chapter 1, some op amp configurations were considered. We will now examine these amplifiers with our present understanding. Op amps usually have values of open-loop voltage gain between 10,000 and 100,000. A triangular symbol is used to represent the entire amplifier as shown in Fig. 13.8. The amplifier is normally an integrated circuit with differential input, as previously discussed. The feedback circuit must always be connected to the inverting input for negative feedback, but the signal may be applied to either terminal, depending upon the required input impedance of the amplifier. When the signal is applied to the noninverting input, as shown in Fig. 13.8, the amplifier input impedance is very high at both input terminals because the feedback voltage is nearly equal to the input voltage. The two voltages appear as a common-mode input signal which sees a very high input impedance because of the very high common-mode impedance in the emitter circuit. The difference between the signal source and feedback signals is the differential signal which is effective in driving the amplifier. In other words, the feedback arrangement of Fig. 13.8 is the series input connection, and the open-loop differential input impedance of the amplifier is multiplied by the factor $(1 + K\mathscr{F})$ as a result of the feedback, as discussed in Chapter 12. Since the effective amplifier impedance is very high, the voltage feedback factor is

$$\mathscr{F}_v = \frac{R_1}{R_1 + R_F} \tag{13.3}$$

and the reference gain of the amplifier is

$$K_{vf} = \frac{1}{\mathscr{F}} = \frac{R_1 + R_F}{R_1} \tag{13.4}$$

assuming the open-loop gain is much higher than the closed-loop gain. Therefore $K\mathscr{F} \gg 1$, as discussed in Chapter 12. The amplifier may require compensation if small values of closed-loop gain are used. However, some integrated circuits have built-in compensation which provides stable operation for all values of closed-loop gain down to unity. Compensation techniques will be discussed in Section 13.6.

Sometimes it is desirable to have very low impedance and very small voltage at the amplifier input terminals. These characteristics are obtained when the input and feedback signals are both applied to the inverting input terminal of the amplifier, as shown in Fig. 13.9. We note that this is the *parallel* input connection discussed in Chapter 12. The non-inverting input terminal is maintained at ground potential. The inverting input signal is the differential input voltage and is in the order of microvolts or so in magnitude. Then, as noted in Chapter 1, the input impedance of the amplifier with feedback is R_1 and the closed loop gain is

$$K_{vf} = \frac{v_o}{v_i} = \frac{R_F}{R_1} \tag{13.5}$$

In addition, the voltage gain from the input source to the output is

$$K'_{vf} = \frac{v_o}{v_s} = \frac{R_F}{R_1 + R_s} \tag{13.6}$$

Example 13.2 | An example will illustrate the use of the relationships above and demonstrate the small error introduced by the approximations which were used. We will assume that an amplifier is needed which will provide an input impedance of 10 kΩ and a voltage gain $v_o / v_i = 100$.

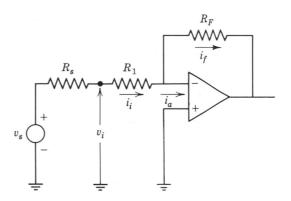

Fig. 13.9. Operational amplifier with the input signal applied to the inverting input terminal.

The driving-source resistance is 1 kΩ. An operational amplifier with open-loop voltage gain = 5×10^4 and input resistance = 100 kΩ will be used. The resistor $R_1 = 10$ kΩ (Fig. 13.9) is used to provide the desired input resistance and $R_F = 100R_1 = 1$ MΩ will provide the desired gain. The voltage gain from voltage source to output is $v_o/v_s = 10^6/(1.1 \times 10^4) = 91$. We will now determine the errors resulting from the assumption that $i_a \simeq 0$ and $v_{ia} \simeq 0$, assuming that the output signal is 5.0 V. The amplifier input voltage is then $v_{ia} = 5/(5 \times 10^4) = 10^{-4}$ V, or 100 microvolts, and the signal input current to the amplifier is 10^{-4} V$/10^5$ Ω $= 10^{-9}$ A, or 1 nA. The feedback current $i_f \simeq 5$ V$/10^6$ Ω $= 5$ μA which is 5,000 times as large as the amplifier input current i_a. Since the current from the driving source flowing through R_1 can differ from i_f only by the amount i_a, this difference can be only one part in 5,000 or 0.02 percent. Similarly, the driving-source voltage is 5 V$/91 = .055$ V, which is 550 times as large as the $v_{ia} \simeq 10^{-4}$ V which we neglected. Therefore, the approximations are extremely good and the gain of the amplifier is essentially as stable as the resistance values of R_1 and R_F.

**13.4
INPUT OFFSET
AND OUTPUT
OFFSET
VOLTAGES**

The bias currents for the differential input stage of the operational amplifier normally flow through the external resistors connected to the input terminals. Therefore, the dc resistance from each of the two input terminals and ground should be approximately the same. Otherwise, the $I_B R$ drop across the two resistors will be different and a dc *offset* voltage will appear between the two input terminals. This offset voltage is multiplied by the closed-loop gain of the amplifier and appears in the output as an offset voltage.

Even though the resistances from the two input terminals to ground are precisely the same, an input offset voltage may exist for two reasons:

1. The values of h_{FE} for the two transistors in the input differential amplifier are different, so the input bias currents are different when the collector currents are equal. The maximum difference between the two bias currents for a given IC is usually listed by the manufacturer as the *maximum* input offset current.

2. The quiescent bias voltages V_{BE} of the input differential transistors are not precisely the same when the input currents are equal. This voltage discrepancy is caused by inaccuracies in the manufacturing process and and is known as *input offset voltage*. The *maximum input offset voltage* for a given IC is usually listed by the manufacturer.

The three components of input offset voltage are therefore:

1. Input bias current times the difference in bias resistance ($I_B \, \Delta R_b$).

2. Input offset current times the nominal (average) bias resistance ($\Delta I_B\, R_b$).

3. Input offset voltage ΔV_{BE}.

These components may have different polarities and tend to cancel, but the circuit designer must be pessimistic and assume that they will all be at their maximum values and will all add directly. This *worst case* input offset voltage is

$$\text{Maximum input offset voltage} = I_B\,\Delta R_b + \Delta I_B R_b + \Delta V_{BE} \quad (13.7)$$

The maximum *output offset voltage* is the maximum input offset voltage multiplied by the closed loop gain of the amplifier, as mentioned before. When an attempt is made to operate an *op amp* without feedback, the output offset voltage is usually large enough to saturate the output. That is, the output potential will be near either $+V_{CC}$ or $-V_{CC}$, frequently known as the *rail* voltages.

The resistance between the two inputs of the op amp and ground may be adjusted to minimize the input offset voltage. That is, ΔR_b in Eq. 13.7 may be either negative or positive and is adjustable. However, it is not always possible to null the input offset voltage by this technique, particularly over a large temperature range, so many op amps have an offset nulling circuit, as noted in Chapter 1 and in Fig. 13.10. The circuit application, the closed loop gain, and the resistance R_b values at the amplifier input determine whether special attention need be given to nulling the input offset voltage. If it is essential to maintain near-zero offset voltage, the IC should be carefully selected in this regard.

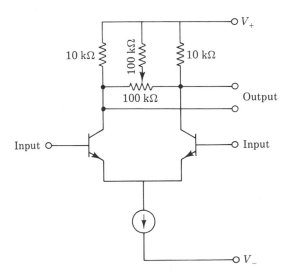

Fig. 13.10. Typical offset-voltage null adjustment.

**13.5
THE SUMMING
AND
INTEGRATOR
CIRCUIT**

Two very useful op amp configurations were developed and used widely in analog computers. One of these configurations is known as the *summing amplifier* or *summer*. The op amp is connected as shown in Fig. 13.11. The equation for the output of this summer is

$$v_o = v_1 \frac{R_F}{R_1} + v_2 \frac{R_F}{R_2} + v_3 \frac{R_F}{R_3} \tag{13.8}$$

If a simple summing or adding operation is desired, then $R_1 = R_2 = R_3 = R_F$, and the accuracy of the operation depends on the precision of the resistors. However, the voltage at any of the inputs may be amplified by the resistance ratios indicated by Eq. 13.8. Of course, any desired number of inputs may be used in a summer configuration.

The resistance R_4 (Fig. 13.11) should be equal to the parallel combination of R_1, R_2, R_3, and R_F, or it may be adjusted, to minimize the input offset voltage. It may appear that R_4 should be bypassed for ac signals since signal currents flow through it and we have neglected the signal voltage across it. However, we have appropriately neglected the input current and voltage of the op amp, and since the amplifier input current flows through R_4 we can neglect signal voltage across R_4 providing its resistance is less than the input resistance of the op amp, which is typically a megohm or higher. The effect of not bypassing R_4 is to slightly reduce the effective open-loop voltage gain of the op amp. The reduction is the ratio $R_{in} / (R_4 + R_{in})$.

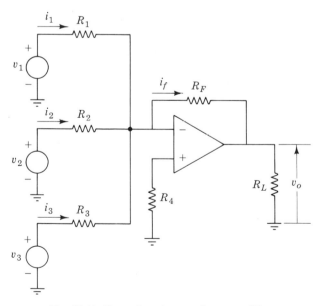

Fig. 13.11. Three-input summing amplifier.

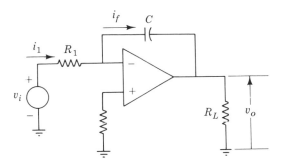

Fig. 13.12. Basic integrator circuit.

The other op amp configuration used in analog computers is the *integrator* circuit shown in Fig. 13.12. The output voltage from this circuit is

$$v_o = \frac{1}{R_1 C} \int v_i \, dt + V_o \qquad (13.9)$$

where V_o is the initial output voltage.

The $1/R_1 C$ factor is known as the gain factor of the integrator. If the time constant $R_1 C$ is unity, the output voltage is a simple integration of the input voltage. When the input voltage contains a dc component, the output voltage continually rises until it is limited by the supply voltage. The integrator is then disabled. Therefore, when this dc component is present, an additional circuit samples the output voltage before saturation occurs and then resets the integrator to zero. The length of time required to reach this reset level is proportional to the dc component present. This circuit, including the integrator, is known as an *integrate and dump* circuit.

When there is no dc input except the usual input offset voltage, a resistor R_F is usually connected in parallel with the capacitor C to limit the dc gain and thus prevent the amplifier from going into saturation without the use of a dump. This type of integrator is known as a *running average integrator* and is usually used as a low-pass filter to remove unwanted high frequency components from the input. The low-frequency cutoff of the filter in r/s is $1/R_F C$.

13.6 AMPLIFIER COMPEN- SATION
Some compensation techniques which will provide stable operation and acceptable transient response of an amplifier with feedback were discussed in Chapter 12. The ideas developed there are very briefly reviewed here with the aid of Fig. 13.13 where curve *a* is the straight-line approximation of the open-loop frequency response curve of a

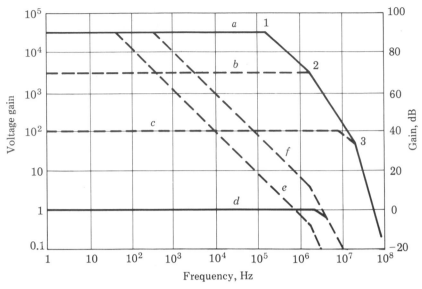

Fig. 13.13. Frequency response curves.

representative operational amplifier with 30,000 or 89 dB reference gain and break frequencies at 150 kHz, 1.5 MHz, and 15 MHz. You may recall from Chapter 12 that approximately 45° or more phase margin will be maintained if the straight-line approximation of the closed-loop response curve intersects the open-loop response curve at a point where the difference in their slopes is only 6 dB per octave or 20 dB per decade. Therefore, no compensation is required in either the amplifier or the feedback network if the closed-loop reference gain is 3,000 or higher, as shown by curve b. This gain corresponds to 20 dB of feedback $(1 + K\mathscr{F} = 10)$. Curve c shows that the closed-loop gain can be reduced to approximately 100 if phase-lead compensation is used in the feedback network with $R_F C_F \simeq 1/6\pi \times 10^6$. $(1 + K\mathscr{F} = 100$, or 40 dB of feedback.) Sometimes a lower closed-loop gain such as 10 is desired, or even a gain of 1, perhaps, which is common for a summing amplifier. Then phase-lag compensation must be used to achieve adequate stability. This type of compensation is achieved by adding capacitance to one of the amplifier stages, as discussed in Chapter 12. If wide bandwidth is not important, no compensation is required in the feedback network if the compensating capacitor produces a sufficiently low break-frequency so that the 6 dB/octave slope of the open-loop gain curve of the compensated amplifier intersects the closed-loop gain curve before the second break-frequency is encountered. For example, in Fig. 13.13 the desired closed-loop gain curve d is drawn at unity gain. The compensating capacitor is added to the amplifier stage which had the break-frequency at 150 kHz, which is reduced to about 40 Hz by the compensation. The open-loop gain curve *e* then decreases at 20 dB per decade from 30,000 gain at 40 Hz to unity gain at 1.2 MHz, which is less than the second break-frequency at 1.5 MHz. Thus, the amplifier is stable with an uncompensated feedback network. Some operational amplifiers have *built-in* compensation of this type and are therefore stable for any amount of uncompensated feedback to unity gain. Bandwidth is sacrificed in this type of amplifier, however, because additional bandwidth could be obtained by using phase-lead compensation in the feedback network, as shown by curve *f*, where the point of intersection of *f* with the closed-loop gain curve can occur at a frequency higher than the second break-frequency. You may recall that the upper cutoff frequency of the amplifier occurs approximately at the first break-frequency in the closed-loop response. This break-frequency is located at the intersection with the open-loop response curve when no phase-lead compensation is used in the feedback network, or it is the break-frequency produced by the compensating network when phase-lead compensation is used.

Observe from Fig. 13.13 the large amount of bandwidth that is sacrificed when an operational amplifier with built-in compensation is used with small amounts of feedback which produce high closed-loop gains.

The op amp manufacturers usually give recommended compensating component values for their externally compensated amplifiers. These components usually include a capacitor and a series resistor to provide a high-frequency zero for improved phase characteristics, as discussed in Chapter 12. The values are given for a variety of voltage gains, and interpolation may be used to obtain compensation component values for gains not listed. Compensation techniques are discussed in more detail in Section 13.8.

13.7
SLEW RATE

The *slew rate* of an amplifier is the rate at which the output voltage of the amplifier rises toward the supply voltage when some or all of the stages of the amplifier are either cut off or are in saturation. In other words, the amplifier is not operating in its normal linear mode. Slew rate is usually expressed in volts/μs. This slew rate becomes important when fast-rising rectangular pulses are applied to the input of an amplifier which has high open-loop gain, such as an operational amplifier, but has a large feedback factor which reduces the gain to a small value. Although the input pulse may seemingly be too small to overload the amplifier, the delay time and rise time of the amplifier prevent the feedback pulse from providing adequate cancellation of the leading edge of the input pulse. Thus, a large, sharp, disabling spike is applied as a differential input signal to the amplifier, as shown in Fig. 13.14. The output voltage does not then rise at the rate predicted on the basis

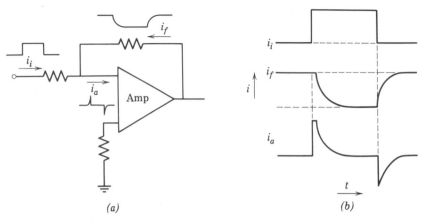

Fig. 13.14. Illustration of the large input spikes that result from fast-rising pulses, large amounts of feedback, and amplifier delay and rise time: (a) circuit diagram; (b) current waveform.

of linear amplifier theory because the amplifier is not entirely operative, but rises at the slower slew rate instead. Thus, slew rate becomes an important parameter in amplifier design. Compensating capacitors can greatly decrease or *slow* the slew rate when they are placed near the output of an amplifier because the lower level stages are then cut off and the output rises slowly due to the large time constants caused by the large compensating capacitance. On the other hand, a compensating capacitance near the input of the amplifier may not be as detrimental because the following amplifiers increase the rate of rise of the slewing voltage.

The relationship between slew rate and bandwidth in determining the rise time of an op amp is illustrated in Fig. 13.15. The op amp is assumed to have a voltage gain = 10, bandwidth = 1 MHz, and slew rate = 4 V/μs. The input signal is assumed to be a rectangular voltage pulse of variable amplitude. When the input pulse is small enough so slewing does not occur, the rise time is independent of signal amplitude at the bandwidth limited value $t_r = 2.2 / \omega_h = 0.32$ μs. When the output increases to (slew rate $\times t_r$) = (4 V/μs) (0.32 μs) = 1.28 V, the amplifier begins to slew. As the output voltage increases beyond this value, the 100 percent rise time is the product of the slew rate and the steady-state output voltage.

The slew rate is the maximum rate of change of voltage with time in the output of an amplifier. Therefore, sinusoidal signals or frequency components will be distorted if their rate of change in amplitude exceeds the slew rate because the amplifier will not be able to reproduce the steepest slopes in the input waveform. For example, let us assume that a sinusoidal signal is applied to an op amp and the output voltage

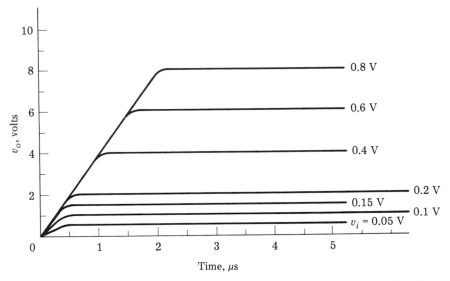

Fig. 13.15. Response time to various input signals of an op amp with 1.0 mHz bandwidth and 4 V/μs slew rate.

is $V_p \sin \omega t$, where V_p is the peak voltage of the sinusoid. We may determine the maximum permissible value of this peak output voltage by equating the maximum slope to the slew rate. The slope of the signal is

$$\frac{d(V_p \sin \omega t)}{dt} = V_p \omega \cos \omega t \qquad (13.10)$$

The maximum slope occurs when $\cos \omega t$ takes on its maximum value of 1. Therefore,

$$V_p \omega = \text{slew rate} \qquad (13.11)$$

and the maximum permissible value of V_p in the amplifier output is

$$V_p = \frac{\text{slew rate}}{\omega} \qquad (13.12)$$

For example, if an op amp with 1 MHz bandwidth and 4 V/μs slew rate is used to amplify a 1 MHz sinusoidal voltage, the peak signal amplitude obtainable in the output without distortion is $V_p = (4 \times 10^6$ V/s)/$(6.28 \times 10^6) = 0.637$ V. If an attempt is made to increase the output amplitude above this value, the sinusoids transform into triangular waves having upward and downward slopes equal to the slew rate. A sketch of available peak output voltage as a function of frequency is given in Fig. 13.16 for the op amp having a 1 MHz bandwidth and 4

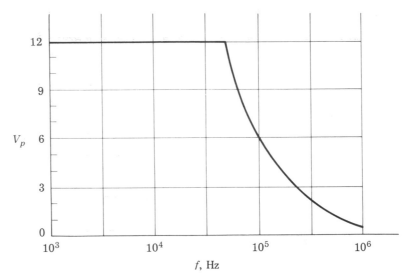

Fig. 13.16. Maximum permissible V_p as a function of frequency in the output
of an op amp having bandwidth = 1 MHz, slew rate = 4 V/μs, and
supply voltages = ±14 V.

V/μs slew rate. The power supply voltages were assumed to be ±14 V.
So a peak voltage of 12 volts was available at low frequencies at which
slew rate limiting did not occur.

**13.8
OP AMPS
HAVING FAST
RISE TIME**

The need frequently arises for an op amp having a very fast rise time.
High-speed pulse-coded systems usually have this requirement. Since
wide bandwidth and high slew rate are both required if the output
signal is to maintain fast rise time over the permissible range of the
amplifier, a broadband amplifier must first be selected and then a compensation technique that achieves high slew rate must be devised. For
example, if the rise time of the output signal must be no greater than 50
ns for any rectangular pulse having an amplitude up to 5 V, the bandwidth of the amplifier must be at least $B = 2.2/6.28 \times 5 \times 10^{-8} = 7$ MHz
and the slew rate must be at least $5 \text{ V}/5 \times 10^{-8} = 10^8 \text{ V/s}$ or 100 V/μs.
The following example illustrates how the op amp selection and com-
pensation may be accomplished.

Example 13.3

Let us assume that an op amp having 50 ns rise time, for output
pulses up to 5 V amplitude, and having a voltage gain of 10 must be
constructed. The first task is to search through the manufacturers'
data sheets to find an op amp with an open-loop gain curve that will
intersect the closed-loop gain of 10 curve at a frequency higher than

7 MHz. The Fairchild μA715 appears to be a good candidate because this intersection occurs at about 20 MHz, as may be seen in Fig. 13.17. The commercial version (μA715C) was chosen because its temperature specifications are adequate for our assumed moderate environment and its cost is significantly lower than the μA715.

Observe from Fig. 13.17 that there are two poles in the open-loop response curve to the left of the intersection with no compensation, which occurs at the third pole frequency. As a result of our previous experience (Chapter 12), we have some confidence that the μA715C may be compensated by shifting the lowest-frequency pole downward with phase-lag compensation and by eliminating the effects of the middle-frequency pole at about 1 MHz with a zero produced by a resistor in series with the phase-lag capacitor. The closed-loop response curve would then intersect the compensated open-loop response curve at a frequency below that of the highest frequency pole at 20 MHz. Let us choose this point of intersection at 10 MHz which allows some safety factor on the phase margin, since there may be additional poles above 20 MHz. Also, this point of intersection allows some safety factor on the required 7 MHz bandwidth. The desired location of the compensated lowest-frequency pole is then 10

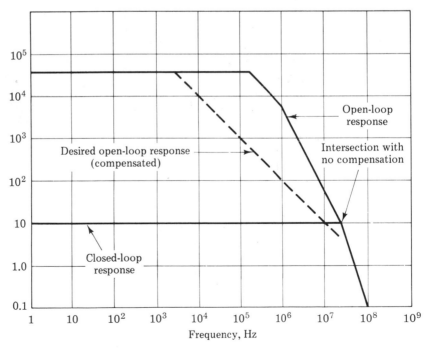

Fig. 13.17. Frequency response curves for the μA715C.

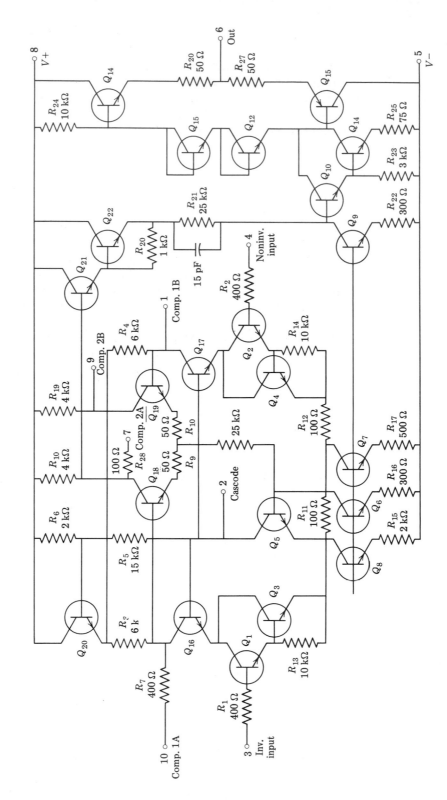

Fig. 13.18. Schematic diagram of the μA715 C. (Courtesy of the Fairchild Semiconductor Company)

MHz divided by the ratio of open loop gain to closed loop gain (30,000/10) which is 3.3 kHz, as shown in Fig. 13.17.

Our next task is to compensate the μA715C in a manner that will provide the required 100 V/μs slew rate. The schematic diagram of the μA715C is given in Fig. 13.18 to assist us. Observe that the input stage is a differential cascode amplifier and the second stage is a simple balanced differential amplifier. This stage is apparently the one having the pole at 110 kHz because compensation terminals are brought out from both bases and both collectors of this stage. This stage is followed by a Darlington-connected emitter follower to avoid both resistive and capacitive loading of the differential amplifier and to provide a low impedance driving source for the Darlington-connected voltage amplifiers Q_{10} and Q_{14}. This voltage amplifier drives a complementary-symmetry emitter follower output stage. The resistors R_{20} and R_{27} probably protect the output transistors in the event the output becomes shorted. The second stage may be compensated by adding capacitance between the two bases, between the two collectors, or between the base and collector of both transistors in the differential pair. This latter technique is preferable because the required compensating capacitance is reduced by the voltage gain of the stage due to the Miller effect, and this smaller capacitance will yield a higher slew rate. The compensating capacitors and the series resistors required to produce the zero are shown in the circuit diagram of the second stage given in Fig. 13.19. Observe that the com-

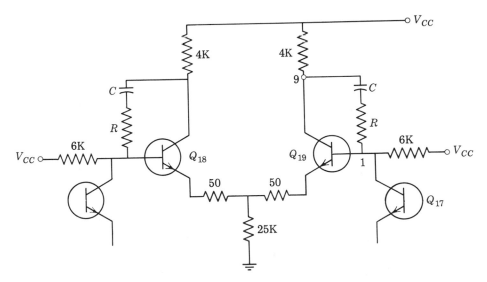

Fig. 13.19. Circuit diagram of the second stage of the μA715C showing the compensating elements C and R.

pensating network C and R appears as a feedback network and the values for C and R could be determined by using the feedback equation and the desired pole and zero frequencies. However, we will use the Miller effect because it is easier.

In order to determine the value of the compensating capacitor, we need the reference voltage gain and the input resistance of the compensated stage (Fig. 13.19). These values are not given by the manufacturer, so we must either determine or estimate them. We may assume that h_{fe} of the transistor is at least 100 and the voltage gain is approximately the ratio of the load resistance (4 kΩ) to the total unbypassed emitter resistance 50 Ω + r_e , as discussed in Chapter 8. Since $r_e \simeq 25/I_E$ (mA), we must estimate I_E . The clue for this estimation is the 25 kΩ emitter circuit resistor in this differential amplifier. A quick analysis of the dc drops and potentials in Fig. 13.18 indicate that about 75 percent of the total supply voltage is dropped across this 25 kΩ resistor. Therefore, if the total supply voltage ($+V_{CC}$ to $-V_{CC}$) is about 32 V, the drop across this resistor would be about 24 V and the current through it would be about 1 mA. The emitter current of each differential transistor is then approximately 0.5 mA and $r_e = 25/0.5 = 50$ Ω. The voltage gain of each side of the differential amplifier is then 4 kΩ/100 Ω = 40, approximately, and the input resistance of each transistor is essentially h_{fe} $(50 + r_e) = 100 (100) = 10$ kΩ. The resistance between base and ground is this 10 kΩ in parallel with the 6 kΩ driving source resistance, which is about 4 kΩ. The time constant of this second stage due to the Miller effect of the compensating capacitor C, neglecting the effect of R (Fig. 13.19), is

$$\tau = (R_s \| R_{it}) (1 + K_v) C \qquad (13.13)$$

The pole frequency, or upper half-power frequency, in r/s is the reciprocal of this time constant, as discussed in Chapter 9.

$$\omega_p = \frac{1}{(R_s \| R_{it}) (1 + K_v) C} \qquad (13.14)$$

Solving for C explicitly in this amplifier to provide a pole frequency of 3.3 kHz,

$$C = 1/[4 \times 10^3 \, (41) \, (6.28 \times 3.3 \times 10^3)] = 294 \text{ pF} \qquad (13.15)$$

We will use the nearest stock size $C = 300$ pF.

The resistor R was included to provide a zero in the vicinity of the amplifier middle-frequency pole at 1 MHz. Since this zero frequency in r/s is the reciprocal of the time constant RC, $R = 1/6.28 \times 10^6 \, (3 \times 10^{-10}) = 530$ Ω. Observe that 500 Ω is built into the μA715 chip (Fig. 13.20) and will appear in series with the compensating capacitor connected to pins 7 and 10. A similar resistance does not appear on the other half of the differential amplifier in series with pins 9 and 1. Therefore, all the resistance must be added externally on this side.

We may estimate the slew rate of the amplifier as determined by the compensating capacitor. When a large positive-going rectangular pulse is applied to the inverting input, transistor Q_{16} (Fig. 13.19) may be driven into saturation, or its collector driven sufficiently negative to cut off Q_{18}. The collector of Q_{18}, which drives the following amplifier, then rises toward V_{CC} at an initial rate equal to $V_{Q_{18}}/\tau$, where τ is the time constant of the circuit consisting of the compensating capacitor C and the resistance facing it, and $V_{Q_{18}}$ is the dc supply voltage for this stage (Q_{18}). From Fig. 13.19, the maximum value the resistance facing C can have is 10.53 kΩ and the supply voltage for Q_{18} from its base to V_{CC} is about 3 V. Then, the initial rate of rise of the output voltage at the collector of Q_{18} is about $3/10.53 \times 10^3 \, (3 \times 10^{-10}) \simeq 10^6 \, \text{V/s} = 1 \, \text{V/}\mu\text{s}$. The slew rate which we calculated must be at least 100 V/μs. The final slew rate is this initial rise rate times the voltage gain of the following amplifiers. Therefore, our specifications will not be met unless the Darlington connected amplifier Q_{10} and Q_{14} of Fig. 13.18 (which is the only amplifier following Q_{18} with a voltage gain greater than 1) has a voltage gain equal to 100 or greater. We may see from Fig. 13.18 that the collector load resistance for this Darlington amplifier is 10 kΩ in parallel with the input of the complementary-symmetry emitter follower which will normally be much higher than 10 kΩ. The resistance in the emitter circuit of Q_{14} is 75 Ω and the quiescent emitter current of Q_{14} is about 1.5 mA for V_{CC} = 15 V. So the voltage gain of this stage is slightly less

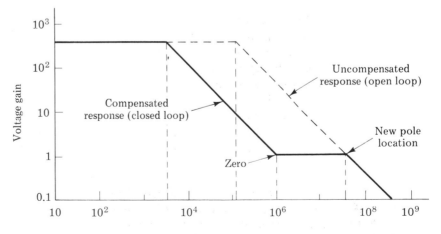

Fig. 13.20. Compensated and uncompensated response curves for the second stage showing the location of the newly created pole.

than $10^4 / (75 + r_e) = 10^4 / (75 + 17) = 109$. Therefore, this stage has the required voltage gain; so the slew rate, and hence the rise time, should meet the specifications.

One remaining item that needs checking is the location of the additional high-frequency pole that was created along with the zero when we added the resistance R in series with the compensating capacitor. Since the order of the denominator of the impedance of an RC network is equal to or greater than the order of the numerator, these high-frequency poles essentially always occur. We can find this pole location by observing the compensated-response curve of the second stage *only* given in Fig. 13.20. The compensating capacitor C causes the response to roll off at 20 dB/decade above the 3.3 kHz corner frequency, as planned. The rolloff continues until $X_C = R$ at 1 MHz, which is the frequency of the zero, beyond which the amplifier appears to have resistive feedback, and hence uniform response. This flat response continues until the closed loop (compensated) response curve intersects the open loop (uncompensated) response curve at 33 MHz where another corner, or pole, is created. Observe that the frequency ratio of this pole to the zero at 1 MHz is the same as the ratio of the original pole at 110 kHz to the compensated pole at 3.3 kHz. This new pole frequency at 33 MHz is only 3.3 times as high as the intersection frequency of the closed- and open-loop response curves of the μA715C, and will therefore reduce the phase margin by about 17 degrees. However, we have allowed for some safety factors, so the phase margin may still be adequate. If it is not, we may still add phase-lead compensation in the feedback path with the zero frequency at about 33 MHz to provide adequate phase margin without sacrificing bandwidth.

We were able, in the example above, to achieve the desired rise time with an amplifier voltage gain of 10. However, op amps with unity gain are often needed for summing amplifiers, voltage followers (to be discussed later), and other applications. If we were to use the same compensation technique for the μA715C as in Example 15.3, but for a unity gain amplifier, the lowest frequency pole after compensation would be at 330 Hz instead of 3.3 kHz, and the compensating capacitor C would need to be ten times as large. Thus the slew rate would be reduced by a factor of 10 and would cause a proportionate increase in rise time.

**13.9
AMPLIFIER
INPUT
COMPENSA-
TION**

A useful technique to produce a low-frequency pole without degrading the slew rate is to use an RC network at the front end of the op amp to produce the lowest frequency pole. Then an extra resistor may be added to create a zero which will cancel another pole in the amplifier. This technique is illustrated in the following example.

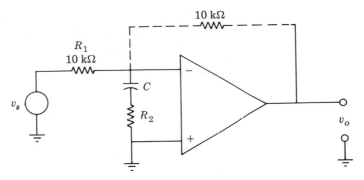

Fig. 13.21. Front-end compensation used to achieve fast rise time in a unity gain amplifier.

Example 13.4

Let us compensate the μA715C for unity gain by adding a compensating network at the input terminals of the op amp shown in Fig. 13.21. The total driving-source resistance, as seen by the op amp, is assumed to be 10 kΩ. The proper pole and zero frequencies for the front-end compensation network may be determined from Fig. 13.22 which shows the open-loop response curve of the μA715C compensated for a gain of 10 in addition to the needed open-loop response

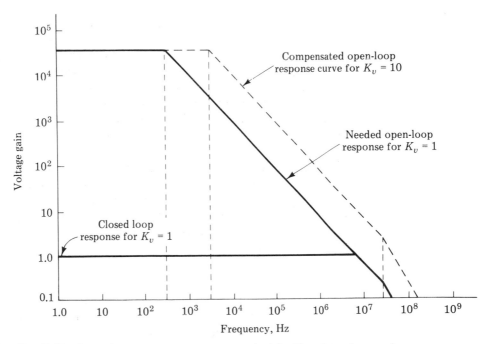

Fig. 13.22. Open-loop response curve needed for $K_v = 1$ and open-loop response curve obtained by compensation of Example 13.4.

curve compensated for unity voltage gain. We can readily see that the front-end compensation needs to provide a zero at 3.3 kHz to cancel the effects of the pole at that frequency. The 3.3 kHz pole will then be effectively moved to 330 Hz without changing the compensation capacitance of the op amp itself. Since the pole frequency ω_p in r/s is the reciprocal of the time constant of the capacitor C and the resistance facing it, without the feedback resistor connected,

$$\omega_p = \frac{1}{(R_1 + R_2)\,C} \tag{13.16}$$

The frequency ω_z of the zero is

$$\omega_z = \frac{1}{R_2 C} \tag{13.17}$$

Then, since $\omega_z = 10\,\omega_p$

$$\frac{10}{(R_1 + R_2)\,C} = \frac{1}{R_2 C} \tag{13.18}$$

Therefore, $R_1 = 9\,R_2$ or $R_2 = R_1/9$. Since R, the total source resistance is 10 kΩ, $R_2 = 1.1$ kΩ and using Eq. 13.17, $C = 1/6.28\ (3.3 \times 10^3) \times (1.1 \times 10^3) = .043$ μF.

The front-end compensation does not degrade the slew rate, as may be seen from Fig. 13.23. The compensation circuit reduces the spike voltage amplitude by a factor of 10 compared with the spike resulting from feedback delay that would be applied to the op amp without front-end compensation. Thus the spike is the same amplitude as for the gain-of-ten amplifier, which had only one-tenth as large an input. The second-stage compensation was not changed so the slew rate is the same as for the amplifier of Example 13.3.

The circuit diagram of the compensated unity-gain μA715C amplifier is shown in Fig. 13.24a. Stock size 10 percent tolerance resistors were used, and decoupling capacitors are shown at the power-supply terminals. The amplifier was tested and an oscillogram of the output voltage, when a rectangular pulse was applied to the input, is given in Fig. 13.24b.

13.10
THE VOLTAGE
FOLLOWER

An op amp version of the emitter follower is known as a *voltage follower*. The desirable characteristics of the voltage follower, like the emitter follower, include extremely high input impedance, very low output impedance, and unity voltage gain. The current gain must be

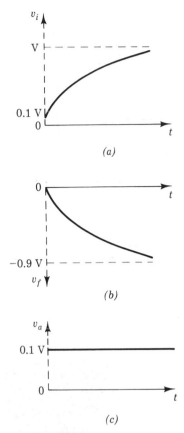

Fig. 13.23. Waveforms resulting from a rectangular driving source voltage when front-end compensation is used: (a) amplifier input voltage v_i without feedback; (b) feedback voltage v_f, showing slight delay behind input voltage; (c) op amp input voltage $v_i - v_f$ showing the input spike.

high, of course, to accomplish this impedance transformation without a loss in voltage. These characteristics require that the input signal be applied to the noninverting input, as shown in Fig. 13.8. The voltage gain with this connection is $(R_1 + R_F)/R_1$, as given in Eq. 13.4. Therefore, either R_1 must be infinite or R_F must be equal to zero, or both, in order to obtain unity voltage gain. A voltage follower with $R_1 = \infty$ is shown in Fig. 13.25. R_F may be zero, but the input bias current must flow through R_F; so R_F may be either chosen or adjusted to provide minimum offset voltage. For example, if R_s is quite large, $R_F = R_s$ may be a good choice. Otherwise, $R_F = 0$ may be preferable. Of course the op amp must be compensated for unity voltage gain.

Some integrated circuits are specifically designed as voltage fol-

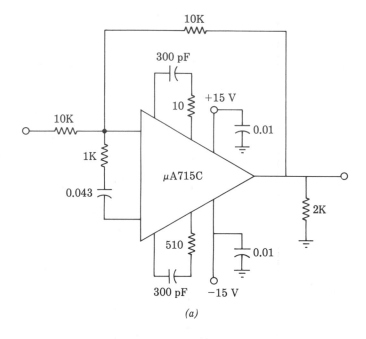

(a)

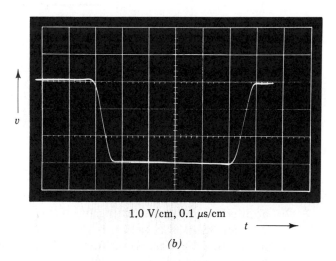

1.0 V/cm, 0.1 μs/cm

$t \longrightarrow$

(b)

Fig. 13.24. Circuit diagram of the unity-gain amplifier using front-end compensation. (b) Oscillogram of the output voltage when a positive-going 3-volt rectangular pulse was applied to the input.

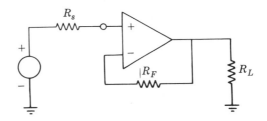

Fig. 13.25. Op amp used as a voltage follower.

lowers. These ICs normally require no external feedback or compensation and they have all the desirable features including wide bandwidth and very high slew rate. The circuit diagram of a typical voltage-follower IC is shown in Fig. 13.26. Observe that the input stage of this voltage follower is an emitter-coupled cascode amplifier with a current source Q_3 as a load to provide very high voltage gain. This stage is followed by a Darlington-connected emitter follower output stage. Transistor Q_7 limits the output current to protect the IC. When the current through R_{25} becomes large enough to turn on Q_7, this transistor robs the base current of Q_5, thus limiting the output current. The feedback resistor is R_7 which applies unity feedback to the inverting side of the emitter-coupled amplifier. The 10 pF capacitor C_1 provides the compensation. The input Darlington-connected emitter follower has its collectors bootstrapped to the output, apparently to reduce the effective input capacitance which is only 1.5 pF. Other typical characteristics of this circuit with ±15 V supply are:

voltage gain = 0.9999
input resistance = 10^{12} Ω
output reistance = 0.75 Ω
slew rate = 30 V / μs
bandwidth = 15 MHz with driving source resistance = 10 kΩ
input bias current = 2 nA
input offset voltage = 2.5 mV

**13.11
VOLTAGE
COMPARATORS**

A final application of an operational amplifier that we will consider in this chapter is the *voltage comparator*. The purpose of this circuit, shown in Fig. 13.27, is to compare the input voltage to a reference voltage, as the name implies. With the input voltage applied to the inverting terminal, the output voltage is at its maximum positive value whenever the input voltage is about 1 mV or more below the reference voltage. Conversely, the output voltage is at its maximum negative value when the input voltage is about 1 mV or more positive with respect to the reference voltage. Thus the comparator indicates whether

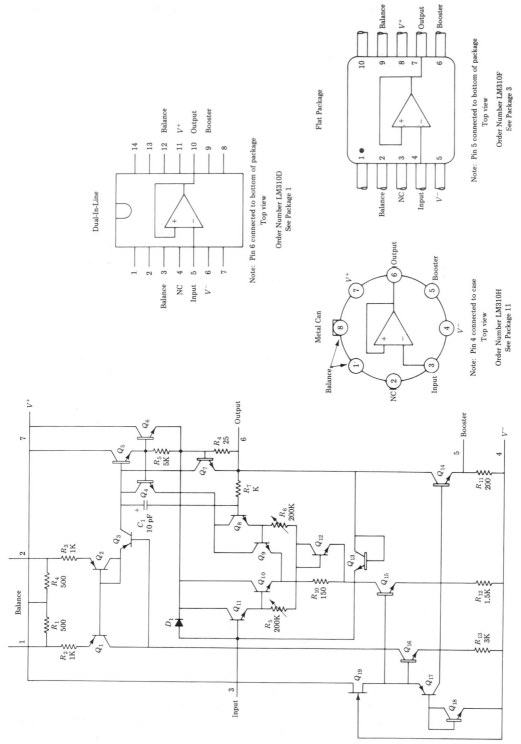

Fig. 13.26. Schematic diagram and circuit connections for the LM310 voltage follower. (Courtesy of the National Semiconductor Corporation)

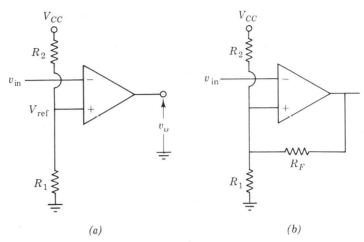

Fig. 13.27. (a) Op amp used as a voltage comparator. (b) Voltage comparator with hysteresis.

the input voltage is above or below the reference voltage. The rate at which the comparator output can change from one state to the other depends upon the open-loop bandwidth of the op amp, since there is no negative feedback.

In normal use, the output of the voltage comparator is a series of rectangular pulses. Therefore, the comparator is useful for cleaning up noisy or sloppy rectangular pulses, changing sine waves into square waves, and many other applications. Several voltage comparators may be used in conjunction with other circuits to convert analog voltages to digital voltages or vice versa.

One problem that arises in a voltage comparator circuit is the noise in the output voltage that results from input noise, as shown in Fig. 13.28a. Observe that input voltages within a millivolt or so of the reference voltage are amplified by the full open-loop gain of the amplifier. Therefore, while the signal voltage is passing through this range, the noise is greatly amplified and may cause several zero crossings in the output. This output noise may be eliminated by applying a small amount of positive feedback to the noninverting V_{ref} terminal of the comparator, as shown in Fig. 13.27b. This feedback voltage shifts the reference voltage by the amount

$$\text{Hysteresis voltage} = \mathscr{F} v_o = \frac{R_1}{R_1 + R_F} \, v_o \tag{13.19}$$

A hysteresis loop is introduced in the v_O versus v_I curve of the voltage comparator. Then, a higher voltage is required to shift the output low than that required to shift the output high, as shown in Fig. 13.29.

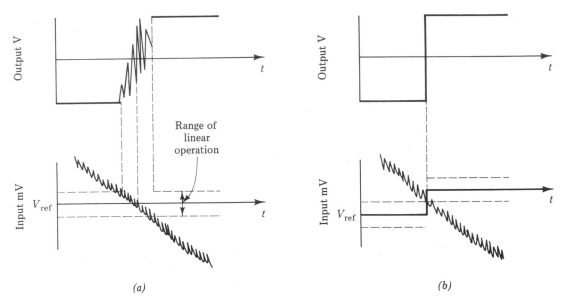

(a)　　　　　*(b)*

Fig. 13.28. (a) Comparator output noise resulting from input noise.
(b) Elimination of output noise by hysteresis.

Therefore, as soon as the input voltage enters the linear region from the positive side toward the negative, the output goes positive and the hysteresis voltage adds to the reference voltage to shift the reference voltage sufficiently above the input voltage to immediately turn off the amplifier and cause the output voltage to rise to its maximum value, as shown in Fig. 13.28b. Thus, the change of reference voltage essentially eliminates the time during which the comparator acts as a linear ampli-

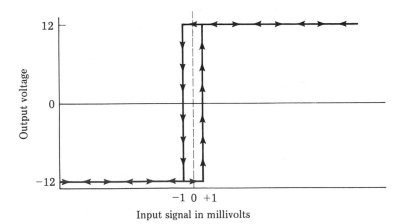

Fig. 13.29. Voltage characteristics of a comparator with positive feedback.

fier, thus eliminating the output noise. As the input voltage returns from its negative excursion toward the positive, the opposite happens, of course, and the reference voltage shifts in the negative direction by the amount $\mathcal{F}\Delta v_O$, where Δv_O is the change in output voltage. The hysteresis voltage in Fig. 13.29b was assumed to be equal to the range of voltage over which the comparator operates as a linear amplifier. This hysteresis was adequate because the peak noise voltage was assumed to be less than the linear input range of the amplifier. However, as the peak noise voltage becomes larger than this linear input range, the hysteresis voltage needs to be increased until it is equal to the peak noise voltage, as may be seen from Fig. 13.28b. Comparators having hysteresis are often known as *Schmitt Triggers*.

Although ordinary op amps make satisfactory voltage comparators when moderate switching speeds are adequate, special op amps which do not allow their transistors to go into saturation are designated as *comparators*. These ICs normally have faster switching speeds than ordinary op amps and some of them are designed to work at logic-level signal and supply voltages.

The schematic diagram and the transfer characteristics of a typical comparator are given in Fig. 13.30. The differential input stage Q_1 and Q_2 is followed by a second differential stage Q_3 and Q_4, which in turn drives a single emitter follower Q_7. As the noninverting input goes positive with respect to the inverting input, the base and emitter of Q_7 go positive until the base of the diode-connected transistor Q_6 becomes positive with respect to its emitter. Then resistors R_4 and R_5 are clamped at almost the same potential at their lower ends, and the collector currents of both Q_3 and Q_4 flow through these two resistors. But since Q_3 and Q_4 are the two halves of a differential amplifier, the sum of their collector currents is constant and the drops across R_4 and R_5 are constant. Thus, the output voltage cannot rise above about 3 V, as shown by the transfer characteristics of Fig. 13.30, which is $V+$ minus the sum of the drop across R_5 plus V_{BE7} plus the 6.2 V reference voltage. When the noninverting input goes negative with respect to the inverting input, the transistor Q_4 is driven to saturation. This also places Q_8 approximately in saturation because the output voltage must be $(6.2 \text{ V} + V_{BE7})$ negative with respect to the collector of Q_4, which is only $(6.2 + V_{CEsat})$ positive with respect to ground. Therefore, the output voltage is clamped at approximately -0.5 V which is $(V_{BE7} - V_{CEsat})$ while the input is negative, as shown by the voltage transfer characteristic in Fig. 13.30. Note that approximately 1 mV input signal is required to drive the output to its clamped voltage in either direction. The biasing circuits control the currents so that the transistors are not driven appreciably into saturation during normal operation.

The voltage comparator is used in digital voltmeters or analog-to-

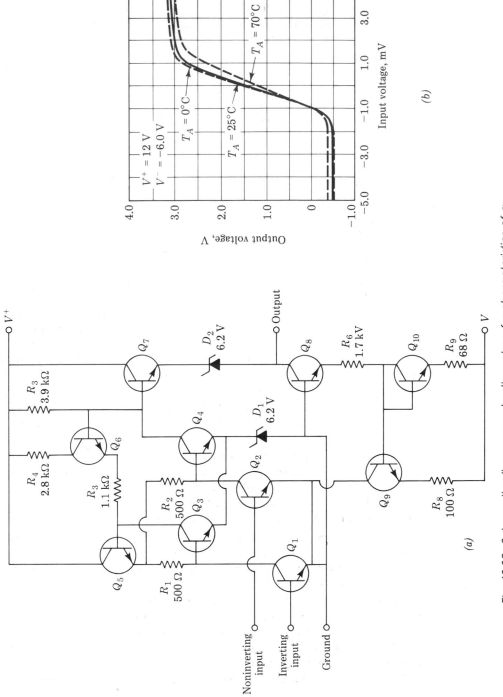

Fig. 13.30. Schematic diagram and voltage transfer characteristics of a typical voltage comparator, µA710. (Courtesy of the Fairchild Semiconductor Company)

digital converters where a stable, precise reference voltage is applied to the inverting input. The output voltage operates some type of read-out when the reference voltage is exceeded. The comparator is also convenient for changing sine waves to square waves, cleaning up a noisy binary signal, or performing many other functions described in the literature.

A careful study of the manufacturer's data and literature is almost essential prior to the choosing and the application of a linear IC to a specific circuit problem. The study of this chapter will hopefully help you understand the literature and data sheets. Numerous additional examples of the use of integrated circuits will be given in the remaining chapters of this book.

Problems   *Section 13.2*

13.1   Continue Example 13.1 and determine both I_5 and I_6 assuming $V_{BE} = 0.5$ V and $V_+ + 8$ V.

Section 13.4

13.2   An IC having input bias current $= 1$ μA, maximum input offset current $= 50$ nA and maximum input offset voltage $= 10$ mV is used in a circuit having 10 kΩ dc resistance between the inverting input, to which the signal is applied, and ground. The amplifier voltage gain is 100.

    a.   Determine the worst case output offset voltage when a 10 kΩ 20 percent tolerance resistor is connected from the noninverting input terminal to ground.

    b.   Can the offset voltage be nulled by adjusting the value of resistance between the noninverting input terminal and ground?

Section 13.5

13.3   Use the circuit of Fig. 13.11 to add the signals from three microphones which have output voltages of 2 mV, 4 mV, and 10 mV, respectively, at normal voice levels of 1 microbar acoustic pressure. It is desired that each microphone produce 0.1 V at the output of the amplifier, and the load resistance for any of the microphones should be 5 kΩ or higher. Design the mixer-circuit using an operational amplifier with an open-loop gain of 5×10^4 with $R_L \geq 2$ kΩ. Assume $R_s \ll 5$ kΩ.

13.4   A given integrator has $R_1 = 10$ kΩ and $C = 0.1$ μF.

    a. What is the gain factor of the integrator?

    b. What is the maximum permissible time between dumps or resets if the dc component in the input is 10 mV and the output saturation voltage is 12 V?

    c. What value of R_F is needed to convert the integrator to a running average integrator having a low-frequency cutoff = 5 Hz?

    d. What is the permissible input-offset of the running average integrator if the dc component in the output is not to exceed 0.1 V?

Section 13.6

13.5   Determine the approximate bandwidth of the amplifier of Fig. 13.13 if the closed-loop gain is 3,000 and the amplifier is:

    a. Internally compensated.

    b. Uncompensated.

Section 13.7

13.6   A type 741 internally compensated op amp has a bandwidth of 100 kHz when its voltage gain is 10. The slew rate is approximately 1 V/μs. Determine the maximum value of peak signal in the output when the frequency is 20 kHz. The supply voltages are ± 14 V. What is the peak amplitude available at 100 kHz?

Section 13.9

13.7   A given IC operational amplifier has the same equivalent circuit as the μA715C except the 25 kΩ resistance in the second differential stage is reduced to 12 kΩ; the open-loop voltage gain is 5×10^4, the three break-frequencies in the open-loop frequency-response curve occur at 150 kHz, 2.0 MHz, and 20 MHz; and there is no built-in resistance in series with the compensation terminals. Design a compensation circuit using $R_F = R_1 = 20$ kΩ for unity gain. Obtain as wide a frequency response as possible.

Section 13.10

13.8 A given internally-compensated op amp has an open-loop voltage gain = 50,000, input resistance = 1 MΩ, output resistance = 150 Ω, and slew rate = 1 V / μs. When feedback is applied to provide unity voltage gain, the bandwidth is 1 MHz. This amplifier is to be used as a voltage follower where the driving source resistance is 10 kΩ and the load resistance is 1 kΩ.

    a. Draw the circuit diagram and determine the expected voltage gain, input resistance, and output resistance of the voltage follower.

    b. Determine the maximum amplitude of a 100 kHz signal that the voltage follower can handle without distortion.

    c. What input impedance does the voltage follower present to a 100 kHz signal if the input capacitance of the op amp is 15 pF?

Section 13.11

13.9 The comparator of Fig. 13.30 is used to produce a square wave from a 1.0 V peak 400 Hz sine wave.

    a. What will be the approximate peak-to-peak magnitude and rise time of the square wave?

    b. Assume that the sine wave has 2 mV peak noise. Devise a hysteresis circuit that will suppress the noise. What influence will the hysteresis circuit have on the rise time?

14

Active Filters

Electrical filters are circuits which pass certain bands of frequencies and reject other frequencies. According to this definition, *all* circuits are filters because they have limited high-frequency response due to their unavoidable shunt capacitance. They thus become either low-pass or band-pass filters, depending upon whether or not series capacitors have been included in the circuit to block dc and discriminate against low frequencies. However, in the previous work, we have been primarily concerned with maximizing the pass band rather than restricting it. In this chapter, we are concerned with selecting a specified range of frequencies and rejecting all others.

We are aware that capacitors and/or inductors in conjunction with resistors form frequency selective networks. Inductors and capacitors are commonly used in resonant circuits to select a radio frequency signal from amongst the myriad of signals being broadcast. The inductors are small, inexpensive, and low-loss at these high frequencies. These narrow-band high frequency filters are treated at considerable depth in the literature.[1] However, the inductors required for filters in the audio-frequency range and below normally have iron cores and are bulky, expensive, and have high loss factor or low Q. This frequency range is also the one in which operational amplifiers perform very well. Therefore, a natural wedding of capacitors, resistors, and op amps has produced a wide variety of inductorless filter circuits having small size, low cost, and superior performance in the frequency range up to about

---

[1] See Chapter 9, Small Signal Tuned Amplifiers, C. L. Alley and K. W. Atwood, *Electronic Engineering*, Third Edition 1973, John Wiley & Sons, New York NY.

0.5 MHz. Some examples of these filters are discussed in this chapter. Smoothing filters for power supplies are discussed in Chapter 16.

**14.1
FILTERS
UTILIZING THE
INVERTING
OP AMP CON-
FIGURATION**

We learned in Chapter 13 that the voltage gain v_o/v_s of an op amp in the inverting mode is R_F/R_1 where R_F is the feedback resistance and R_1 is the driving source resistance as seen by the inverting input terminal of the op amp. The resistive impedances R_F and R_1 were used only because broad bandwidth or uniform frequency response was desired from the amplifier. The transfer function or gain equation is generalized by writing

$$G_v = \frac{Z_F}{Z_1} = \frac{Y_1}{Y_F} \tag{14.1}$$

where $Y_1 = 1/Z_1$ and $Y_F = 1/Z_F$.

When Y_1 and/or Y_F are complex or, in other words, contain poles and zeros, the frequency response characteristics of the amplifier are determined by the location of the poles and zeros of these elements. Therefore, the desired filter characteristics may be chosen, or synthesized, by selecting the proper pole and zero frequencies in the input and feedback networks.

Equation 14.1 does not distinguish between driving point and transfer admittances and is therefore limited to simple series or parallel elements. A more general relationship is developed from Fig. 14.1. Since the op amp input current is negligible, $I_3 = -I_2$. But $I_2 = Y_{21a}V_s$ and $I_3 = Y_{12b}V_o$. Therefore,

$$G_v = \frac{V_o}{V_s} = -\frac{Y_{21a}}{Y_{12b}} \tag{14.2}$$

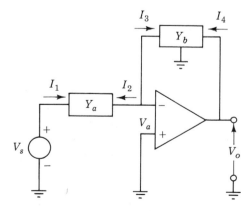

Fig. 14.1. Active filter utilizing an op amp in the inverting mode.

The negative sign results from the polarity reversal of the op amp. The transfer admittances of some commonly used filter networks are given in Table 14.1. The driving point admittances are also given, so we can determine the effects of the driving source resistance.

**14.2
ACTIVE FILTER
DESIGN**

One approach to designing active filters is first to write the transfer function or gain equation in the s-domain that contains the poles and zeros required to provide the desired filter response. This equation is then synthesized by substituting appropriate transfer admittances from Table 14.1, or elsewhere, into Eq. 14.2 to provide the desired transfer function. Finally, the networks which yield these transfer admittances are connected into the op amp circuit with due regard to offset voltages and stability. The following example illustrates this procedure.

Example 14.1

A student desires to construct a color organ. He wishes to have a red light with the light intensity proportional to the amplitude of the low-frequency component of an audio signal. The intensity of a yellow light is to be controlled by the amplitude of the midfrequency component of the audio signal. Finally, a blue light is to be controlled by the amplitude of the high-frequency component of the audio signal. In order to separate the three audio components, the student decides he must have a low-pass filter which will pass frequencies from zero to 1,000 rad/s (about 160 Hz) to activate the red light. He also needs a band-pass filter from 1,000 rad/s to 10,000 rad/s (about 1,600 Hz) to activate the yellow light. Finally, he needs a high-pass filter from 10,000 rad/s to the upper end of the audio spectrum to activate the blue light.

The desired characteristics for the three filters are given in Fig. 14.2. We will design the band-pass filter. From Fig. 14.3, we note that the transfer function of this filter is

$$\frac{V_o}{V_i} = \frac{As}{(s + 10^3)(s + 10^4)} \tag{14.3}$$

Notice that Eq. 14.3 is identical to that of an RC-coupled amplifier with $\omega_\ell = 10^3$ rad/s and $\omega_h = 10^4$ rad/s. Then, the constant A must be equal to $K\omega_h$ where K is the midband gain of the amplifier. The desired midband gain of our filter is +20 dB or a voltage gain of 10. Then $A = 10 \times 10^4 = 10^5$. From Eq. 14.2, we have

$$\frac{V_o}{V_i} = -\frac{y_{21a}}{y_{12b}} = \frac{10^5 s}{(s + 10^3)(s + 10^4)} \tag{14.4}$$

Table 14.1. Typical RC circuits and their y-parameters.

Circuit Configuration	y-parameters
1 	$y_{11} = y_{22} = \dfrac{1}{R_1} \dfrac{s}{\left(s + \dfrac{1}{R_1 C_1}\right)}$ $y_{12} = y_{21} = -\dfrac{1}{R_1} \dfrac{s}{\left(s + \dfrac{1}{R_1 C_1}\right)}$
2	$y_{11} = y_{22} = C\left(s + \dfrac{1}{RC}\right)$ $y_{12} = y_{21} = -C\left(s + \dfrac{1}{RC}\right)$
3 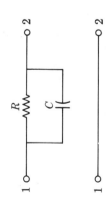	$y_{11} = y_{22} = C_2\dfrac{s^2 + s\left(\dfrac{1}{R_1 C_1} + \dfrac{1}{R_2 C_2} + \dfrac{1}{R_1 C_2}\right) + \dfrac{1}{C_1 C_2 R_1 R_2}}{\left(s + \dfrac{1}{R_1 C_1}\right)}$ $y_{12} = y_{21} = -C_2\dfrac{s^2 + s\left(\dfrac{1}{R_1 C_1} + \dfrac{1}{R_2 C_2} + \dfrac{1}{R_1 C_2}\right) + \dfrac{1}{C_1 C_2 R_1 R_2}}{\left(s + \dfrac{1}{R_1 C_1}\right)}$

4

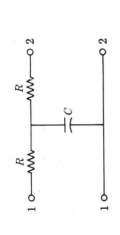

$$y_{11} = y_{22} = \frac{1}{R}\,\frac{\left(s + \dfrac{1}{RC}\right)}{\left(s + \dfrac{2}{RC}\right)}$$

$$y_{12} = y_{21} = -\frac{1}{R^2C}\,\frac{1}{\left(s + \dfrac{2}{RC}\right)}$$

5

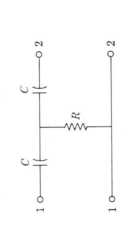

$$y_{11} = y_{22} = \frac{C}{2}\,\frac{s\left(s + \dfrac{1}{RC}\right)}{\left(s + \dfrac{1}{2RC}\right)}$$

$$y_{12} = y_{21} = -\frac{C}{2}\,\frac{s^2}{\left(s + \dfrac{1}{2RC}\right)}$$

6

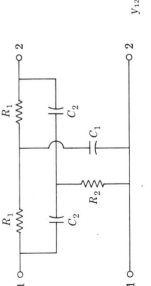
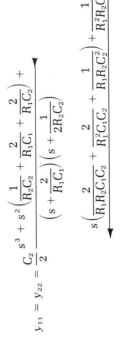

$$y_{11} = y_{22} = \frac{C_2}{2}\,\frac{s^3 + s^2\left(\dfrac{1}{R_2C_2} + \dfrac{2}{R_1C_1} + \dfrac{2}{R_1C_2}\right) + s\left(\dfrac{2}{R_1R_2C_1C_2} + \dfrac{2}{R_1^2C_1C_2} + \dfrac{1}{R_1R_2C_2^2}\right) + \dfrac{1}{R_1^2R_2C_1C_2^2}}{\left(s + \dfrac{2}{R_1C_1}\right)\left(s + \dfrac{1}{2R_2C_2}\right)}$$

$$y_{12} = y_{21} = -\frac{C_2}{2}\,\frac{s^3 + s^2\,\dfrac{2}{R_1C_1} + s\left(\dfrac{2}{R_1^2C_1C_2} + \dfrac{1}{R_1R_2C_2^2} + \dfrac{2}{R_1^2R_2C_1C_2^2}\right) + \dfrac{1}{R_1^2R_2C_1C_2^2}}{\left(s + \dfrac{2}{R_1C_1}\right)\left(s + \dfrac{1}{2R_2C_2}\right)}$$

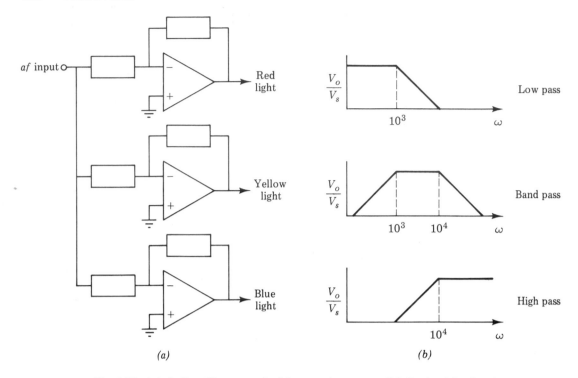

Fig. 14.2. (a) Active filters needed for a color organ. (b) Desired Bode plots
for the respective filters.

We note that the first circuit in Table 14.1 will provide a zero and
one of the poles. Let us use this circuit for Y_a. Then,

$$y_{21a} = -\frac{1}{R_1} \frac{s}{[s + (1/R_1C_1)]} \tag{14.5}$$

At higher frequencies, the impedance of C_1 is very small, so the input
impedance to the amplifier is essentially equal to R_1. In order to
maintain a high impedance to the filter (so we will not load down the
signal source), let us choose $R_1 = 10^5\ \Omega$. If y_{21a} is to provide the pole
at $\omega = -10^3$ rad/s, then $1/R_1C_1 = 10^3$ or $C_1 = 1/(10^3 \times R_1) = 1/10^8 =
10^{-8}$ F or $C_1 = 0.01\ \mu$F. When these values are substituted back into
Eq. 14.5, y_{21a} is

$$y_{21a} = -10^{-5} \frac{s}{(s + 10^3)} \tag{14.6}$$

Now,

$$\frac{y_{21a}}{y_{12b}} = -10^{-5} \frac{s}{(s + 10^3)\, y_{12b}} \tag{14.7}$$

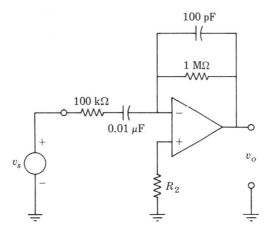

Fig. 14.3. Band-pass filter of Example 14.1.

However, an expression for $-y_{21a}/y_{12b}$ is given in Eq. 14.4. Consequently,

$$\frac{10^{-5}s}{(s+10^3)\,y_{12b}} = \frac{10^5 s}{(s+10^3)\,(s+10^4)} \tag{14.8}$$

or

$$y_{12b} = 10^{-10}(s+10^4) \tag{14.9}$$

The second circuit in Table 14.1 has y_{12} of the form given by Eq. 14.9, so this configuration will be used for Y_b. The actual y_{12} has a negative sign, so V_o of Eq. 14.4 will be inverted with respect to V_1. Since we are applying the signal to the inverting input, this behavior would be expected. Now, from Table 14.1,

$$y_{12} = -C\left(s + \frac{1}{RC}\right) \tag{14.10}$$

Comparing Eq. 14.9 and Eq. 14.10, $C = 10^{-10}$ F or 100 pF. (Notice that C could be made larger if R_1 of circuit Y_a is made smaller.) Then $1/RC = 10^4$ so $R = 1/(10^4 \times C) = 1/10^{-6} = 1$ MΩ. Our band-pass filter would be as shown in Fig. 14.3. There is no dc path to the input terminals, but there is a path through the 1 MΩ resistor to the output terminals. Therefore, if the offset voltage is to be maintained near zero, R_2 should be 1 MΩ also.

**14.3
STABILITY OF
ACTIVE FILTERS**

Of course, each active filter should be checked for stability. The methods outlined in Chapters 12 and 13 can be used to determine the circuit stability.

First, the op amp is compensated by creating a dominant low-frequency pole, so the response drops off at 6 dB/octave (or 20 dB/decade) until the gain is reduced to one. An internally compensated op amp may be used if desired. If the feedback were purely resistive, this step would be sufficient to ensure stable operation for any amount of feedback. Unfortunately, since the feedback networks of filters may be quite complex, the foregoing procedure *does not* ensure freedom from oscillation for the filter circuit. However, if the op amp is compensated so the response drops off at 6 dB/octave, the gain equation of the op amp (over its useful frequency range) is

$$G_a = \frac{-K\omega_h}{s + \omega_h} \tag{14.11}$$

where K is the dc open-loop gain of the amplifier and ω_h is the open-loop -3 dB break-frequency for the compensated amplifier.

The closed-loop gain of the amplifier is

$$G_f = \frac{G_a}{1 + G_a \mathscr{F}} \tag{14.12}$$

where $\mathscr{F}$ is the voltage feedback factor v_a/v_o of the filter when the source voltage v_s is turned off (Fig. 14.1). Referring to Fig. 14.1, the current flowing into the input network a resulting from v_a is

$$I_2 = Y_{22a} V_a \tag{14.13}$$

Also, the current I_3 flowing into the feedback network is

$$I_3 = Y_{11b} V_a + Y_{12b} V_o \tag{14.14}$$

However, the op amp input current is negligible, so $I_3 \simeq -I_2$ and

$$-Y_{22a} V_a = Y_{11b} V_a + Y_{12b} V_o \tag{14.15}$$

Therefore, the voltage feedback factor is

$$\mathscr{F} = \frac{V_a}{V_o} = -\frac{Y_{12b}}{Y_{11b} + Y_{22a}} \tag{14.16}$$

The stability of the filter may be determined from the loop transfer function, using Eqs. 14.16 and 14.11,

$$G_a \mathscr{F} = \frac{K\omega_h Y_{12b}}{(s + \omega_h)(Y_{11b} + Y_{22a})} \tag{14.17}$$

Either a Bode plot or a root locus plot could be used to check the stability. However, the denominator term $(Y_{11b} + Y_{22a})$ must first be factored. Let us use the example of the band-pass filter (Example 14.1) to illustrate this point.

Example 14.2

When the expressions for Y_{11b} and Y_{22a} given in Table 14.1 are substituted into Eq. 14.16, this equation becomes

$$\mathscr{F} = \frac{C\left(s + \dfrac{1}{RC}\right)}{C\left(s + \dfrac{1}{RC}\right) + \dfrac{s}{R_1\left(s + \dfrac{1}{R_1C_1}\right)}} \tag{14.18}$$

After the actual component values are substituted into this equation, we have

$$\mathscr{F} = \frac{10^{-10}(s + 10^4)}{10^{-10}(s + 10^4) + \dfrac{10^{-5}s}{s + 10^3}} \tag{14.19}$$

When both numerator and denominator of Eq. 14.19 are multiplied by $(s + 10^3)$, the value of $\mathscr{F}$ is

$$\mathscr{F} = \frac{10^{-10}(s + 10^4)(s + 10^3)}{10^{-10}(s + 10^4)(s + 10^3) + 10^{-5}s} \tag{14.20}$$

or

$$\mathscr{F} = \frac{(s + 10^4)(s + 10^3)}{s^2 + 1.11 \times 10^5 s + 10^7} \tag{14.21}$$

The denominator of Eq. 14.21 can be factored to yield

$$\mathscr{F} = \frac{(s + 10^4)(s + 10^3)}{(s + 1.11 \times 10^5)(s + 90)} \tag{14.22}$$

Using the expression for G_a in Eq. 14.11,

$$G_a\mathscr{F} = \frac{-K\omega_h(s + 10^4)(s + 10^3)}{(s + \omega_h)(s + 1.11 \times 10^5)(s + 90)} \tag{14.23}$$

If we assume $\omega_h \simeq 2{,}000$, the root-locus plot for G_f will be as shown in Fig. 14.4. Notice that this plot indicates that the filter will be stable for all values of amplifier gain since the root locus branches all lie on the negative real axis.

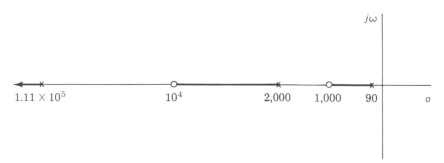

Fig. 14.4. Root locus plot for Eq. 14.23 with $\omega_h = 2,000$.

**14.4
HIGH-ORDER
FILTERS**

The filters considered to this point have not discriminated sharply against the unwanted frequencies. The slopes of their response curves have been only 6 dB/octave or 20 dB/decade in the *stop* ranges. Although this *rolloff* may be adequately sharp for a color organ, it is unsuitable for many other applications. One approach is to search Table 14.1 and other sources to find circuits with more storage elements and thus y-parameters having more poles and zeros than the simple circuits previously used. However, an inspection of Table 14.1 reveals that these higher-order equations are rather complicated, and a good deal of ingenuity is required to find the proper networks and then determine all the component values required to provide the desired filter response. An easier approach is to cascade two or more simple filters to provide the additional poles and the steeper rolloffs. Example 14.3 illustrates this procedure.

Example 14.3

The filter of Example 14.1 with a pass band from 1,000 rad/s to 10,000 rad/s does not have sufficient rejection of a signal near the pass band. Consequently we desire the response to drop off at 12 dB/octave (or 40 dB/decade) outside of the pass band.

The equation for the transfer function of the filter must have two zeros at (or very near) zero and two poles to produce the low-frequency breakpoint. Then two additional poles are required to produce the high-frequency breakpoint. The transfer function will have the following form

$$\frac{V_o}{V_i} = \frac{As^2}{(s + \omega_\ell)^2(s + \omega_h)^2} \qquad (14.24)$$

We are tempted to let ω_ℓ be 10^3 rad/s, but then our gain will be down 6 dB at 10^3 rad/s. Consequently, we must use the concepts of Chapter 11. Then, from Eq. 11.15, or from Table 11.1, we find ω_ℓ must be 643 rad / s. Similarly, ω_h must be 1.556×10^4 rad / s. If we desire the mid-band gain to be 10, $A = K\omega_h^2 = 10 \times (1.556 \times 10^4)^2 = 2.42 \times 10^9$. Thus the transfer function is

$$G_f = \frac{2.42 \times 10^9 \; s^2}{(s + 643)^2 (s + 1.556 \times 10^4)^2} \tag{14.25}$$

A study of Table 14.1 reveals that this function cannot be achieved by using any combination of the given circuits and a single op amp. In fact, the high- and low-frequency drop-off of 12 dB / decade indicate that the circuit may either oscillate or have poor transient response if the proper circuit configuration *could* be found. Therefore let us divide the transfer function into two identical parts

$$G_f = \frac{4.92 \times 10^4 \; s}{(s + 643) \, (s + 1.556 \times 10^4)} \cdot \frac{4.92 \times 10^4 \; s}{(s + 643) \, (s + 1.556 \times 10^4)} \tag{14.26}$$

Each part of this function now has the form of Eq. 14.4. Thus, the required filter can be constructed from two filters with the form shown in Fig. 14.3 connected in cascade. The total filter configuration would be as shown in Fig. 14.5.

In Example 14.2, we found that a circuit similar to each unit in Fig. 14.5 was stable. Thus, since there is negligible interaction between units, the total configuration is also stable.

It is possible to purchase two op amps in a single case, so the cost and size of this filter need not be twice as much as the filter in Example 14.2.

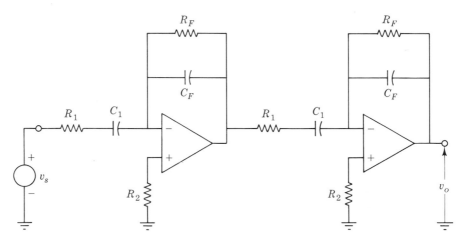

Fig. 14.5. Band-pass filter of Example 14.3.

**14.5
A
DOUBLE-POLE
LOW-PASS
FILTER
UTILIZING A
VOLTAGE
FOLLOWER** A double-pole low-pass filter with controllable roll-off characteristics may be built using a voltage follower. The circuit diagram is shown in Fig. 14.6. The voltage follower is assumed to have extremely high input impedance and unity gain.

The transfer function of the filter may be found by summing the currents into node a. Using $G_1 = 1/R_1$ and $G_2 = 1/R_2$,

$$(v_s - v_a)\,G_1 + (v_o - v_a)\,G_2 + (v_o - v_a)\,s\,C_1 = 0 \qquad (14.27)$$

Collecting terms

$$(v_s - v_a)\,G_1 + (v_o - v_a)\,(G_2 + s\,C_1) = 0 \qquad (14.28)$$

Since the relationship between v_a and the v_o at the input of the amplifier is simply a voltage divider,

$$v_o = v_a \frac{\dfrac{1}{sC_2}}{R_2 + \dfrac{1}{sC_2}} = \frac{v_a}{R_2 C_2 s + 1} \qquad (14.29)$$

Solving for v_a,

$$v_a = v_o \, R_2 C_2 \left(s + \frac{1}{R_2 C_2} \right) \qquad (14.30)$$

Substituting this expression for v_a into Eq. 14.28,

$$\left[v_s - v_o \, R_2 C_2 \left(s + \frac{1}{R_2 C_2} \right) \right] G_1 +$$

$$v_o \left[1 - R_2 C_2 \left(s + \frac{1}{R_2 C_2} \right) \right] (G_2 + s\,C_1) = 0 \qquad (14.31)$$

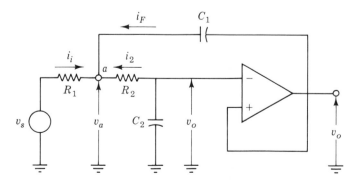

Fig. 14.6. Two-pole low-pass filter.

Solving Eq. 14.31 explicitly for v_o as a function of v_s,

$$v_o = \frac{v_s\, G_1}{R_2 C_2 G_1 \left(s + \dfrac{1}{R_2 C_2}\right) + \left[R_2 C_2 \left(s + \dfrac{1}{R_2 C_2}\right) - 1\right](G_2 + s\, C_1)} \tag{14.32}$$

Expanding the product terms in the denominator and collecting terms with equal powers of s, the transfer function is

$$\frac{v_o}{v_s} = \frac{G_1}{R_2 C_2 C_1 s^2 + (R_2 G_1 C_2 + C_2)\, s + G_1} \tag{14.33}$$

Let us divide both numerator and denominator by $R_2 C_2 C_1$. Then, recognizing that $G_1 = 1/R_1$, Eq. 14.33 becomes

$$\frac{v_o}{v_s} = \frac{1/(R_1 R_2 C_1 C_2)}{s^2 + \dfrac{(R_1 + R_2)\, C_2 s}{R_1 R_2 C_1 C_2} + \dfrac{1}{R_1 R_2 C_1 C_2}} \tag{14.34}$$

The standard form for a second-order transfer function is

$$\frac{v_o}{v_s} = \frac{A}{s^2 + 2\, \zeta \omega_n s + \omega_n^2} \tag{14.35}$$

Therefore, comparing Eq. 14.34 with Eq. 14.35,

$$\omega_n = \frac{1}{\sqrt{R_1 R_2 C_1 C_2}} \tag{14.36}$$

and

$$\zeta = \frac{(R_1 + R_2)\, C_2}{2\, R_1 R_2 C_1 C_2\, \omega_n} = \frac{\omega_n}{2}\, (R_1 + R_2)\, C_2 \tag{14.37}$$

The numerator of Eq. 14.34 is seen to be ω_n^2. Therefore, the low-frequency gain $(s \ll \omega_n)$ is one.

Designing the filter consists of determining the proper component values. The following procedure might be used:

a. Choose R_1, which is the minimum (high-frequency) impedance load on the driving source.

b. Select a ratio of R_2/R_1, perhaps $R_2 = R_1$.

c. Select ω_n and ζ to give the desired pass band and frequency response or time response for the filter. The theoretical response curves for several values of ζ are given in Fig. 14.7.

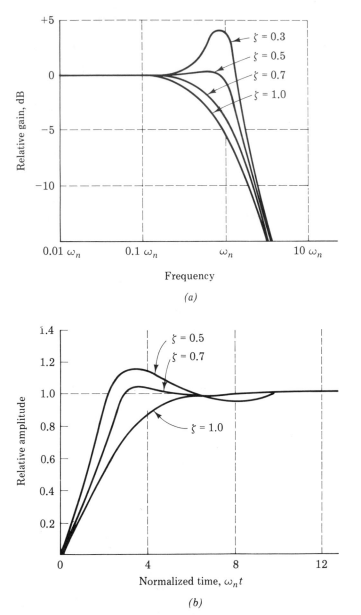

Fig. 14.7. (a) Frequency- and (b) time-response of a low-pass
double-pole filter.

d. Determine C_2 from Eq. 14.37.

$$C_2 = 2\,\zeta/(R_1 + R_2)\,\omega_n \qquad (14.38)$$

e. Determine C_1 from Eq. 14.36.

$$C_1 = \frac{1}{R_1 R_2 C_2 \,\omega_n^2} \qquad (14.39)$$

If either C_2 or C_1 are not convenient values, a different ratio of R_2 to R_1 may be used.

14.6 A DOUBLE-POLE HIGH-PASS FILTER   A double-pole high-pass filter may be obtained by interchanging the Rs and the Cs in the low-pass filter (shown in Fig. 14.6) as shown in Fig. 14.8. When this circuit is solved for the transfer function V_o/V_s as before, it is found that

$$\frac{V_o}{V_s} = \frac{s\left(s + \dfrac{1}{R_2 C_c}\right)}{s^2 + \dfrac{C_1 + C_2}{C_1 C_2 R_2}\,s + \dfrac{1}{R_1 R_2 C_1 C_2}} \qquad (14.40)$$

When we compare Eq. 14.40 with Eq. 14.35, we see, as before, that

$$\omega_n = \frac{1}{\sqrt{R_1 R_2 C_1 C_2}} \qquad (14.41)$$

and

$$2\,\zeta\omega_n = \left(\frac{C_1 + C_2}{C_1 C_2}\right)\frac{1}{R_2} = \left(\frac{1}{C_1} + \frac{1}{C_2}\right)\frac{1}{R_2} \qquad (14.42)$$

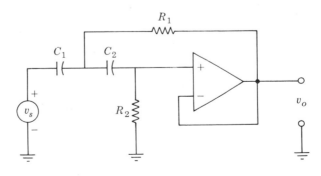

Fig. 14.8. Double-pole high-pass filter.

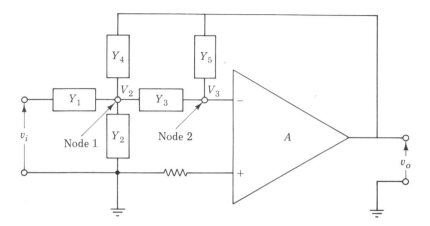

Fig. 14.9. Multiple-feedback filter.

Therefore,

$$\zeta = \frac{1}{2\omega_n} \left(\frac{1}{C_1} + \frac{1}{C_2} \right) G_2 \qquad (14.43)$$

The design procedure suggested for the low-pass filter may also be used for the high-pass filter except the minimum impedance (at high frequencies) is equal to R_2 instead of R_1.

The filters incorporating the voltage follower are stable when the op amp is properly compensated for unity gain, so the voltage follower gain K cannot exceed unity at any frequency.

14.7 MULTIPLE-FEEDBACK FILTERS The foregoing filters have used feedback to a single node in the input networks. An interesting and fairly common type of filter employs feedback to more than one node of the input network. This configuration is known as a *multiple-feedback network*[2] and is shown in Fig. 14.9. If we note that $V_o = -K V_3$, we can solve this network by using two nodal equations. The nodal equations at nodes 1 and 2 are:

$$0 = -V_i Y_1 + V_2 (Y_1 + Y_2 + Y_3 + Y_4) - V_3 Y_3 - V_o Y_4 \qquad (14.44)$$

$$0 = 0 - V_2 Y_3 + V_3 (Y_3 + Y_5) - V_o Y_5 \qquad (14.45)$$

We have assumed the input impedance to the amplifier is so high that the current into the amplifier is negligible. Now, if we substitute $-V_o/K$ for V_3, these equations can be written as follows:

---

[2] L. P. Huelsman, *Active Filters: Lumped, Distributed, Integrated, Digital, and Parametric*, McGraw-Hill Book Company, Inc., New York, 1970.

$$V_i Y_1 = V_2 (Y_1 + Y_2 + Y_3 + Y_4) - V_o \left(\frac{Y_3}{K} + Y_4 \right) \tag{14.46}$$

$$0 = -V_2 Y_3 + V_o \left(-\frac{Y_3}{K} - \frac{Y_5}{K} - Y_5 \right) \tag{14.47}$$

These two equations can be solved for V_o where

$$V_o =$$

$$\frac{-V_i Y_1 Y_3}{Y_5(Y_1 + Y_2 + Y_3 + Y_4) + Y_3 Y_4 + \frac{1}{K} [(Y_3 + Y_5)(Y_1 + Y_2 + Y_3 + Y_4) - Y_3^2]}$$

$$\tag{14.48}$$

If K is assumed to be very large, the term $1/K$ will become very small, and we can write

$$\frac{V_o}{V_i} \simeq -\frac{Y_1 Y_3}{Y_5 (Y_1 + Y_2 + Y_3 + Y_4) + Y_3 Y_4} \tag{14.49}$$

Now, since Y for a capacitor is sC and Y for a resistor is $1/R$, we can produce a second order equation in the denominator.

The equation for a second-order low-pass filter will have the form

$$\frac{V_o}{V_i} = -\frac{K \omega_n^2}{s^2 + 2\zeta \omega_n s + \omega_n^2} \tag{14.50}$$

In order for the numerator to be real, Y_1 and Y_3 must both be resistors. Since we need a second order equation in the denominator, Y_5 must be a capacitor and either Y_2 or Y_4 must be a capacitor. However, in order to obtain a real term for ω_n^2, Y_4 must be a resistor. Then, Y_2 must be the second capacitor. The low-pass circuit will have the form shown in Fig. 14.10.

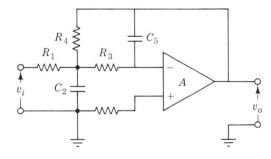

Fig. 14.10. Multiple-feedback low-pass filter.

When the circuit parameters are inserted into Eq. 14.49, we have

$$\frac{V_o}{V_i} = \frac{G_1 G_3}{sC_5(G_1 + sC_2 + G_3 + G_4) + G_3 G_4} \tag{14.51}$$

This relationship can be rewritten as follows:

$$\frac{V_o}{V_i} = \frac{G_1 G_3 / C_2 C_5}{s^2 + s(G_1 + G_3 + G_4)/C_2 + G_3 G_4 / C_2 C_5} \tag{14.52}$$

From this equation,

$$K = \frac{G_1}{G_4} \tag{14.53}$$

$$\zeta = \frac{1}{2}\left(\frac{C_5}{C_2 G_3 G_4}\right)^{1/2} (G_1 + G_3 + G_4) \tag{14.54}$$

$$\omega_n = \left(\frac{G_3 G_4}{C_2 C_5}\right)^{1/2} \tag{14.55}$$

The roll-off for this filter will be 40 dB/decade and ζ will determine the roll-off characteristics near ω_n.

The band-pass filter must have the form

$$\frac{V_o}{V_i} = -\frac{sK\omega_n}{s^2 + 2\zeta\omega_n s + \omega_n^2} \tag{14.56}$$

In this filter, let Y_1, Y_2, and Y_5 be resistors. Then, Y_3 and Y_4 will be capacitors. The filter will have the form shown in Fig. 14.11. Then, Eq. 14.49 will be

$$\frac{V_o}{V_i} = \frac{sG_1 C_3}{G_5(G_1 + G_2 + sC_3 + sC_4) + s^2 C_3 C_4} \tag{14.57}$$

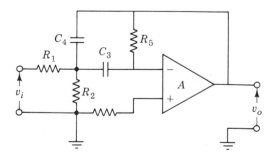

Fig. 14.11. Multiple-feedback band-pass filter.

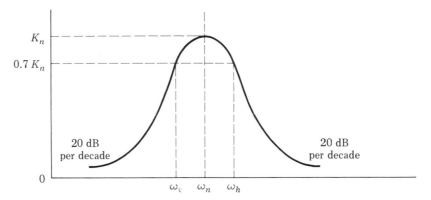

Fig. 14.12. Frequency response of a band-pass filter.

This equation can be rewritten as

$$\frac{V_o}{V_i} = \frac{sG_1/C_4}{s^2 + sG_5(C_3 + C_4)/C_3C_4 + G_5(G_1 + G_2)/C_3C_4}$$ (14.58)

From this equation,

$$K = G_1 \left| \frac{C_3}{C_4G_5(G_1 + G_2)} \right|^{1/2}$$ (14.59)

$$\zeta = \frac{C_3 + C_4}{2} \left[\frac{G_5}{C_3C_4(G_1 + G_2)} \right]^{1/2}$$ (14.60)

$$\omega_n = \left[\frac{G_5(G_1 + G_2)}{C_3C_4} \right]^{1/2}$$ (14.61)

The frequency-response characteristics of the band-pass filter are determined by the choice of ζ, as indicated by Fig. 14.7. However, the roll-off is 20 dB/decade for frequencies well removed from the band edges, both below and above the pass band, because of the s in the numerator of Eq. 14.56. The roll-off may be much steeper at the band edges, depending upon ζ. The frequency response of the filter is sketched in Fig. 14.12. The peak response with voltage gain K_n is obtained at ω_n. (This gain K_n varies with ζ and may be different from K in Eq. 14.56.) It can be shown[3] that the half-power bandwidth in radians per second is

$$B = \omega_h - \omega_\ell = \frac{\omega_n}{Q} = 2\zeta\omega_n$$ (14.62)

---

[3] C. L. Alley and K. W. Atwood, *Electronic Engineering*, Third Edition, Chap. 9, John Wiley & Sons.

where $Q = 1/2\zeta$. Also, the reference gain K_n at ω_n may be obtained by substituting $j\omega_n$ for s in Eq. 14.58.

$$K_n = \frac{j\omega_n\, G_1/C_4}{-\omega_n^2 + j\omega_n\left[\dfrac{G_5(C_3 + C_4)}{C_3 C_4}\right] + \omega_n^2}$$

$$= \frac{G_1 C_3}{G_5(C_3 + C_4)} = \frac{R_5 C_3}{R_1(C_3 + C_4)} \tag{14.63}$$

The filter may now be designed after specifying the values of ω_n, bandwidth ($2\zeta\omega_n$) in rps, and reference gain K_n. The value of K_n should not be greater than 10, or the center frequency and bandwidth may become too sensitive to changes in parameter values and temperature. Since there are five filter components and only three specified parameters, we must choose two components on the basis of other considerations. Since R_1 determines the minimum input impedance, let us select a value for R_1 that will provide a suitable load for the driving source. The design equations will be simplified if we make $C_4 = C_3 = C$. Then, using Eq. 14.63, R_5 may be determined from the desired value of reference gain K_n.

$$K_n = \frac{R_5}{2R_1} \tag{14.64}$$

Observe that R_5 may be used as a gain control. Next, G_2 and thus R_2 may be determined from Eq. 14.60.

$$\zeta = \frac{1}{\sqrt{R_5(G_1 + G_2)}} \tag{14.65}$$

Finally, C may be determined from Eq. 14.61.

$$C = \frac{\sqrt{G_5(G_1 + G_2)}}{\omega_n} \tag{14.66}$$

In this filter, as well as all the others, the bandwidth of the amplifier must be large in comparison with ω_n and the open-loop gain of the op amp must be large in comparison with K_n. These assumptions were made in deriving the equations. The compensated open-loop gain of the op amp must be at least 10 at the highest frequency of interest, perhaps 100 times ω_n. Otherwise, the design will be inaccurate and instability may occur. When very wide bandwidths are needed, the equations can take the form given in Example 14.1.

The equation for a second-order high-pass multiple-feedback filter has the following form:

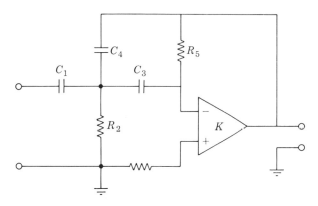

Fig. 14.13. Multiple-feedback high-pass filter.

$$\frac{V_o}{V_i} = -\frac{s^2 K}{s^2 + 2\zeta\omega_n + \omega_n^2} \tag{14.67}$$

In this filter, both Y_1 and Y_3 must be capacitors. In order to have a real term in the denominator, Y_5 and Y_2 must be resistors. Then Y_4 must be a capacitor in order to obtain a second-order equation in the denominator. The high-pass filter will have the configuration shown in Fig. 14.13. For this configuration, Eq. 14.49 becomes:

$$\frac{V_o}{V_i} = \frac{s^2 C_1 C_3}{G_5(sC_1 + G_2 + sC_3 + sC_4) + s^2 C_3 C_4} \tag{14.68}$$

This equation can be rewritten as

$$\frac{V_o}{V_i} = \frac{s^2 C_1 / C_4}{s^2 + sG_5(C_1 + C_3 + C_4)/C_3 C_4 + G_5 G_2 / C_3 C_4} \tag{14.69}$$

From this equation,

$$K = \frac{C_1}{C_4} \tag{14.70}$$

$$\zeta = \frac{C_1 + C_3 + C_4}{2}\left(\frac{R_2}{R_5 C_3 C_4}\right)^{1/2} \tag{14.71}$$

$$\omega_n = \left(\frac{1}{R_2 R_5 C_3 C_4}\right)^{1/2} \tag{14.72}$$

In order to the test the stability of these multiple-feedback networks, the value of $\mathcal{F}$ must be known. This feedback factor can be found from

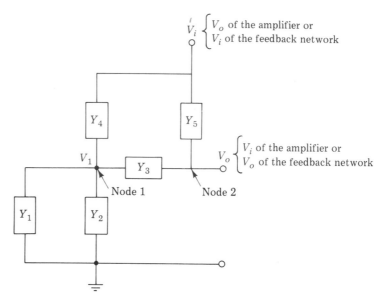

Fig. 14.14. Configuration to find F in Fig. 14.9.

Fig. 14.14. The impedance of the signal source has been assumed to be negligible. If this signal source is not negligible, it must be included as part of Y_1. The node equations for this network are:

$$0 = -V_iY_4 + V_1(Y_1 + Y_2 + Y_3 + Y_4) - V_oY_3 \tag{14.73}$$

$$0 = -V_iY_5 - V_1Y_3 + V_o(Y_3 + Y_5) \tag{14.74}$$

or

$$V_iY_4 = V_1(Y_1 + Y_2 + Y_3 + Y_4) - V_oY_3 \tag{14.75}$$

$$V_iY_5 = -V_1Y_3 + V_o(Y_3 + Y_5) \tag{14.76}$$

When these equations are solved for V_o and simplified,

$$\frac{1}{\mathscr{F}} = \frac{V_o}{V_i} = \frac{Y_5(Y_1 + Y_2 + Y_3 + Y_4) + Y_3Y_4}{(Y_3 + Y_5)(Y_1 + Y_2 + Y_4) + Y_3^2} \tag{14.77}$$

With $\mathscr{F}$ known, the stability can be determined as outlined in Section 14.3.

**14.8
TWIN-T
BAND-STOP
FILTER**
The last circuit in Table 14.1 is known as a twin-T circuit because of its configuration. This circuit may be used in a band-pass configuration, if it is placed in the feedback network, or as a band-stop filter, if placed in the input circuit as shown in Fig. 14.15. In this band-stop filter, an

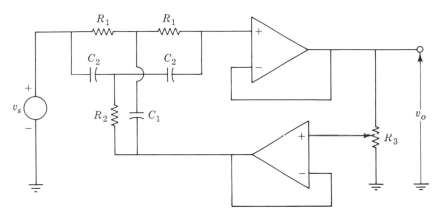

Fig. 14.15. Twin-T band-stop filter with adjustable Q.

adjustable positive feedback voltage has been added through a voltage follower to the common terminal of the filter network in order to control the bandwidth of the filter. Positive feedback, in contrast to negative, decreases the bandwidth and thus increases the Q of the circuit.

The transfer *impedance* of the RC network is needed, rather than the transfer admittance given in Table 14.1, because the voltage follower input acts as an open circuit rather than a short circuit. The transfer function is rather lengthy and tedious to derive because the three RC elements generate a third-order transfer function. As shown in Table 14.1, however,

$$\omega_n = \frac{1}{\sqrt{R_1^2 R_2 C_1 C_2^2}} \tag{14.78}$$

Again, R_1 determines the minimum input resistance, so R_1 should be large enough to avoid loading the source. In our previous filters, the source resistance was lumped with R_1, but since there are two R_1 resistors, the two C_2s must connect in parallel with these R_1s. Then, R_s cannot be included as part of the input R_1. Therefore, R_1 should be about 10 R_s or higher in order to preserve the desired response of the circuit. Then, if R_2 is chosen so that $R_2 = R_1/2$ and $C_1 = 2C_2$, so the time constants of the three RC elements are equal, $\zeta = 1.6$ and $Q = 0.3$ when the damping control R_3 is adjusted for zero feedback. This setting provides maximum width of the stop band. As the feedback is increased, the damping decreases and Q increases until $\zeta = .05$ and $Q = 10$ when 95 percent of R_3 appears between the wiper and ground. The frequency response of the filter is shown for these two positions of the wiper when $\omega_n = 2\pi (60)$ rps in Fig. 14.16.

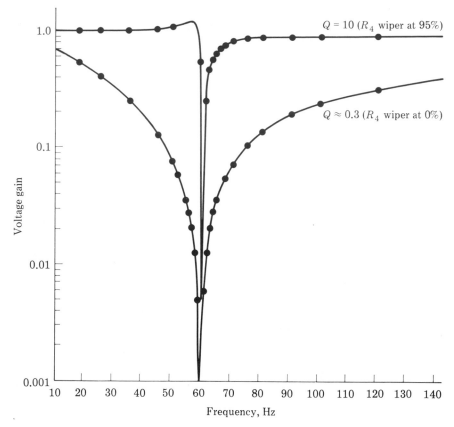

Fig. 14.16. Frequency response of a twin-T band-stop filter.

The integrator circuit can be used to construct a very versatile filter.[4] The block diagram of this filter is shown in Fig. 14.17. Each block contains the transfer function of that portion of the circuit. Consequently, the two integrator configurations (Fig. 13.14) are represented by the blocks which contain $(-\omega/s)$ terms.

From Fig. 14.17, the following relationships can be noted:

$$V_3 = -\frac{\omega_2}{s} V_2 = \frac{\omega_1\omega_2}{s^2} V_1 \qquad (14.79)$$

$$V_2 = -\frac{\omega_1}{s} V_1 \qquad (14.80)$$

[4] Gunnar Hurtig, "Positive Results from Negative Feedback," *Electronics*, Vol. 42, No. 7, March 31, 1969.

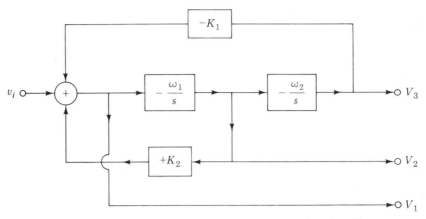

Fig. 14.17. Block diagram of a universal active filter.

Also,

$$V_1 = V_i + K_2 V_2 - K_1 V_3 \qquad (14.81)$$

Then, using Eqs. 14.79 and 14.80,

$$V_1 = V_i - \frac{K_2 \omega_1}{s} V_1 - \frac{K_1 \omega_1 \omega_2}{s^2} V_1 \qquad (14.82)$$

This equation can be arranged as follows:

$$V_1 + \frac{K_2 \omega_1}{s} V_1 + \frac{K_1 \omega_1 \omega_2}{s^2} V_1 = V_i \qquad (14.83)$$

or

$$V_1 \left(\frac{s^2 + s K_2 \omega_1 + K_1 \omega_1 \omega_2}{s^2} \right) = V_i \qquad (14.84)$$

Finally,

$$\frac{V_1}{V_i} = \frac{s^2}{s^2 + s K_2 \omega_1 + K_1 \omega_1 \omega_2} \qquad (14.85)$$

Therefore, V_1 is the output terminal of a second-order *high-pass filter*. Now, from Eq. 14.80, the value of V_2 is $-(\omega_1/s) V_1$. When this relationship is applied to Eq. 14.85, the resulting equation is

$$\frac{V_2}{V_i} = -\frac{s \omega_1}{s^2 + s K_2 \omega_1 + K_1 \omega_1 \omega_2} \qquad (14.86)$$

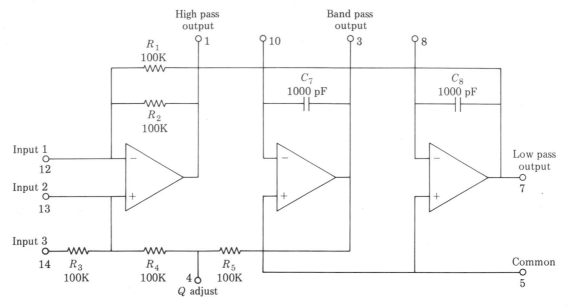

Fig. 14.18. Circuit of the UAF 31, UAF 21/25, and UAF 11/15.

Consequently, V_2 is the output terminal of a second order *band-pass filter.* Finally, from Eq. 14.79, the value of V_3 is $-(\omega_2/s)V_2$, so Eq. 14.86 can be converted to the form

$$\frac{V_3}{V_i} = \frac{\omega_1\omega_2}{s^2 + sK_2\omega_1 + K_1\omega_1\omega_2} \tag{14.87}$$

Thus, V_3 is the output terminal of a second-order *low-pass filter.*

The fact that a single configuration can be used for such a range of filters can lead to standardization of components. For example, Burr-Brown[5] produces several models of universal active filters. Each module contains three op amps connected as shown in Fig. 14.18. By adding external resistors, as shown in Fig. 14.19, the circuit has the form shown in Fig. 14.17. The values of R_8 (external) and C_8 determine ω_2, and the values of R_7 (external) and C_7 give ω_1. The values of resistors R_1 and R_2 cause K_1 to have a value of one. Other gains can be obtained by placing resistors in parallel with R_1 and R_2. The first op amp acts as the summing junction. The fact that R_1 feeds into the inverting input of this op amp cause K_1 to have a negative sign. Voltage V_1 of Fig. 14.17 appears on terminal 1 of Fig. 14.19.

---

[5] *Universal Active Filter Applications,* Burr-Brown Research Corporation, International Airport Industrial Park, Tucson, Arizona 85734.

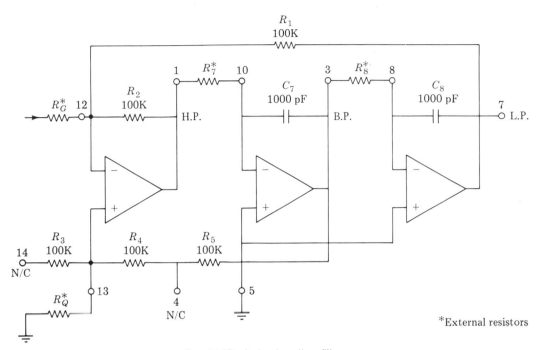

Fig. 14.19. Actual active filter.

The ratio of R_G (external) and R_2 determines the effective magnitude of V_i that is applied to the filter. Of course R_G and R_2 determine the gain of the input op amp as seen from the noninverting terminals. However, K_2 can also be adjusted by the ratio of $(R_4 + R_5)$ and R_Q (external). Since R_4 feeds into the noninverting terminal of the summing amplifier, the sign of K_2 will be positive. Voltage V_2 of Fig. 14.17 appears at terminal 3 of Fig. 14.19 and voltage V_3 of Fig. 14.17 appears at terminal 7 of Fig. 14.19.

Application notes provide additional configurations and specifications. For the configuration shown in Fig. 14.19 and $\omega_n < 2\pi \times 5 \times 10^3$ rad/s, the following equations are given:

$$K_{BP}(2R_G \, 10^{-5} + 1) > 1$$

where K_{BP} is K in Eq. 14.56 for the band-pass filter.

$$R_7 = R_8 = 10^9 / \omega_n$$

$$QK_{LP} = QK_{HP} = K_{BP}$$

$$R_G = 10^5 \, Q / K_{BP}$$

$$R_Q = 2 \times 10^5 / (2K_{BP}R_G \times 10^{-5} + K_{BP} - 1)$$

The value of Q is equal to $1/(2\zeta)$.

While many other filter configurations exist, the filters described in this chapter can meet the majority of modern filter requirements. The fact that cutoff and resonant frequencies are determined by RC combinations permits the use of variable resistors for tunable filters.

Problems   *Section 14.1*

14.1   Verify that the y-parameters given for the first two circuits of Table 14.1 are correct.

14.2   The sixth circuit given in Table 14.1 is known as a *bridged T network*. Verify that the given y-parameters for this network are correct.

Section 14.2

14.3   Design the low-pass filter with the characteristics given in Fig. 14.2.

14.4   Design the high-pass filter with the characteristics given in Fig. 14.2.

Section 14.3

14.5   Use a Bode plot to prove that Eq. 14.23 represents a stable configuration. Assume $\omega_h = 2,000$ rad/s.

14.6   Determine if the circuits which you designed in Problem 14.3 and Problem 14.4 are stable.

Section 14.4

14.7   Use cascaded sections to produce second-order low- and high-pass filters for the color organ of Example 14.1.

Section 14.5

14.8   Design a low-pass double-pole filter having a minimum input impedance of 10 kΩ, upper cutoff frequency of 1 kHz, and $\zeta = 0.5$. Sketch its frequency response.

Section 14.6

14.9   Design a double-pole high-pass filter having a cutoff frequency of 2 kHz, a minimum input impedance of 10 kΩ, and a $\zeta = 0.7$. Sketch its frequency response.

Section 14.7

14.10   Use a multiple-feedback filter to design a low-pass filter with a gain in the pass band of one and a cutoff frequency of 10^3 rad/s. The attenuation outside the pass band should be 40 dB/decade. The gain of the op amp is

$$G_a = \frac{5 \times 10^5}{s + 50}$$

  a.   Draw the circuit diagram.
  b.   Determine the magnitude of each circuit component.
  c.   Determine the minimum phase margin for your circuit.

14.11   Design a band-pass filter having cutoff frequencies of 1.2 kHz and 1.8 kHz and a minimum input impedance of 10 kΩ. Sketch the frequency response of the filter.

14.12   Use a multiple-feedback filter to design the band-pass filter of Example 14.1. Check the stability of your filter.

Section 14.8

14.13   A given signal source having 2 kΩ internal resistance has an undesirable 120 Hz signal induced into it. Design a band-stop filter with variable bandwidth capability that will essentially eliminate this unwanted signal.

Section 14.9

14.14   Use the universal active filter shown in Fig. 14.19 to design the band-pass filter of Example 14.1. Specify the size of all external resistors.

General

14.15   A filter has the configuration shown in Fig. 14.20. Make a Bode plot of the open-loop transfer function for this amplifier

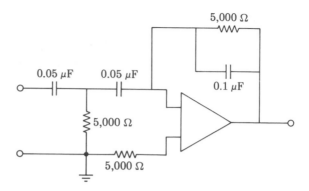

Fig. 14.20. Configuration for Prob. 14.15.

and determine if the system is stable. The transfer function of the operational amplifier is

$$G_a = \frac{5 \times 10^6}{(s + 500)}$$

14.16   Design a filter which has the following transfer function.

$$G = \frac{10^5 s}{(s + 100)(s + 10{,}000)}$$

    a.   Draw the circuit diagram of your filter.
    b.   Find the values of all resistors and capacitors in the circuit if the minimum input impedance is 1,000 Ω.
    c.   Draw the open-loop gain and phase plot for your circuit if

$$G_a = \frac{10^5}{s + 50}$$

    d.   Determine the phase margin of your filter.

14.17   Design a filter that has the following transfer function

$$G_f = \frac{-10(s + 100)(s + 100{,}000)}{(s + 1{,}000)(s + 10{,}000)}$$

Make a gain plot of this response. Determine if your filter is stable.

14.18   Design a band-rejection filter which has the following transfer function:

$$G = \frac{10(s + 8{,}000)(s + 10{,}000)}{(s + 2{,}000)(s + 40{,}000)}$$

Make a gain plot of this response. Determine if your filter is stable.

15

Power Amplifiers

A typical amplifier consists of several stages of amplification. Most of these stages are small-signal, low-power devices. For these stages, efficiency is usually unimportant, distortion is negligible, and the equivalent circuits accurately predict the amplifier behavior. In contrast, the final stage of an amplifier (and in some cases an additional driver stage) is usually required to furnish appreciable signal power to its load. Typical loads include loudspeakers, antennas, positioning devices, and so on. These amplifiers are commonly called power amplifiers. Because of this relatively high-power level, the efficiency of the power amplifier is important. Also, distortion becomes a problem because the amplifier parameters vary appreciably over the signal cycle. Therefore, the equivalent circuits are only rough approximations, and graphical methods assume increased importance. Heat dissipation also becomes a problem. This chapter will discuss these problems which are peculiar to power amplifiers, namely: distortion, efficiency, push-pull configurations, and thermal conduction.

15.1
POWER
OUTPUT AND
EFFICIENCY

We will first discuss a method of determining the power output and the efficiency of an amplifier. The voltage drop and power loss in emitter-circuit stabilizing resistors will be neglected. These losses may be included by increasing V_{CC} to compensate for the IR drop across R_E and then using this higher V_{CC} to calculate the power delivered by the power supply.

Figure 15.1 shows how the power output and efficiency of an amplifier can be determined from the collector characteristics. The maximum power available is of major interest, and that power is obtained when

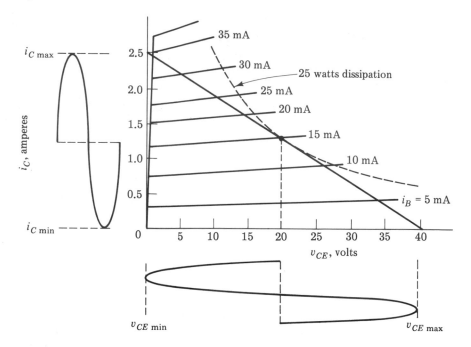

Fig. 15.1. Current, voltage, and power relationships in an amplifier.

the transistor is biased halfway between saturation and collector-current cutoff and is driven from saturation to cutoff. The maximum power output for a sinusoidal driving signal is then $V_c I_c$ where V_c and I_c are the rms values of collector voltage and current. But these rms values can be obtained by dividing the peak-to-peak values by two, to obtain peak values, and then dividing by $\sqrt{2}$ to obtain rms values. Therefore,

$$P_o = \frac{(v_{CEmax} - v_{CEmin})(i_{Cmax} - i_{Cmin})}{8} \qquad (15.1)$$

Observe from Fig. 15.1 that alternate half cycles of the output signals will be dissimilar and thus contain even harmonic distortion unless the characteristic curves are evenly spaced. The peak value of the fundamental component of signal current or voltage in the output may be obtained from $(i_{max} - i_{min})/2$ or $(v_{max} - v_{min})/2$, respectively.

The amplifier of Fig. 15.1 can deliver the same signal power to the load with approximately half as much input power from the power supply if the load is transformer coupled, as shown in Fig. 15.2, rather than RC coupled. The improved efficiency occurs because the dc power loss in the load is eliminated. If the ohmic resistance of the transformer primary is negligible, the dc load line is nearly vertical. The ac signal

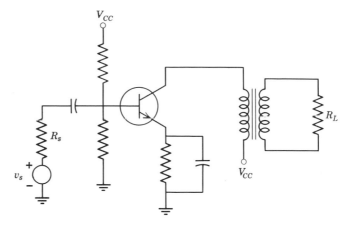

Fig. 15.2. Transformer-coupled power amplifier.

will swing as high above V_{CC} as it swings below V_{CC}. Thus, V_{CC} is now the average value of v_{CE}. An example will be used to illustrate the contrast between the amplifier with R_L in the collector circuit and a transformer-coupled amplifier.

Example 15.1

The collector characteristics of an amplifier with a 16 Ω load is shown in Fig. 15.1. We will determine the collector circuit efficiency if R_L is connected between the collector and V_{CC}, and then compare this efficiency to that of a transformer-coupled amplifier (Fig. 15.2) which operates along the same load line.

From Eq. 15.1, the ac signal power is $P_o = (40 - 1)(2.45 - 0)/8 = 11.9$ W. The collector power input for the amplifier with the resistance in the collector circuit is $P_i = V_{CC} I_{Cave} = 40 \times 1.225 = 49$ W. The collector circuit efficiency is $\eta_c = (P_o/P_i)100 = (11.9/49)100 = 24$ percent.

When the transformer-coupled amplifier (Fig. 15.2) is considered, V_{CE} will swing as high above V_{CC} as it swings below V_{CC}. Then, $V_{CC} = (V_{CEmax} + V_{CEmin})/2 = (40 + 1)/2 = 20.5$ V. The ac power output is the same as above. However, the power input is now $P_i = V_{CC} I_{Cave} = 20.5 \times 1.225 = 25.1$ W. The collector-circuit efficiency is now $\eta_c = (11.9/25.1) = 47.4$ percent, which is essentially twice as great.

The power output would be increased and, hence, the efficiency increased if the collector voltage v_{CE} could be driven to zero on one-half cycle and the collector current i_C could be driven to zero on the next half cycle. Then, using Fig. 15.1 and Eq. 15.1,

$$P_{omax} = \frac{2V_{CC}\,2I_{Bq}}{8} = \frac{V_{CC}\,I_{Cq}}{2} \qquad (15.2)$$

The maximum efficiency is then

$$\text{max eff} = \frac{(V_{CC}\,I_{Cq})/2}{V_{CC}\,I_{Cq}} \times 100 = 50\% \qquad (15.3)$$

This is the maximum theoretical efficiency of a class A amplifier with sinusoidal input. The efficiency can approach 100 percent if the input signal is a square wave. A class A amplifier is defined as an amplifier in which collector current flows during the entire cycle. We have considered only class A amplifiers to this point in our work, except for the complementary-symmetry amplifier in Chapter 10.

The class A amplifier is inefficient because of the relatively large q-point current which flows through the amplifier all the time. The q-point current must be equal to the peak signal current. Otherwise, the collector current is cut off during part of the signal cycle and serious distortion results. Note that the average collector power input to a class A amplifier is essentially constant. However, the signal output power increases with the signal level. The collector dissipation is the collector power input minus the signal power output. Consequently, the collector power dissipation decreases and the collector efficiency increases as the signal level increases.

**15.2
PUSH-PULL
AMPLIFIERS**
A high-efficiency amplifier can be built using two transistors in an arrangement known as push-pull, shown in Fig. 15.3. This amplifier is similar to the balanced amplifier discussed in Chapter 10 except that transformer coupling is used to provide balanced signals of opposite polarity to the two bases. Transformer coupling is also used to add the balanced outputs of the two transistors in a single load R_L. The push-pull amplifier can be operated class A and will provide about twice the power output obtainable from a single transistor under similar circumstances. Of course, the total collector input power also doubles (two stages) so the collector efficiency remains the same. The distortion of the push-pull connection is lower, however, because the even harmonic distortion is cancelled as a result of the balanced arrangement. The positive half cycle of one transistor adds to the negative half cycle of the other transistor to provide an output with similar successive half cycles, as shown in Fig. 15.4.

The main advantage of a push-pull amplifier results from the fact that the q-point currents of the individual transistors can be drastically reduced and the efficiency, therefore, markedly increased. In fact, each transistor can be biased at very nearly cutoff, so each transistor delivers

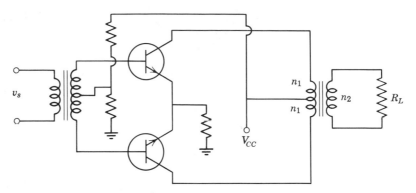

Fig. 15.3. Transformer-coupled push-pull amplifier.

power to the load only during one-half of the signal cycle. Collector current flows approximately half the time in each transistor. This type of operation is known as *class B*.

When each transistor is biased precisely at cutoff, distortion occurs in the output because h_{fe} decreases rapidly with collector current at very small values of collector current. This type of distortion is known as

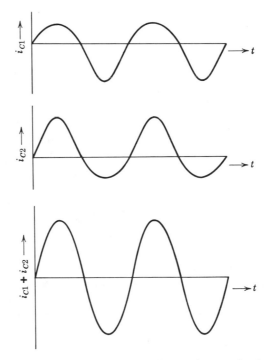

Fig. 15.4. Illustration of the cancellation of even harmonics in the output of a balanced push-pull amplifier.

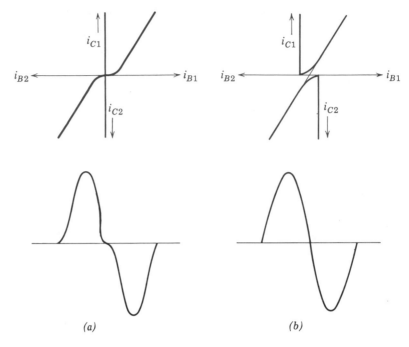

Fig. 15.5. Illustration of the crossover distortion that results from cutoff bias and the reduction of distortion by the use of projected cutoff bias: (a) crossover distortion; (b) projected cutoff bias.

crossover distortion and is illustrated in Fig. 15.5a. In this figure, the transfer characteristic of one transistor is inverted and reversed as compared to the other transistor, so the sum of the two curves is the total transfer curve of the push-pull amplifier. Most of the nonlinearity of the total transfer curve occurs at small values of i_C as shown.

The crossover distortion can be eliminated by providing a small forward bias current, as shown in Fig. 15.5b. The minimum value of bias is determined by extending the relatively straight portion of the transfer curve of one transistor until the extension crosses the $i_C = 0$ axis. The value of i_B found at the intersection is known as *projected-cutoff bias*. If the transfer curve for the other transistor is inverted and reversed and the projected transfer curves are made to coincide at the projected-cutoff bias value of i_B, the total transfer curve is essentially straight and little distortion occurs. Since the magnetic flux in the output transformer, which results from the dc components of collector currents, tends to cancel, the net flux which is effective in producing voltage in the secondary is proportional to the difference between the individual collector currents. Therefore, the total or *composite* transfer characteristic is the difference between the individual transfer characteristics at any point.

A set of composite (or total) collector characteristics can be drawn by inverting one set and reversing it as compared with the other, and vertically aligning the individual q-points. However, the bottom half of such a composite set contains the same information as the top half, so only the top half, as shown in Fig. 15.6, is needed. The composite (total) load line passes through the composite q-point which is at $i_C = 0$ and $v_{CE} = V_{CC}$. The operation of the individual transistor follows the composite load line except during the time when both transistors are conducting. Then the total collector current is the difference between the individual collector currents. Since each transistor works into one-half of the output transformer primary, the composite load line represents the impedance of only one-half of the primary, which is one-fourth of the total primary impedance.

15.3 POWER OUTPUT AND EFFICIENCY OF CLASS *B* AMPLIFIERS

The maximum permissible collector current and the maximum power output for the class *B* push-pull amplifier can be determined as functions of the permissible transistor dissipation with the aid of Fig. 15.6. The signal is assumed to be sinusoidal. The total transistor power dissipation is

$$P_d = P_i - P_o \qquad (15.4)$$

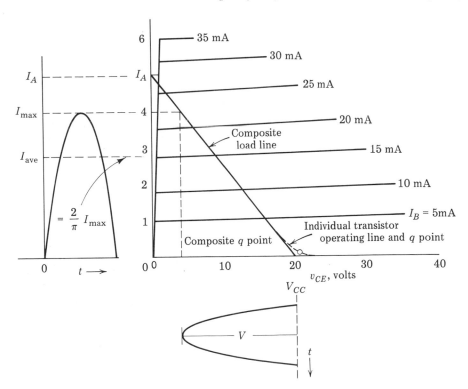

Fig. 15.6. Plot of i_C vs. i_B to find projected cutoff.

where P_i is the collector power input from the power supply and P_o is the signal power output, as previously discussed. But, assuming cutoff bias,

$$P_i = I_{ave} V_{CC} = \frac{2}{\pi} I_{max} V_{CC} \qquad (15.5)$$

and

$$P_o = \frac{I_{max}^2}{2} R_L \qquad (15.6)$$

Then, substituting Eq. 15.6 and Eq. 15.5 into Eq. 15.4,

$$P_d = \frac{2}{\pi} I_{max} V_{CC} - \frac{I_{max}^2}{2} R_L \qquad (15.7)$$

We need to find the value of I_{max} which will give maximum power dissipation P_{dmax}. This can be done by differentiating P_d with respect to I_{max} and equating this derivative to zero.

$$\frac{dP_d}{dI_{max}} = \frac{2}{\pi} V_{CC} - I_{max} R_L = 0 \qquad (15.8)$$

Solving Eq. 15.8 explicitly for the I_{max} which gives maximum power dissipation,

$$I_{max} = \frac{2}{\pi} \frac{V_{CC}}{R_L} = \frac{2}{\pi} I_A \qquad (15.9)$$

where I_A is the current axis intercept (V_{CC}/R_L) of the load line, as shown in Fig. 15.6. Substituting this value of I_{max} (Eq. 15.9) into Eq. 15.7, the maximum power dissipation can be found in terms of I_A and V_{CC}.

$$P_{dmax} = \frac{4}{\pi^2} I_A V_{CC} - \frac{2}{\pi^2} I_A^2 R_L \qquad (15.10)$$

But $I_A R_L = V_{CC}$, so

$$P_{dmax} = \frac{4}{\pi^2} I_A V_{CC} - \frac{2}{\pi^2} I_A V_{CC} = \frac{2}{\pi^2} I_A V_{CC} \qquad (15.11)$$

However, the power dissipation capabilities of the two transistors P_d is usually known or obtainable, and the maximum permissible value of I_A can be found from Eq. 15.11.

$$I_A (max) = \frac{\pi^2}{2} \frac{P_d}{V_{CC}} \simeq \frac{5P_d}{V_{CC}} \qquad (15.12)$$

A method of determining P_d will be discussed in Section 15.8. The i_c axis intercept current I_A must be reduced as the q-point collector current I_{Cq} is increased above cutoff. An empirical alteration of Eq. 15.12 takes the q-point current of the individual transistor into account.

$$I_A(\text{max}) \simeq \frac{5P_d}{V_{CC}} - 3I_{Cq} \qquad (15.13)$$

This P_d is the total dissipation of both transistors because each transistor dissipates only during one-half cycle.

The minimum load resistance, maximum power output, and maximum efficiency can now be determined if V_{CC} and P_d are known. We will demonstrate this computation by an example.

Example 15.2

Let us consider the amplifier with characteristics given in Fig. 15.6 and $V_{CC} = 20$ V. Let us assume that $P_d = 20$ W total for the two transistors and $I_{Cq} = 100$ mA. Then, using Eq. 15.13, $I_A = 5 - 0.3 = 4.7$ A. The minimum value of load resistance is $R_L = V_{CC}/I_A = 20/4.7 = 4.25$ Ω. The maximum power output, assuming the saturation voltage is negligible, is $P_o = V_{CC}I_A/2 = 47$ W. At this power output, the collector power input is $P_i = I_{ave}V_{CC} = [I_A(2/\pi) + I_{Cq}]V_{CC} = (4.7 \times .636 + .1) 20 = 62$ W. The maximum collector circuit efficiency is $(47/62) 100 = 75$ percent.

Figure 15.7a shows the power output, power dissipation, and percent efficiency of a class B amplifier as a function of peak output signal amplitude I_{max}. The class B amplifier is assumed to be biased at cutoff. The power output, power dissipation, and efficiency of a class A ampli-

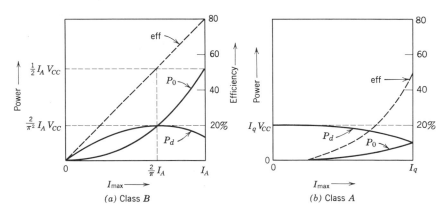

Fig. 15.7. Power output, power dissipation, and efficiency characteristics of class A and class B amplifiers.

fier, either push-pull or single-ended, is given as a function of I_{max} in Fig. 15.7b. Both amplifiers have the same power dissipation capabilities P_{dmax}. Note the increased power output capability of the class B amplifier as compared with the class A. For example, if the power dissipation capability is 10 W in either amplifier, the maximum power output from the class A amplifier is 5 W, and the maximum power output from the class B amplifier is 25 W, neglecting the saturation voltage. Thus, the class B amplifier can deliver five times the power output of the class A amplifier.

The balanced source voltage can be obtained from a phase *inverter* or *splitter* instead of a transformer with some saving in cost and weight. A differential amplifier can be used as a phase inverter, as shown in Fig. 15.8. The collector voltages of the differential pair T_1 and T_2 provide the opposite polarity voltages required for the push-pull amplifier. The diodes in the base circuits of the push-pull amplifier provide projected-cutoff bias and clamp the bases at this bias voltage. A small resistance may be needed in series with the diodes to adjust the bias. Without the diodes, the base-emitter junctions (diodes) would clamp the bias voltage to the peak of the base driving voltage.

15.4 COMPLEMEN-TARY-SYMMETRY AMPLIFIERS

Both the input and output transformers can be eliminated in a push-pull amplifier type known as the complementary-symmetry amplifier that was discussed briefly in Chapter 10. This amplifier uses one n-p-n and one p-n-p transistor with similar characteristics, as shown in Fig. 15.9. The bias resistors R_1, R_2, R_3, and R_4 provide the desired forward

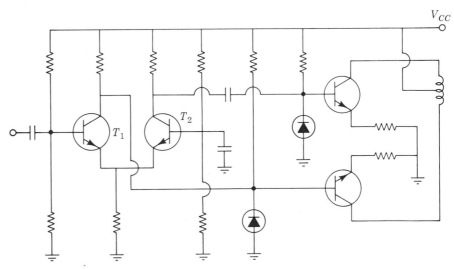

Fig. 15.8. Differential amplifier used as a phase splitter.

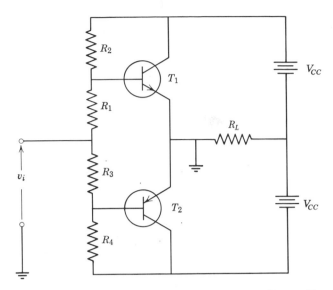

Fig. 15.9. Complementary-symmetry push-pull amplifier.

bias, usually projected-cutoff bias, for the transistors. When the input signal v_i goes through a positive half cycle, the emitter junction of the n-p-n transistor T_1 is forward biased and the emitter junction of the p-n-p transistor is reverse biased, causing the collector current of T_1 to pass through the load resistor R_L from left to right. During the next half cycle when the input voltage is negative, the p-n-p transistor is forward biased and the n-p-n transistor is cut off so the collector current of T_2 flows through the load resistor R_L from right to left. When the signal current is zero, the two transistor currents are nearly equal, so essentially no direct current flows through the load resistor. Note that this amplifier, like any push-pull amplifier, can be operated class A or at any q-point between class A and class B. Operation at q-points between class A and class B is known as class AB. However, high-power amplifiers are usually operated at projected-cutoff class B because of the high efficiency of this mode.

The circuit of Fig. 15.9 has a serious limitation. That is, the power supply which is represented by two batteries does not have a ground point. Therefore, the output signal voltage appears between the power supply and ground, and this power supply would not be usable for other amplifiers in the system. The shunt capacitance between the power supply and ground or chassis would also limit the high-frequency response of the amplifier.

The emitter-follower version of the complementary-symmetry amplifier permits the grounding of the power supply, as shown in Fig. 15.10. However, as compared with the common-emitter configuration, the input impedance is high and the required input voltage v_i is larger. The

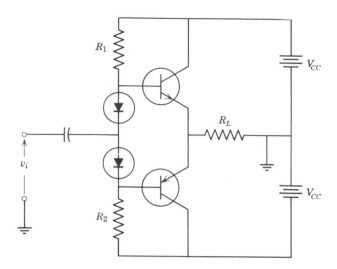

Fig. 15.10. Common-collector configuration of the
complementary-symmetry amplifier.

forward bias for the transistors in Fig. 15.10 is provided by the two
forward-biased diodes. The low dynamic resistance of these diodes as
well as their negative temperature coefficients provide improved thermal stability of the transistors as compared with the circuit of Fig. 15.9.
A small resistance can be connected in either series or parallel with the
diodes to provide precise adjustment of the q-point.

The driver transistor for the complementary-symmetry amplifier may
be directly coupled, as shown in Fig. 15.11. Let us consider the design
of a typical amplifier of this type to improve our understanding of its
operation.

Example 15.3

We will assume that V_{CC} (each supply in Fig. 15.11) = 15 V and
$R_L = 100$ Ω. Then the current axis intercept I_A for either T_1 or T_2 is
$I_A = 15$ V $/100$ Ω $= 150$ mA. From Eq. 15.22 using projected-cutoff
bias, the power dissipation requirement for the two transistors T_1 and
T_2 is $I_A V_{CC}/5 = 450$ mW. The silicon complementary pair 2N3703
and 2N3705 have adequate current capability and power dissipation
and voltage ratings for this application and, therefore, will be used.
These transistors have $\beta_o \simeq 100$ over the collector current range 10
mA to 150 mA. Projected-cutoff bias is approximately $I_C = 1$ mA and
$I_B = 15$ μA. Therefore, the forward-biasing diodes, with perhaps
some additional resistance, should be chosen so these q-point values
are obtained. We should expect that the diodes will be low-current
silicon diodes to match the characteristics of the emitter junctions of
the transistors.

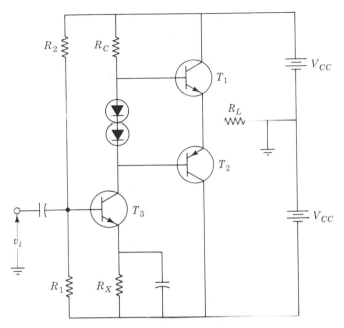

Fig. 15.11. Complementary-symmetry amplifier with a direct coupled driver.

When the input signal v_i is positive, the collector of the driver T_3 goes negative, so the transistor T_1 is cut off and transistor T_2 conducts. Then the emitter end of the load resistor has approximately the same potential as the collector of the driver. When the driver transistor T_3 is driven to saturation, the potential of the emitter end of the load resistor R_L differs from the potential of the negative end of the lower V_{CC} by the drop across R_E plus the saturation voltage of transistor T_3 plus the voltage V_{BE} of transistor T_2. Therefore, the voltage $I_E R_E$ of the driver transistor should be fairly small; we will use one volt.

When the input signal goes negative, the collector of the driver transistor T_3 goes positive, so transistor T_1 conducts and transistor T_2 is cut off. Then the emitter end of the load resistor approaches the potential of the positive end of the upper V_{CC}. However, both the collector current of T_3 and the base current of T_1 must flow through the collector load resistor R_C. Therefore, at the peak negative input voltage v_i, the transistor T_3 is cut off and the maximum-signal base current of transistor T_1 must flow through R_C. Thus, the difference between the maximum positive load potential and the positive power supply potential is the voltage $i_{Bmax} R_C$ plus the voltge V_{BE} of transistor T_1. So the voltage $i_{Bmax} R_C$ should not exceed a volt or so. We will allow 1.5 V. Then with i_C (of T_1) max = 150 mA and $\beta_o = 100$, $i_{Bmax} = 1.5$ mA, and $R_C = 1.5$ V / 1.5 mA = 1 kΩ.

We can now design the driver stage with $R_C = 1$ kΩ. At quiescent conditions, the voltage drop across the driver load R_C is approximately $V_{CC} = 15$ V. Therefore, the q-point collector current is 15 mA, and the value of R_E required for one volt drop is $1/.015 = 67$ Ω. The driver operates as a class A amplifier with $V_{CC} = 30$ V and maximum dissipation $P_d = 15$ V (15 mA) = 225 mW. The 2N3704 transistor, which is similar to the 2N3705 except $\beta_o \simeq 200$, has adequate voltage and dissipation ratings for this application. The bias resistors R_1 and R_2 can be determined by the techniques discussed in Chapter 5. We will assume that a current stability factor of 25 is adequate for this silicon transistor. Then, $R_B = 25 \times 67 = 1.7$ kΩ, $V_B = I_B R_B + V_{BE} - I_E R_E = 1.63$ V, $R_1 = (30/28.37) 1.7$ k = 1.8 kΩ and $R_2 = (30/1.63) 1.7$ k = 32 kΩ. The coupling capacitor C and the emitter bypass capacitor C_E can be determined when the driving-source resistance is known. The bypass capacitor C_E is frequently omitted to provide higher input resistance and better linearity (less distortion) at the expense of lower power gain. The voltage gain with R_E unbypassed is 1 kΩ / 67 Ω = 15, so the required peak input voltage v_i for maximum power output is $15/15 = 1$ V.

You may have observed that the power dissipation rating of the class A driver must be about the same as either of the class B pair. Therefore, a high-power amplifier will require a high-power driver unless a class B intermediate amplifier is used, as shown in Fig. 15.12. This figure also shows that a single power supply can be used if the load resistor R_L is capacitively coupled to the amplifier. The amplifier design can proceed in the same manner as previously discussed, except the effective β of the Darlington pair is treated as $\beta_1 \beta_2$, as previously discussed. The capacitor C_2 is calculated in the usual manner. Since the output resistance of the (emitter follower) amplifier is very low, the coupling capacitance can be determined from the values of f_1 and R_L. For example, if $R_L = 10$ Ω and $f_1 = 32$ Hz, determined by C, $C = 1/\omega_1 R_L = 500$ μF.

15.5 QUASI-COMPLEMENTARY SYMMETRY

The availability of complementary-symmetry pairs of transistors is limited, especially in the high-power types. This problem is resolved by using the *quasi-complementary-symmetry* circuit of Fig. 15.13. Note that this amplifier has two transistors of the same type in the output. However, the lower transistor is driven by the collector current of the driver amplifier instead of its emitter current. In other words, the lower amplifier consists of a p-n-p–n-p-n combination rather than a Darlington connection. The polarity reversal in this arrangement permits the

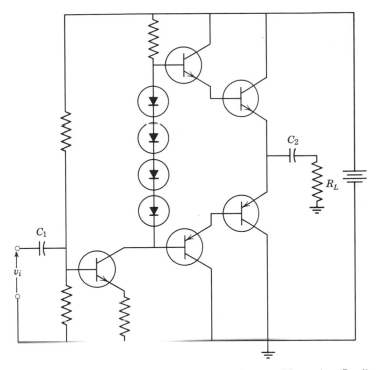

Fig. 15.12. *High-power complementary-symmetry amplifier using Darlington intermediate amplifiers.*

transistor T_2 to be an n-p-n instead of a p-n-p type. At first glance, one might suspect that the input resistance of the driver T_4 is much lower than the input resistance of T_3. This is not true because the output voltage $\Delta i_{C2} R_L$ is between the emitter of T_4 and ground, and the base-to-ground signal voltage v_{B4} of transistor T_4 is essentially equal to the output voltage. Therefore,

$$R_{it} = \frac{\Delta v_{B4}}{\Delta i_{B4}} \simeq \frac{\Delta i_{C2} R_L}{\Delta i_{B4}} = \beta_2 \beta_4 R_L \qquad (15.14)$$

Thus, the input resistance of transistor T_4, with $R_L = 10\ \Omega$, $\beta_2 = 100$, and $\beta_4 = 200$ is $R_{it} = 200{,}000\ \Omega$. The input resistance of transistor T_3 is also approximately $200{,}000\ \Omega$.

The resistors R_1, R_2, R_3, and R_4 (Fig. 15.13) are needed to provide adequate thermal stability when germanium transistors are used. The resistors R_1 and R_2 should be small in comaprison with R_L to avoid serious loss of output power. For example, if $R_L = 8\ \Omega$, R_1 and R_2 should not exceed about $1\ \Omega$. The resistors R_3 and R_4 should be of the same order of magnitude as the input resistance of the power transistors T_1 and T_2, as discussed in the Darlington amplifier section of Chapter 10.

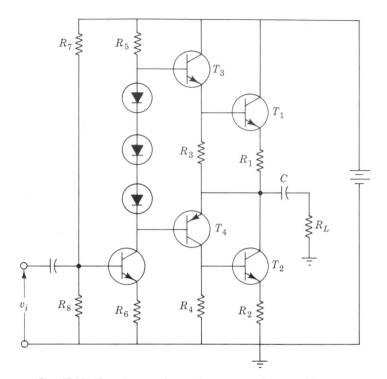

Fig. 15.13. Quasi-complementary-symmetry amplifier.

Thus, these resistors should be of the order of 100 Ω in a high-power amplifier. Note that the input resistance of T_2, as seen from the viewpoint of R_4, is $h_{ie2} + (\beta_2 + 1)R_2$. Although h_{ie2} varies widely over the signal cycle, it is usually small in comparison with $(\beta + 1)R_2$. The same relationship holds for the input resistance of transistor T_1 as seen from the viewpoint of R_3. Note that the output resistances of transistors T_3 and T_4 are in parallel with R_3 and R_4, respectively. The diodes between the bases of T_3 and T_4 cause this output resistance to be quite low, so long as emitter current flows in transistors T_3 and T_4. Thus the current stability factors of transistors T_1 and T_2 are much lower than the ratio R_3/R_1 or R_4/R_2 when current flows in transistors T_3 and T_4.

An improved version of the quasi-complementary-symmetry amplifier is shown in Fig. 15.14. This amplifier is similar to the one shown in Fig. 15.13, previously discussed, except for the following changes:

1. An additional amplifier T_1 has been added to increase the voltage gain so ample negative feedback may be applied to stabilize the q-point voltages and reduce the distortion.

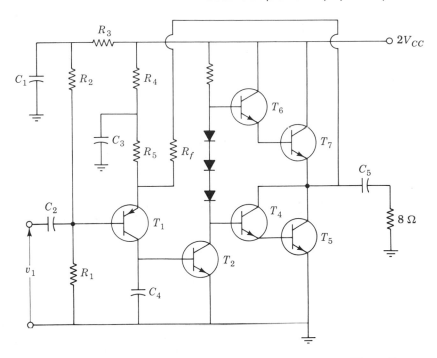

Fig. 15.14. Quasi-complementary-symmetry audio amplifier with negative feedback.

2. One-half of the total supply voltage is dropped across the emitter-circuit resistance $(R_4 + R_5)$ of T_1. The resistor R_4 is bypassed by C_3. R_5 only provides signal feedback while $(R_4 + R_5)$ provide dc feedback to stabilize the q-point voltage at the output terminal.

3. The resistor R_3 and capacitor C_1 constitute a low-pass filter which filters the power supply ripple voltage from the base-bias circuit of transistor T_1 to provide low hum level. The resistors R_1, R_2, and R_3 should be chosen so that the current through them is several times as large as the base current of T_1, and so that the emitter potential of T_1 is one-half the supply voltage potenital.

4. The capacitor C_4 provides phase-lag compenstion to the input stage in order to stabilize the feedback system. This capacitor should not be larger than is necessary to provide adequate phase margin. Excessive compensation will degrade the slew rate and thus make the amplifier susceptible to distortion of the high frequencies at high output levels, as discussed in Chapter 13.

All the other components in the circuit of Fig. 15.14 have been discussed previously.

The complementary-symmetry power amplifier can be arranged into a common-emitter configuration, as shown in Fig. 15.15. The basic amplifier does not include either R_3 or R_F. Then the capacitor C_E holds the emitters of the driver transistors T_3 and T_4 at signal ground potential so that the entire amplifier acts as a complementary-symmetry two-stage amplifier with both stages in the common-emitter configuration. However, the q-point stability of this amplifier is unsatisfactory because of its high dc gain. The q-point collector voltage of the output transistors T_1 and T_2 may be effectively stabilized by adding the feedback resistor R_F between the output and the emitters of the driving transistors, as shown in Fig. 15.15. A large dc feedback factor (approaching unity) may be obtained from this resistor R_F because the input impedance looking into emitters of the drivers is high while the signal is being applied to their bases, as discussed in Chapter 12, while R_F may be as low as 10 R_L. The distortion of the amplifier may be reduced, at the price of reduced voltage gain, by adding the resistor R_3 in series with the capacitor C_E, as shown in Fig. 15.15, to provide negative feedback over the signal frequency range. Then the reactance of C_E should be small in comparison with the resistance of R_3 at the lower edge of the desired pass band. The resistance of R_1 and R_2 (Fig. 15.15) should be small enough to allow a bleeder current that is several times as large as the base current of either T_3 or T_4. Then, if $R_1 = R_2$, the emitter potential of T_3 and T_4 will be approximately midway between the power supply potentials, as desired.

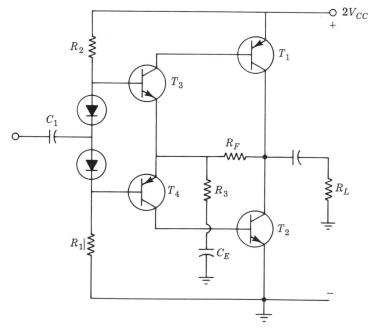

Fig. 15.15. Common-emitter complementary-symmetry amplifier.

**15.6
IC POWER
AMPLIFIERS**

Power amplifiers are available in integrated form. Two representative types are given here. The first type is the RCA HC2500 power hybrid circuit shown in Fig. 15.16. It is a hybrid circuit because a monolithic circuit is not capable of dissipating the heat generated by high-power output transistors. Therefore, these transistors are attached to the chip in a manner that will allow adequate heat dissipation, to be discussed later. This circuit is similar to the quasi-complementary-symmetry circuit shown in Fig. 15.14. The essential differences are:

1. The HC2500 uses a differential input amplifier consisting of transistors Q_1, Q_2, and the emitter current source Q_3, whereas the amplifier of Fig. 15.14 used a single transistor with a high, bypassed, emitter resistance to control its emitter current.

2. The driving transistor Q_5 for the complementary pair uses a current source Q_4 in place of the usual load resistor R_C. In addition to providing the proper q-point current for Q_5, this circuit source provides a very large load impedance, and consequently high voltage gain, for this driver.

3. The thermal stabilization resistors are included in the power output stage so the circuit will operate over its specified junction-temperature range of $-55°C$ to $150°C$.

The open-loop voltage gain of this amplifier is typically 70 dB and the amplifier is self compensated with the dominant pole at 30 kHz.

The second representative integrated power amplifier is the LM379 monolithic dual 6 watt (maximum) amplifier shown in Fig. 15.17. This amplifier is manufactured by National Semiconductor Corporation and is intended for stereo applications using 8 Ω or 16 Ω loudspeakers. This amplifier is low cost and has adequate power for home use. Some of the special features of the quasi-complementary-symmetry circuit shown in Fig. 15.17a follow.

1. The differential amplifier input amplifier uses two transistors, Q_2 and Q_3, with two (split) collectors. One pair of these collectors adds current to the emitters of the emitter followers Q_1 and Q_4 to increase the transconductance and bandwidth of these transistors. The second pair of split collectors provide reduced current, and consequently higher output resistance, to the following stage Q_7. The higher output resistance is desirable because this stage is compensated by capacitor C_1 which may therefore be smaller for the same required time constant. This compensating circuit provides stable operation for closed-loop voltage gains down to 10.

2. The unusual arrangement involving Q_5 and Q_6 is known as a *current mirror* and is used to double the gain of the differential amplifier. Since the bases and emitters of these two identical transistors are each tied together, the base currents of Q_6 and Q_5

HC2000H, HC2500
Multi-Purpose 7-Ampere Operational Amplifiers

Linear Amplifiers for Applications in Industrial and Commercial Equipment

The RCA-HC2000H and HC2500 hybrid-circuit operational amplifiers are designed for operation from either single or split power supplies at output currents up to 7 amperes and power outputs up to 100 watts. These versatile amplifiers are recommended for servoamplifiers, audio power amplifiers, driven inverters, power operational amplifiers, deflection amplifiers, solenoid drivers, voltage regulators, and similar linear-amplifier power applications. They are supplied in a metal hermetic package.

The HC2000H and HC2500 employ a quasi-complementary-symmetry output stage with hometaxial-base output transistors. They feature low distortion, with a maximum total harmonic distortion of 0.5 per cent over a bandwidth of 30 kHz at a power output of 60 watts and a typical intermodulation distortion of less than 1 per cent at rms power outputs from 0.2 to 70 watts. At an rms output of 50 milliwatts, the HC2500 has an exceptionally low typical intermodulation distortion of only 0.06 per cent.

The HC2000H includes a load-line limiting network that provides protection against short-circuit loads and against high-energy transients when the amplifier is used to drive inductive loads. Both circuits also feature adjustable idling current and direct coupling to the load.

Features:
- **Bandwidth: 30 kHz at 60 W**
- **High power output: up to 100 W(rms)**
- **High output current: 7A (peak)**
- **Low IMD and THD**
- **Adjustable idling current**
- **Stability with resistive or reactive loads**
- **Single or split power supply**
  (30 to 75 V, single, ± 15 to ±37.5, split)
- **Class AB output stage (HC2500)**
  Class B output state (HC2000H)
- **Direct coupling to load**
- **Built-in load-line-limiting circuit to protect amplifiers from accidentally short-circuited output terminals (HC2000H)**
- **Reactive-load fault protection (HC2000H)**
- **Socket available**
- **Rugged package with heavy leads**
- **Light weight: 100 grams**

MAXIMUM RATINGS, *Absolute-Maximum Values:*

		HC2000H HC2500
V_S:		
Between leads 1 & 10		75 V
I_{OM}:		7 A
P_T:		
Per Output Device		See Figs. 3 & 4
T_{stg}		−55 to +125°C
T_J		−55 to +150°C
T_L (During Soldering):		
At distance ≥ 1/8 in. (3.17 mm) from case for 10 s max.		235°C
ϕL (Min):		
At distance ≥ 0.075 (1.91 mm) from case		0.04 in. (1.02 mm)

TERMINAL DESIGNATION

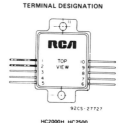

92CS-27727

HC2000H, HC2500

(See dimensional outline "AA".)

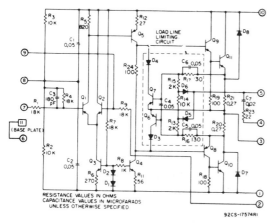

Fig. 1 — Schematic diagram of type HC2000H operational amplifier.

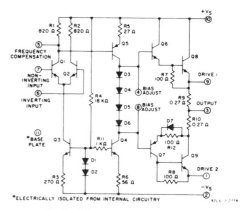

Fig. 2 — Schematic diagram of type HC2500 operational amplifier.

Fig. 15.16. Circuit diagrams and specifications of RCA types HC2000 and HC2500 hybrid power integrated circuits. (Courtesy of Radio Corporation of America)

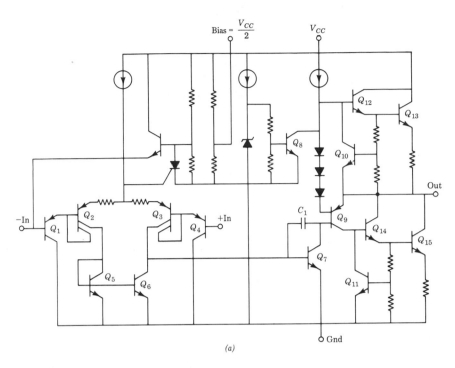

(a)

Electrical characteristics

$V_S = 20$ V, $T_{TAB} = 25°$C, $R_L = 8$ Ω, $A_V = 50$ (34 dB), unless otherwise specified.

Parameter	Conditions	Min.	Typ.	Max.	Units
Total supply current	$P_{OUT} = 0$ W		15	65	mA
	$P_{OUT} = 1.5$ W/channel		430		mA
dc output level			14		V
Supply voltage		10			V
Output power	T.H.D. = 5%		6		W
	T.H.D. = 10%	6	7		W
T.H.D.	$P_{OUT} = 1$ W/channel, $f = 1$ kHz		0.07	1	%
	$P_{OUT} = 4$ W/channel, $f = 1$ kHz		0.2		%
Offset voltage			15		mV
Input bias current			100		nA
Input impedance		3			MΩ
Open loop gain	$R_S = 0$ Ω	66	90		dB
Channel separation	$C_F = 250$ μF, $f = 1$ kHz	50	70		dB
Ripple rejection	$f = 120$ Hz, $C_F = 250$ μF		70		dB
Current limit			1.5		A
Slew rate			1.4		V/μs
Equivalent input noise voltage	$R_S = 600$ Ω, 100 Hz–10 kHz		3		μVrms

Note 1: For operation at ambient temperatures greater than 25°C the LM379 must be derated based on a maximum 150°C junction temperature using a thermal resistance which depends upon device mounting techniques.

(b)

Fig. 15.17. (a) Simplified circuit diagram. (b) Electrical characteristics of the LM379 dual 6 watt power amplifier. (Courtesy of the National Semiconductor Corporation)

are equal, and therefore their collector currents are equal under quiescent conditions. This is not unusual for differential transistors. However, while the input differential signal is causing the collector current of Q_2 and consequently the base currents of both Q_5 and Q_6 to increase, the collector current of Q_3 is decreasing. Both of these effects tend to reduce the potential of the base of Q_7. The effects have equal weight. Therefore, the voltage gain is doubled. In fact, Q_5 and Q_3 appear to be functioning as a balanced class A complementary-symmetry pair since one is driven on while the other is driven off.

3. The circuit elements in the upper center of Fig. 15.17a are designed to maintain the quiescent voltages at both the input and output terminals of the amplifier at $V_{CC}/2$ for any supply voltage ≥ 10 V. This circuit also provides $V_{CC}/2$ bias for external use.

4. The lateral p-n-p transistor which provides the complementary drive to Q_{14} and Q_{15} has a unique symbol to indicate that it is a *field-assisted* lateral p-n-p. This means that it has a *built in* electric field in the base, similar to the electric field in a conventional graded-base transistor. The main advantage to this field is the improved frequency response, or increased f_T, of the p-n-p transistor. This is an important advantage, since the low f_T of the lateral p-n-p would otherwise cause poor balance in the complementary-symmetry amplifier at high frequencies. It would also cause instability on negative half cycles at small closed-loop gains because of the inadequate spacing between the poles of the compensated differential amplifier and the bottom half of the output amplifier.

5. Transistors Q_{10} and Q_{11} limit the output current, and thus protect the chip, by shunting the base current of either Q_{12} or Q_{14} whenever the voltage drop across the emitter resistors of either Q_{13} or Q_{15} is sufficiently high to turn on either Q_{10} or Q_{11}.

External circuit terminals and recommended external components as well as additional data and specifications are available in the manufacturer's data sheets for these and other integrated power amplifier circuits.

15.7 VMOS POWER AMPLIFIERS

The conventional FETs and MOSFETs discussed in Chapter 8 do not have sufficiently large current capacities to be attractive for power amplifier use. This is because the short channel lengths and wide channel widths required to yield drain currents in the ampere range were not practical or economical.

A power MOSFET known as *VMOS*, which is short for *vertical MOSFET* has been developed. A cross-sectional view of the VMOS is shown in Fig. 15.18a, and a similar view of a conventional MOSFET is shown in Fig. 15.18b for comparison. The $n-$ doped substrate of the VMOS serves as a drain. A lightly-doped epitaxial layer also appears as part of the drain, but this layer becomes the depletion region between the drain and the channel when the drain potential is a few volts positive with respect to the body, as it is in normal operation. The $n+$ and $n-$ symbolism indicates only that the $n-$ material is lightly doped in comparison with the $n+$ material. The p doped body and $n+$ doped source as well as the $n+$ doped substrate (drain) are actually diffused into the lightly-doped epitaxial material. The purpose of the epitaxial material is to provide a comparatively wide, constant width depletion region which yields a high drain-to-source breakdown voltage and small drain-to-gate capacitance, since the gate overlaps the epitaxial region rather then the $n+$ drain. A V-shaped groove is etched through the $n+$ source and the p body into the epitaxial layer. Then a silicon dioxide insulating layer is formed in an oxidizing atmosphere and finally a layer of aluminum is deposited over the oxide in the groove to form the V-shaped metal gate. A positive potential on the gate with respect to the source, and the base region connected to it, induces an n channel on both sides of the V, and therefore the channel width is doubled. The channel length is very short because of the vertical construction. Thus the width-to-length ratio of the VMOS is large compared with the conventional MOSFET, as may be seen in Fig. 15.18.

The only commercially-available VMOS at this writing is commonly

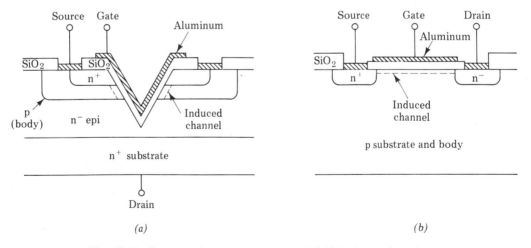

Fig. 15.18. Cross-sectional view of (a) a VMOS channel and (b) a conventional MOSFET.

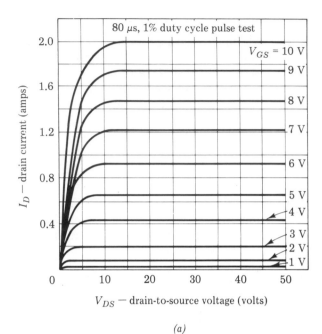

(a)

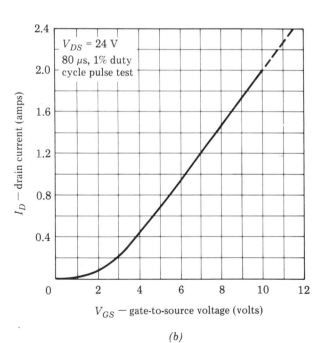

(b)

Fig. 15.19. (a) Drain characteristics and (b) transfer characteristics
of the VMP1 VMOS.

known as a VMP1. The drain and the transfer characteristics of this device are shown in Fig. 15.19. You may observe that the drain characteristics are unusually flat for $V_{DS} \geq V_T + V_{GS}$ where the threshold voltage V_T is about 1 V. This flatness results from the epitaxial layer which contains the depletion region, and thus isolates the channel from changes in drain voltage once pinchoff has been achieved. The other striking characteristic that may be observed from either the drain or transfer characteristics is the linearity of drain current with respect to V_{GS} for values of drain current above 400 mA. All the previously studied amplifying devices have exhibited either a square-law or exponential relationship between output current and input voltage. The cause of the square-law relationship in the conventioanl MOSFET or FET is the acceleration of the charge carriers as they pass through the channel. However, in the VMOS transistor, the electrons reach a limiting velocity known as velocity saturation when V_{GS} exceeds about 0.4 V because of the high velocities achieved in the short channel and the resulting high-loss collisions between the electrons and other particles in the channel. The transductance achieved in this linear region is about 260 mS, as determined from the characteristics given in Fig. 15.19.

The basic VMP1 is available in several different case styles and drain-voltage ratings which have either 2N- or VN- series designation. The 2N6656 to 2N6658 series uses the TO-3 case which is most suitable for audio-frequency power amplifiers because of its minimum thermal resistance, to be discussed in Section 15.8. The *absolute maximum ratings* of this series are given in addition to the case drawing in Fig. 15.20.

A VMOS audio amplifier capable of 30 watts sinusoidal power output is used as an example of VMOS amplifier design. The circuit diagram of this amplifier is shown in Fig. 15.21. The 2N6658 VMOS output transistors operate push-pull in the source-follower configuration and are driven by the common emitter transistors T_3 and T_4. These transistors receive their signals from the input differential amplifier which serves as a phase splitter.

The VMOS power amplifier does not appear to be a push-pull source follower because the lower transistor appears to be in the common source mode. This arrangement is not unlike the quasi-complementary-symmetry amplifier previously discussed. However, voltage drive instead of current drive is required for the VMOS, so complementary-symmetry bipolars do not serve well as drivers. Therefore, another scheme was devised to provide the same voltage gain and output resistance for the lower VMOS and its driver as for the upper source-follower-connected VMOS and its driver. This scheme is illustrated in Fig. 15.22 where the drive systems for both output transistors are shown.

The VMOS in Fig. 15.22a is seen to be operating as a conventional

TO-3

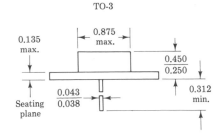

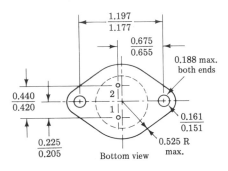

All dimensions in inches

Absolute Maximum Ratings

*Maximum drain-source voltage
2N6656 . . . . 35 V
2N6657 . . . . 60 V
2N6658 . . . . 90 V

*Maximum drain-gate voltage
2N6656 . . . . 35 V
2N6657 . . . . 60 V
2N6658 . . . . 90 V

*Maximum continuous drain current . . . . 2.0 A

Maximum pulsed drain current . . . . 3.0 A

*Maximum continuous forward gate current . . . . 2.0 mA[1]

*Maximum pulsed forward gate current . . . . 100 mA

*Maximum continuous reverse gate current . . . . 100 mA

*Maximum forward gate-source (zener) voltage . . . . 15 V

*Maximum reverse gate-source voltage . . . . 0.3 V

*Maximum dissipation at 25°C case temperature . . . . 25 W

*Linear derating factor . . . . 200 mW/°C

*Temperature (operating and storage) . . . . −55 to +150°C

*Lead temperature
(1/16″ from case for 10 sec.) . . . . 300°C

*Indicates JEDEC registered data

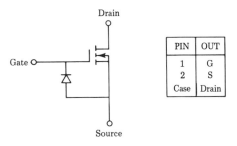

PIN	OUT
1	G
2	S
Case	Drain

Fig. 15.20. Case drawing and maximum ratings of the 2N6656 series VMOS.
(Courtesy of Siliconix Incorporated)

source follower. The voltage gain of the driving transistor is very nearly R_C/R_E and, using the unity feedback concept, the voltage gain of the VMOS source follower is $g_m R_L/(1 + g_m R_L)$. Therefore, the total reference voltage gain of this circuit is

$$K_V = \frac{g_m R_L R_C}{(1 + g_m R_L) R_E} \qquad (15.15)$$

The output resistance of the source follower is $1/g_m$, as discussed in Chapter 8.

The VMOS amplifier of Fig. 15.22b appears as a common-source amplifier with R_C acting as a feedback resistor. The voltage gain of the

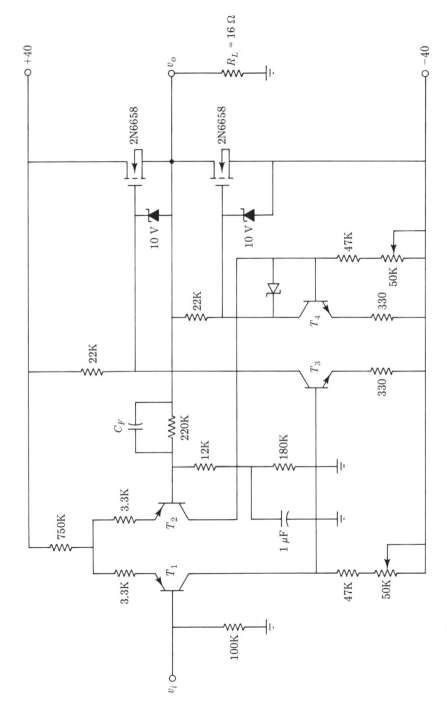

Fig. 15.21. Circuit diagram of a 30 watt VMOS amplifier.

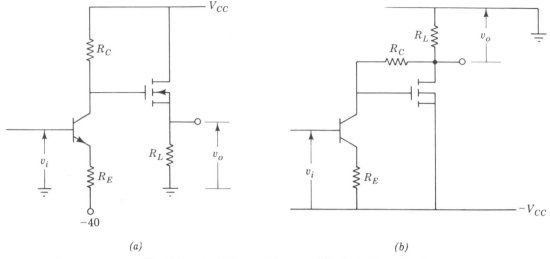

(a)   *(b)*

Fig. 15.22. VMOS amplifiers and their driving circuits:
(a) conventional source follower amplifier;
(b) common-source amplifier with feedback to the driver.

VMOS is $g_m R_L$ and the voltage gain of the driving transistor is approximately $R_C / R_E (1 + g_m R_L)$ because the load resistance of the driver is $R_C / (1 + g_m R_L)$ due to the Miller effect acting on R_C. Therefore, the total voltage gain of this circuit is

$$K_V = \frac{g_m R_L R_C}{R_E (1 + g_m R_L)} \qquad (15.16)$$

This voltage gain is identical to that of the source follower configuration, as seen by comparing Eqs. 15.15 and 15.16.

The output resistance of Fig. 15.22b is obtained by applying a voltage v_o to the drain terminal, as shown, and determining the current that flows into the drain. Both the output resistance of the driving transistors and the input resistance of the VMOS are very large in comparison with R_C, so essentially all of v_o is applied to the gate of the VMOS. Therefore, the drain current of the VMOS is essentially $g_m v_o$ and the output resistance is $v_o / v_o g_m = 1/g_m$, which is the same as the source follower.

Referring to Fig. 15.21, the 10 V reference diodes are connected between each gate and source of the VMOS transistors to protect the VMOS from excessive gate-source voltage. These zeners are actually built into the 2N6656 series, as shown in Fig. 15.20, to protect the VMOS from static charges as well as excessive driving voltages. The Schottky diode (Fig. 15.21) prevents the driving transistor T_4 from being driven appreciably into saturation. Because of diffusion capacitance

(Sec. 3.11), a saturated bipolar transistor cannot recover immediately to normal operation. Therefore, if T_4 were allowed to saturate, increased distortion would result at high frequencies where the recovery time may be an appreciable part of the period. The Schottky diode conducts with a small value of forward bias voltage compared with a silicon diode, and thus diverts current from the base of the transistor to its collector before the base-collector junction becomes appreciably forward biased, which is the condition for saturation. The 50 kΩ potentiometers in the collector circuits of T_1 and T_2 (Fig. 15.21) are used to adjust the bias of the VMOS transistors to projected cutoff (about 2 V) and to equalize the gains of the two halves of the amplifier. All of the other components and features of the amplifier have been discussed previously in other circuits.

The VMOS transistor is capable of amplifying very-high-frequency signals (several hundred MHz). Therefore, amplifiers designed for audio-frequency use may have *parasitic oscillations*. These oscillations are caused by connecting leads whose inductances resonate with the stray capacitances in the circuit at frequencies within the pass band of the amplifier, and thus cause high power gain at these frequencies. Feedback is also required for oscillation, but at these very high frequencies, adequate feedback is obtained through either C_{gd} or between connecting leads. These parasitic oscillations may be eliminated, if they occur, by adding resistance in series with the leads to the VMOS electrodes. The resistors should be connected to the VMOS terminals with minimum lead length. The gate resistor should be several hundred ohms, but the drain and source resistances may be only a few ohms. A few turns of insulated wire may be wound around the drain and source resistors and connected in parallel with them to eliminate the power loss of these resistors in the audio-frequency range. These resistors and coils are known as *parasitic suppressors*.

15.8 THERMAL CONDUCTION AND THERMAL RUNAWAY

As noted in Chapter 4, small-signal transistors rely on air convection and connecting-lead conduction to remove heat from the transistor to the surrounding atmosphere. Therefore, the heat dissipation rating is based on the ambient temperature, usually 25°C, and this rating must be reduced as the ambient temperature is increased above the reference temperature. You may recall that the reciprocal of the derating factor is known as *thermal resistance*.

Power transistors (above about 1 watt) are usually fastened securely to a large metal plate or chassis which serves as a heat sink. The collector of the power transistor is usually connected electrically and mechanically to the transistor case. Maximum thermal conductivity is obtained when the transistor case is electically and mechanically fastened to the heat sink. However, an insulator is usually required to

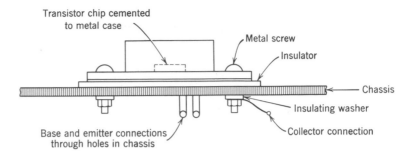

Fig. 15.23. Heat sink arrangement for a typical power transistor.

electrically isolate the collector from the heat sink, as shown in Fig. 15.23.

Every transistor has a maximum permissible junction temperature $T_{j\max}$ which ranges from about 85°C to 110°C for germanium and about 130°C to 175°C for silicon. Thermal resistance θ_T is defined as the ratio of temperature rise in degrees centigrade (or Kelvin) to the power conducted in watts. Therefore, the junction temperature can be related to the power being dissipated P_d, the thermal resistance θ_T, and the ambient temperature T_a by the following equation, previously given in Chapter 7.

$$T_j = \theta_T P_d + T_a \tag{15.17}$$

The thermal resistance θ_T consists of three parts:

1. The thermal resistance between the collector junction and the transistor case θ_{jc}.

2. The thermal resistance between the transistor case and the heat sink θ_{cs}.

3. The thermal resistance between the heat sink and the ambient surroundings θ_{sa}.

Heat conduction is comparable to electrical conduction. Thus, the equivalent circuit of Fig. 15.24 represents the *thermal relationships* of the transistor and its heat sink. The capacitors shown represent the thermal capacitance of the three parts of the circuit. The thermal capacitance C_j of the collector junction is very small because the mass of the transistor chip, and hence its heat capacity, is very small. Therefore, the thermal time constant $\theta_{jc} C_j$ is of the order of milliseconds. The significance of this time constant is that the load line can pass through the area

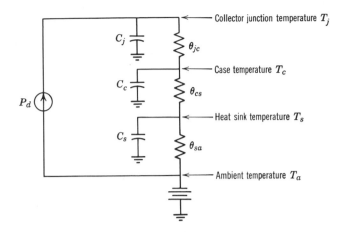

Fig. 15.24. An equivalent thermal circuit.

above the maximum dissipation curve, providing the *excessive dissipation does not continue for more than a few milliseconds.* This situation may occur in a class B or AB amplifier. The thermal capacitance C_c of the case is much greater than that of the junction. Therefore, a transistor amplifier which is designed to operate with a heat sink can operate at least several seconds without the heat sink before the transistor is damaged. Similarly, the thermal capacity of the heat sink is usually much greater than that of the transistor case, so the amplifier may operate for several minutes with an inadequate heat sink.

The thermal resistance θ_{jc} from the collector junction to the case is usually given by the manufacturer for each type of power transistor. Also, thermal resistance θ_{cs} data are usually available for the various transistor mounting systems.[1] The thermal resistance θ_{sa} of a ⅛-thick sheet of bright aluminum is given as a function of area (both sides) in Fig. 15.25. When the heat sink is horizontal, the thermal resistance increases by about 10 percent because of the reduced convection. On the other hand, black painting or anodizing of the heat sink will lower its thermal resistance because of increased heat radiation.

We can now determine the dissipation capability of a power transistor with a specific heat sink system by applying "Ohm's law" for thermal cirucits wherein power dissipation P_d is comparable to current, temperature difference is analogous to potential difference, or voltage, and thermal resistance compares with electrical resistance. The following example illustrates this procedure.

---

[1] See *Motorola Power Transistor Handbook*, First Edition, Motorola Semiconductor Products Division, Phoenix, Arizona, p. 23.

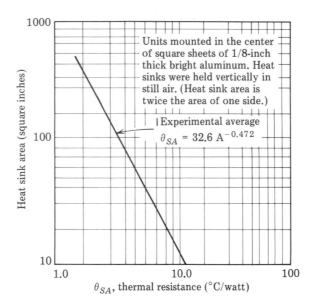

Fig. 15.25. Heat sink versus thermal resistance. (Courtesy of the Motorola
Semiconductor Products Division)

Example 15.4

Let us consider a transistor with $\theta_{jc} = 1°C/W$ mounted on a ⅛-inch aluminum 10 inch by 10 inch sheet. A mica washer coated with silicone grease is used to insulate the transistor from the chassis. The thermal resistance of this washer is 0.5°C/W. The thermal resistance of the aluminum heat sink is $\theta_{sa} = 2.5°C/W$, as seen in Fig. 15.25, if mounted vertically, and 2.75°C/W if mounted horizontally. The total thermal resistance from junction to ambient is $\theta_T = 1.0 + 0.5 + 2.5 = 4°C/W$ with vertical heat sink. If the maximum junction temperature of the transistor is 175°C and the ambient temperature is 25°C, the maximum permissible transistor dissipation is $P_{dmax} = (175 - 25)/4 = 37.5$ W. However, if the maximum expected ambient temperature is 75°C, the maximum safe transistor dissipation is $P_{dmax} = (175 - 75)/4 = 25$ W. The maximum power output from this transistor in a class A amplifier with $T_a = 75°C$ is $P_o = 25/2 = 12.5$ W. However, if two of these transistors are mounted on the same heat sink with the same type washers, the total thermal resistance is $\theta_T = (1.5/2) + 2.5 = 3.25°C/W$. The maximum dissipation of both transistors at $T_a = 75°C$ is $P_{dmax} = 100/3.25 \approx 30$ W. The maximum sinusoidal power output from these two transistors when operated class B at 75°C ambient temperature is $P_o = 2.5 \times 30 = 75$ W, as can be noted in Fig. 15.7a.

A power transistor is often given a power dissipation rating with the case temperature held at 25°C. For example, the transistor in the example above can dissipate $(175 - 25)/1 = 150$ W at 25°C case temperature. This case temperature can be maintained by cooling the cases with circulating water. Normally, this rating is intended to indicate only the thermal resistance between junction and case. For example, if a transistor is rated as having $T_{jmax} = 100$°C and $P_{dmax} = 150$ W at 25°C case temperature, we may calculate that $\theta_{jc} = (100 - 25)/150 = 0.5$°C/W.

A bipolar transistor is not completely protected when the heat sink is adequate for the required dissipation; the transistor can be destroyed by *thermal runaway*. This thermal runaway results from the increasing collector current caused by the increasing I_{CO}. You may recall from Chapter 6 that $\Delta I_C = S_I \Delta I_{CO}$, where S_I is the current stability factor. We were not concerned about thermal runaway in the RC-coupled amplifier because the dc load line did not cross the maximum dissipation curve, so the transistor could not be destroyed by heat. In fact, the power dissipation decreases with increasing collector current when the q-point is to the left of the load line center. However, in a transformer-coupled or complementary-symmetry amplifier, the dc load line is almost vertical, as shown in Fig. 15.26. Therefore, the dc collector voltage V_{CE} is almost independent of the collector current, at the value V_{CC}, and

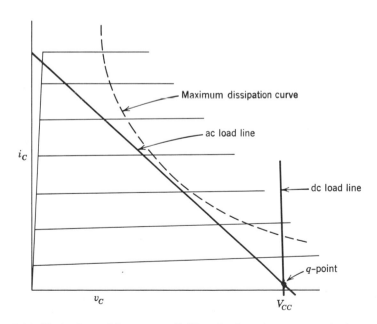

Fig. 15.26. *Illustration of the susceptibility of a transformer-coupled amplifier or complementary-symmetry amplifier to thermal runaway.*

the power dissipation is essentially proportional to the average collector current.

Thermal runaway occurs when the rate of increase of collector junction temperature exceeds the ability of the heat sink to remove the heat. The portion of heat dissipation which results from I_{CO} is

$$P_d = S_I I_{CO} V_{CC} \qquad (15.18)$$

Then the rate of increase of this power dissipation with temperature is

$$\frac{dP_d}{dT} = S_I V_{CC} \frac{dI_{CO}}{dT} \qquad (15.19)$$

The rate of heat conduction away from the collector junction to the ambient surroundings is the reciprocal of the thermal resistance, or $1/\theta_T$. Thus, in order to prevent a continual build-up of heat, the following inequality must hold,

$$\frac{1}{\theta_T} > S_I V_{CC} \frac{dI_{CO}}{dT} \qquad (15.20)$$

If I_{CO} doubles for each 10°C temperature increment, the I_{CO} increases by about 7 percent per °C. Then,

$$\frac{dI_{CO}}{dT} = 0.07 \, I_{CO} \qquad (15.21)$$

and

$$\frac{1}{\theta_T} > S_I V_{CC} (0.07 \, I_{CO}) \qquad (15.22)$$

Note that the current stability factor S_I must be controlled in order to prevent thermal runaway. Using Eq. 15.22 to obtain S_I explicitly,

$$S_I < \frac{14.3}{\theta_T V_{CC} I_{CO}} \qquad (15.23)$$

The thermal current I_{CO} must be determined at the maximum expected junction temperature.

Example 15.5

Let us consider a class B amplifier with $V_{CC} = 16$ V, $R_L = 8 \, \Omega$ and $I_{CO} = 0.1$ mA at 25°C, and doubles for each 10°C increase. Then $I_A = 16/8 = 2.0$ A and the maximum dissipation occurs when $I_{Cmax} = 0.636 \, I_A = 1.27$ A and $I_{ave} = 0.636 \, (1.27) = 0.81$ A. Therefore, the maximum dc power input is $V_{CC} I_{ave} = 16 \times 0.81 = 13$ W and the maximum dissipation is $13/2 = 6.5$ W since the amplifier is 50 percent efficient at this signal level (Fig. 15.7). We will assume that θ_T is $4.0°C/W$ as in the preceding example and that the maximum ambient temperature is 40°C. Then, from Eq. 15.17, $T_{jmax} = 6.5 \times 4 + 40 = 66°C$ and I_{CO} at this temperature is

$$I_{CO} = 0.1 \text{ mA } (2)^{(66-25)/10} = 0.1 \text{ mA } (2)^{4.1} \simeq 1.8 \text{ mA}$$

The thermal-stability factor must be less than $S_I = 14.3 / (4 \times 16 \times .0018) = 123$ to ensure that thermal runaway will not occur.

You may recall from Chapter 7 that the increasing channel resistance of an FET with temperature tends to decrease the drain current as the temperature increases. This negative temperature coefficient eliminates the problem of thermal runaway in VMOS amplifiers.

**15.9
CONTROL
CIRCUITS FOR
POWER
AMPLIFIERS**

Provision must be made for selecting the input signal and controlling the output level of the amplifier. These controls should be inserted at about the 1 V signal level for the following reasons:

1. Potentiometers and switches are often noisier than fixed resistors, and their connecting leads are usually long because the control is usually mounted on the front panel. Therefore, the signal level needs to be fairly high at their insertion point in order to avoid degrading the signal-to-noise ratio.

2. All signals ahead of the level, or *volume*, control must be of sufficiently low level to avoid distortion or clipping in these circuits.

A typical switching and level control circuit is shown in Fig. 15.27a. The phono and tape inputs normally come from the outputs of preamplifiers. The capacitors C_1 and C_3 are used to prevent dc from flowing through the potentiometer P because dc makes the control noisy. These blocking capacitors may be included in the preamps and the main amplifier. The components R and C_2 are not essential to the level control but are often used to compensate for the characteristics of the ear in music systems designed for homes where the desired sound levels are usually much lower than the levels in the concert hall or recording studio. Under these conditions, the ear perceives a reduced bass, or low frequency, response because of the lower sensitivity of the ear for low frequencies at low listening levels. This ear characteristic is illustrated by the constant *perceived-loudness* contours, shown in Fig. 15.27b. These curves are often known as *Fletcher-Munson* curves in honor of

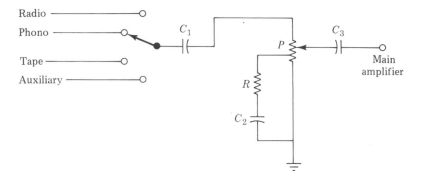

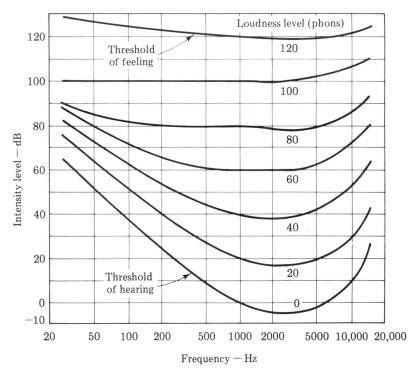

Fig. 15.27. (a) Typical input-switching and level control circuit.
(b) Equal-loudness contours of the ear.

their developers. The potentiometer P, which normally has a logarithmic or *audio* taper, is tapped at about the midpoint of its resistance.
The resistance R is then chosen to be about 10 percent of the potentiometer resistance between the tap point and ground. Finally, C_2 is
chosen so that $X_{C2} = R$ at about 300 Hz. This type of level, or *volume*,

control is often known as a *loudness* control. Sometimes a second potentiometer, having about the same resistance value as *P*, is connected in parallel with C_2 to control the amount of low-frequency compensation.

Problems *Section 15.1*

15.1   The transistor whose characteristics are given in Fig. 15.1 is to be operated with $V_{CC} = 40$ V and R_L (connected from collector to V_{CC}) = 20 Ω. Determine the maximum value of ac power output, collector power input, and collector efficiency.

15.2   A circuit is connected as shown in Fig. 15.2. The characteristics of the transistor are given in Fig. 15.1. The maximum collector voltage is 40 V and the ac impedance of the transformer primary is 20 Ω. Determine the maximum value of ac power output, collector power input, and collector efficiency, neglecting the ohmic resistance of the transformer.

Section 15.2

15.3   The transfer characteristic for a 2N2147 power transistor is given in Fig. 15.28. Determine the projected-cutoff bias for this transistor.

Section 15.3

15.4   Two transistors with a total power dissipation capability of 15 W are used in the class *B* push-pull amplifier with $I_{Cq} = 50$

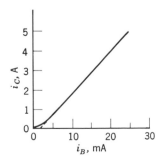

Fig. 15.28. Current transfer characteristics for the 2N 2147.

mA and $V_{CC} = 25$ V. Determine the minimum permissible value of ac load resistance and the maximum power output for this amplifier, assuming sinusoidal signals. Neglect the transistor saturation voltage.

15.5   Determine the power output and the collector circuit efficiency of the amplifier of Example 15.2 without neglecting the transistor saturation voltage.

Section 15.4

15.6   Repeat Example 15.3 but let $R_L = 8$ ohms. Assume that the β_o of your transistors is 75. Find the values of all the resistors in the circuit and the required transistor characteristics.

Section 15.5

15.7   Use the circuit of Fig. 15.15 to design an amplifier. Assume that $R_L = 8$ Ω, $2V_{CC} = 32$ V, all transistors have $h_{fe} = 100$ over their required current range, and the distortion of the amplifier at maximum power output is ten percent when signal feedback is *not* used. Devise a feedback circuit that will reduce the distortion to one percent and determine all the component values, the maximum sinusoidal power output, neglecting saturation voltage, and the required input voltage assuming $f_\ell = 30$ Hz.

Section 15.6

15.8   The LM379 integrated amplifier has 90 dB open-loop voltage gain, 10 MHz gain bandwidth product, and is internally compensated for stable operation at closed-loop voltage gains down to 10, as previously stated.

a. Would this amplifier be suitable for amplifying the audio signal from a microphone having a -54 dB output rating with reference to 1 V. Open circuit? Why?

b. Would this amplifier be suitable for amplifying the signal from a magnetic phonograph pickup having 10 mV output signal in the midfrequency range? Why?

Section 15.7

15.9   Select suitable bipolar transistors for the circuit of Fig. 15.21 and check the appropriateness of the component values listed. Pay particular attention to the voltage rating of transistor T_3. Verify the power-output rating of the amplifier.

Section 15.8

15.10   If the supply voltage V_{CC} is raised to 20 V and the thermal resistance θ_T is increased to 5°C/W, all other conditions remaining the same as in Example 15.5, determine the maximum value of the stability factor S_I to ensure thermal stability.

Section 15.9

15.11   A given potentiometer P has 100 kΩ resistance and is tapped at the center point. Determine suitable values for R and C_2 and sketch the expected frequency response at the output of the potentiometer when the slider is at the tap point, assuming negligible load on the potentiometer. Which loudness contour does this setting most nearly match? Does the control adequately compensate the ear characteristics for very low listening levels?

15.12   Design a transistor power amplifier using an emitter-follower type complementary-symmetry arrangement which will provide 20 W power output to an 8 Ω speaker. The driving source is a radio tuner with 0.5 V rms output voltage, open circuit, and 2.0 kΩ internal resistance R_s. The distortion should not exceed 0.5 percent. Use transistors of your choice, including the characteristics given in Appendix A. Draw a circuit diagram. Calculate the maximum permissible stability factor using the heat sink of your choice.

15.13   Use the common-emitter, complementary-symmetry arrangement for the ouptut stage instead of the emitter-follower arrangement in the design of the amplifier of Problem 15.12.

15.14   Design the amplifier of Problem 15.12 using a quasi-complementary-symmetry arrangement for the power amplifier instead of the complementary-symmetry configuration.

15.15   Imagine that you are hired to design the audio amplifier for a new-model TV set. The desired power output is about 2 W, the available signal is 0.5 V rms, the loudspeaker has 3.2 Ω resistance, and the distortion should not exceed 5 percent. Cost and reliability are important factors. A bandwidth of 100 Hz to 10 kHz is adequate. Select the component types, assuming the cost of bipolar and FET types are about the same, and design the amplifier. Specify the heat sink area or thermal resistance.

15.16   Design a high-fidelity 40 W amplifier using VMOS in the output with an 8 Ω loudspeaker as a load. Two or more VMOS may be paralleled if needed. The input voltage is assumed to be 0.5 V rms. Specify the required heat sink thermal resistance or area.

16

Power Supplies

Essentially all of the electronic devices discussed in the preceding chapters of this book require a dc power source for their operation. Up to this point, little attention has been given to the problem of obtaining this dc power. In fact, the inference may have been made that one or more batteries will supply adequate power for any device. This is possible, but in most instances the battery is neither the most convenient nor the most economical means of obtaining dc power. It is usually much more convenient and economical to obtain dc power from the ac power line by the use of rectifiers and appropriate filters. Some of the commonly used rectifying and filtering systems are discussed in this chapter.

**16.1
POWER SUPPLY
CHARACTER-
ISTICS**

The requirements of a power supply differ widely among the various electronic devices. The primary characteristics which need to be considered in the design of a power supply follow.

1. The dc voltage or voltages required by the device, which is known as the load on the supply, is of primary importance.

2. The power supply must be able to furnish the maximum current requirement of the load.

3. The variation of the dc output voltage with change in load current may be important. The voltage regulation of a power supply is defined as the change in output voltage, when the current is changed from no load to full load, divided by the full load voltage and expressed as a percent, as shown in the following equation.

447

$$\% \text{ regulation} = \frac{\text{no load voltage} - \text{full load voltage}}{\text{full load voltage}} \times 100 \quad (16.1)$$

4. The rapid variations of the output voltage which result from imperfect filtering must be considered. These voltage variations have a fundamental frequency which is related to the ac power line frequency and are called ripple voltage, or simply *ripple*. The ripple is defined as the ratio of the rms value of the ripple voltage to full load dc voltage. This ripple is usually expressed in percent as indicated below.

$$\% \text{ ripple} = \frac{\text{rms ripple voltage}}{\text{full load dc voltage}} \times 100 \quad (16.2)$$

A commonly encountered term in power supply design is the *peak-to-peak ripple ratio (pprr)* which is defined as $(V_{omax} - V_{omin}) / V_{oave}$. Power supply characteristics in addition to those listed are also considered in power supply design. These characteristics will be considered as the need arises.

16.2
CAPACITOR
INPUT FILTERS

Some basic rectifier circuits were considered in Chapter 3. These circuits included both half-wave and full-wave rectifiers. Also, the capacitor was considered as a filtering element in Chapter 3. Whenever the filter capacitor immediately follows the rectifier, the filter is known as a capacitor input filter. Three basic rectifier circuits are shown in Fig. 16.1. These circuits were introduced in Chapter 3.

Equations that express the approximate relationships between peak input voltage, average output voltage, peak-to-peak ripple ratio, load resistance, and filter capacitance were developed in Chapter 3. However, those equations were based on the following assumptions.

1. The peak-to-peak ripple ratio (pprr) is small (about 10 percent or less).

2. The resistance of the transformer windings is negligible.

3. The ohmic, or bulk, resistance of the rectifier diodes is negligible.

Unfortunately, these constraints are not always met in actual practice. In fact, the *iR* drops in the transformer and the rectifiers may be significant, even though the resistances of these elements are small, because the peak currents may be very high in comparison with the average current in the load. Therefore, J. Phillip Stringham, a systems design

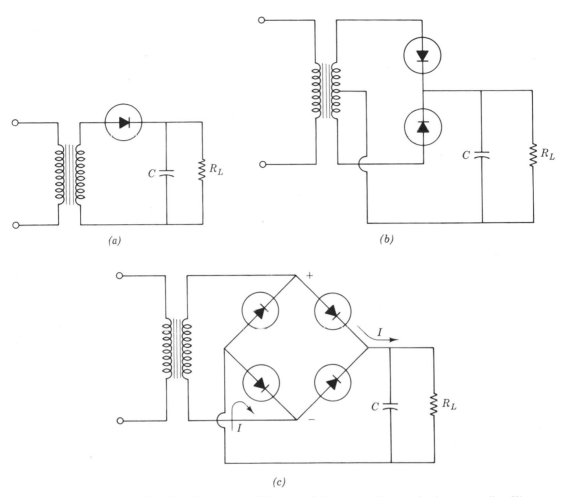

Fig. 16.1. *Typical rectifier circuits incorporating a single capacitor filter:*
(a) half-wave rectifier; (b) full-wave rectifier using a tapped transformer;
(c) bridge rectifier circuit.

engineer, developed a computer program and obtained a solution for
the full-wave rectifier with a capacitive filter which is *not* based on the
above assumptions. Stringham prepared a monograph from his solution, which provides the needed relationships among the parameters
with accuracy limited only by the tolerances of the components and the
preciseness of reading the nomograph. The nomograph parameters are
illustrated in Fig. 16.2, and the nomograph is given in Fig. 16.3. A
second nomograph prepared by Stringham, which can be used to determine the peak diode currents, is given in Fig. 16.4.

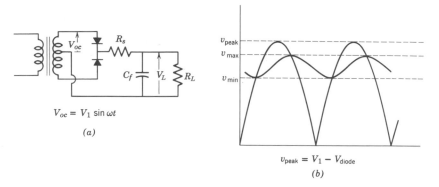

$$V_{oc} = V_1 \sin \omega t$$

(a)

$$v_{peak} = V_1 - V_{diode}$$

(b)

Fig. 16.2. Full-wave rectifier circuit: (a) circuit diagram; (b) voltages defined.

16.3
CIRCUIT AND
VOLTAGES FOR
THE FULL-WAVE
RECTIFIER

Figure 16.2a shows the circuit diagram of the full-wave rectifier with the open circuit (no load) voltage V_{OC} across one-half of the secondary. The resistance R_s includes the dynamic resistance of the diode plus the ac resistance looking into one-half of the secondary of the transformer. The value of R_s could be experimentally determined by disconnecting the filter capacitor and plotting the voltage V_L across R_L as a function of $I_L = V_L / R_L$. Then R_s is the slope of the curve, $\Delta V_L / \Delta I_L$. Typical values of R_s for some representative power supply ratings are given in Table 16.1. This source resistance is included in the nomograph of Fig. 16.3 by *defining* the parameter R_{eq} (which is actually a resistance ratio) as

$$R_{eq} = R_L / R_s \tag{16.3}$$

Also, this nomograph can be used for any primary power frequency f by *defining* C_{eq} as

$$C_{eq} = R_s C_f (f / 60) \tag{16.4}$$

The nomograph gives v_{Lmax} and v_{Lmin} as percentages of V_{Lpeak}, where V_{Lpeak} is the peak voltage across the load when the load resistance is very high, or when R_L approaches infinity. Thus V_{Lpeak} can be obtained from the relationship

Table 16.1. Typical R_s as a function of secondary voltage and current ratings.

	10 V	25 V	50 V	100 V	200 V
0.02 A	70.0 Ω	120.0 Ω	220.0 Ω	420.0 Ω	820.0 Ω
0.2 A	7.0 Ω	12.0 Ω	22.0 Ω	42.0 Ω	82.0 Ω
1.0 A	1.4 Ω	2.4 Ω	4.4 Ω	8.4 Ω	16.4 Ω
2.0 A	0.7 Ω	1.2 Ω	2.2 Ω	4.2 Ω	8.2 Ω
5.0 A	0.3 Ω	0.5 Ω	0.9 Ω	1.7 Ω	3.3 Ω

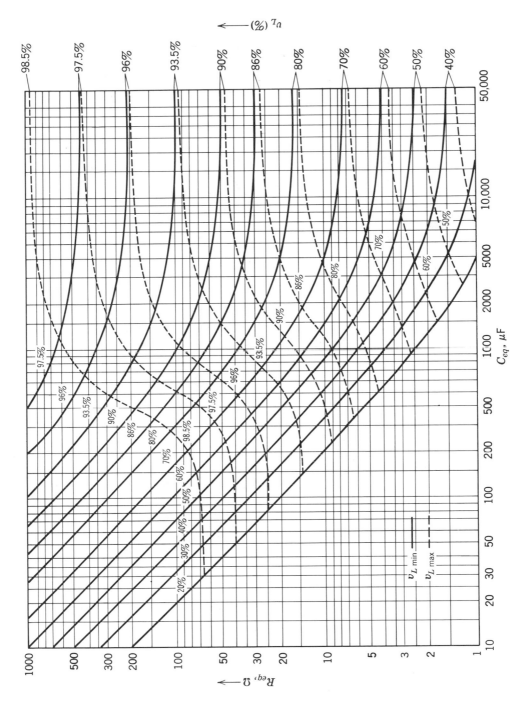

Fig. 16.3. Nomograph to determine the circuit parameters of Fig 16.2.

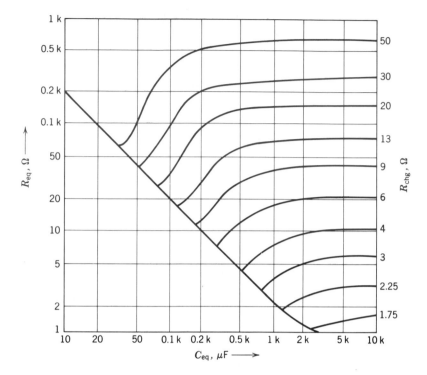

Fig. 16.4. Nomograph to determine the peak diode currents.

$$V_{Lpeak} = V_1 - V_{diode} \tag{16.5}$$

where V_{diode} is the forward voltage drop across the diode when the diode current is small (perhaps 10 percent of the average value).

Finally, after the parameters R_{eq} and C_{eq} have been determined (Eqs. 16.3 and 16.4), a *defined* parameter known as R_{chg} can be found from the nomograph given in Fig. 16.4. Then the peak diode current I_{peak} can be determined from the following equation.

$$I_{peak} = \frac{V_{Lpeak}}{R_s R_{chg}} \tag{16.6}$$

The proper use of the nomographs is illustrated by the following examples.

Example 16.1

We want to determine the minimum output voltage v_L, the peak-to-peak ripple voltage, and the peak diode current of a full-wave circuit such as Fig. 16.2a when $V_{OC}(\frac{1}{2}$ secondary$) = 35$ V, $R_L = 10$ Ω, $R_s = 0.35$ Ω, $C_f = 5{,}000$ μF, $f = 60$ Hz, and $V_{diode} = 0.5$ V.

1. $R_{eq} = R_L / R_s = 10/0.35 = 28.6$ (from Eq. 16.3), $C_{eq} = R_s C_f f / 60 = 0.35 \times 5{,}000 \times 60/60 = 1750 \ \mu F$ (Eq. 16.4).

2. Using these values of R_{eq} and C_{eq}, read v_{Lmax} (%) and v_{Lmin} (%) from Fig. 16.3. Then compute the V_{peak} and the minimum and maximum load voltages.

$$v_{Lmax} \ (\%) \simeq 90\% \ \text{(dashed line)}$$

$$v_{Lmin} \ (\%) \simeq 80.5\% \ \text{(solid line)}$$

$$V_{peak} = (1.414 \times 35 - 0.5) = 49 \ V$$

$$v_{Lmax} = 49 \times 0.90 = 44.1 \ V$$

$$v_{Lmin} = 49 \times 0.805 = 39.4 \ V$$

Peak-to-peak ripple voltage $= 44.1 - 39.4 = 4.7$ V.

3. The parameter R_{chg} can be obtained from Fig. 16.4.

$$R_{chg} \simeq 8.5 \ \Omega$$

$$I_{peak} = V_{peak} / R_s R_{chg} - 49/0.35 \times 8.5 = 16.5 \ A$$

Example 16.2

The transformer and diodes of Example 16.1 are used in the circuit of Fig. 16.2a, but the frequency is increased to 400 Hz and the load resistance is increased to 12 Ω. A voltage regulator will follow the rectifier and filter. The voltage input to the regulator must not fall below 41 V. We need to determine the required filter capacitance C_f and the peak diode currents I_{peak}.

1. $R_{eq} = R_L / R_s = 12/0.35 = 34 \ \Omega$

   $v_{Lmin} \ (\%) = 41 \times 100/49 = 83.6\%$

   $C_{eq} \simeq 2500 \ \mu F$ (Fig. 16.3)

   $C_f = C_{eq} \times 60/R_s f = 2500 \times 60/0.35 \times 400 = 1070 \ \mu F$ (Eq. 16.4)

2. Using Fig. 16.4,

   $R_{chg} \simeq 8 \ \Omega$

   $I_{peak} = V_{Lpeak} / R_s R_{chg} = 49/0.35 \times 8 = 17.5 \ A$ (Eq. 16.6)

The nomographs can be used for the bridge rectifier, but R_s includes the dynamic resistance of two diodes in series and $V_{peak} = V_1 - 2V_{diode}$.

The power supply discussed above has rather poor regulation unless $R_L C \gg T$ because the output voltage is a strong function of the rate of discharge of the filter capacitor through the load. This drop adds to the diode drop and the voltage drop in the transformer windings. A sketch of V_o as a function of I_L is given in Fig. 16.5 for two different values of capacitance in a typical rectifier circuit.

Power supply *output resistance* is a common and useful term to express the relationship between output voltage and load current. This dynamic output resistance is defined as

$$r_o = \frac{\Delta V_O}{\Delta I_L} \tag{16.7}$$

For example, the power supply that has the output characteristics given in Fig. 16.5 has an average value of output resistance $r_o = 9 \text{ V} / 1 \text{ A} = 9$ Ω when the filter capacitance is 1,000 μF.

A full-wave *bridge* rectifier was discussed in Chapter 3 and is shown in Fig. 16.6 for convenience. Observe that this rectifier does not require a center-tapped transformer. Therefore, the transformer is required only for voltage transformation and may be eliminated if this function is not required. Note that the conductor C charges to the maximum value of the *full secondary* voltage minus the drop across two forward-biased diodes. Therefore, the bridge rectifier provides twice as much dc output voltage for the same full secondary voltage as the circuit of Fig. 16.2. The peak inverse voltage across the diodes is the same in both circuits, however, so the ratio of output voltage to peak inverse voltage is twice

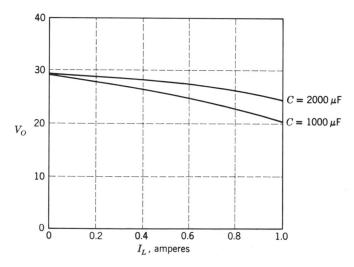

Fig. 16.5. Output voltages as a function of output current for two different values of filter capacitance.

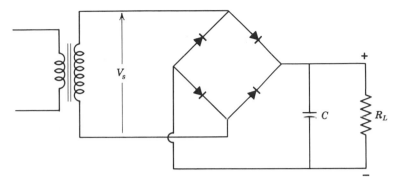

Fig. 16.6. Full-wave bridge rectifier circuit.

as high in the bridge circuit. The bridge rectifier has the disadvantage of requiring four diodes, and in very low voltage circuits the rectifier efficiency may be seriously reduced because two diodes are in series with the load.

6.4 EMITTER-FOLLOWER REGULATORS

Reference diodes (Zener diodes) were discussed in Chapter 3, and a simple circuit that will maintain a fairly constant voltage across a load resistor was analyzed. A circuit of this type can be used to improve the voltage regulation and reduce the ripple in a power supply. However, the load currents usually encountered require high-power dissipation capability in the reference diode, and the power supply efficiency is low because of the power loss in this diode. One or more transistors can be used in conjunction with the reference diode, however, to greatly increase the efficiency of the regulator by reducing the current through the reference diode. A typical circuit of this type is shown in Fig. 16.7. This circuit is known as an *emitter-follower regulator* because the output voltage follows (is nearly equal to) the reference voltage, the difference being the base emitter voltage V_{BE}.

The emitter-follower regulator operates in the same manner as the simple Zener regulator except that the currents through the reference diode are reduced by the factor $(\beta + 1)$ because of the transistor. To illustrate this behavior, let us assume both the input voltage and the reference-diode voltage to be constant; then the current I through R_B is constant. A reduction in load current I_L will reduce the base current I_B since $I_B = I_L/(h_{FE} + 1)$, but the reference diode current I_D will increase because the base voltage will tend to rise. Thus the sum of the currents $(I_B + I_D)$ is essentially constant. The diode voltage does change by the amount $\Delta V_B = \Delta I_D r_d$ where r_d is the dynamic resistance of the diode, discussed in Chapter 3. The load voltage V_o increases more than the base voltage because of the reduced V_{BE}. If the change in load current

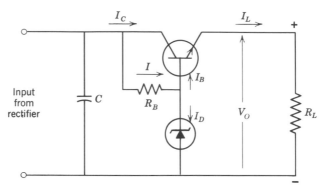

Fig. 16.7. Emitter-follower regulator.

is small so that r_d and r_e are essentially constant over the current range, the change in output voltage is

$$\Delta V_O = \Delta I_L r_e + \Delta I_D r_d \qquad (16.8)$$

But $\Delta I_D \simeq \Delta I_B = \Delta I_L / (h_{FE} + 1)$, so

$$\Delta V_O \simeq \Delta I_L \left(r_e + \frac{r_d}{h_{FE} + 1} \right) \qquad (16.9)$$

and the output resistance of the regulator is

$$r_o = \frac{\Delta V_O}{\Delta I_L} \simeq r_e + \frac{r_d}{h_{FE} + 1} \qquad (16.10)$$

Of course, if the input voltage does not remain constant, the current I through R_B changes with input voltage, and this current change must be absorbed by the reference diode.

Example 16.3

An example will illustrate the design of an emitter-follower regulator. Let us assume that the required output voltage is 20 V and the maximum load current is 1.0 A. Let us also assume that the chosen transistor is silicon with $h_{FE} = 50$ and $V_{BE} = 0.6$ V at $I_E = 1.0$ A. The reference diode current is minimum when the load current is maximum, and we shall choose this minimum diode current to be 5 mA. Then, the reference diode voltage should be $(V_O + V_{BE}) = 20.6$ V at $I_D = 5$ mA. We must choose the minimum voltage that can be tolerated across R_B, keeping in mind that the higher this voltage is, the higher the required input voltage and hence the lower the efficiency of the power supply. However, we shall see shortly that low values of R_B, which result from choosing a low minimum voltage, result in high diode dissipation and inferior regulator characteristics. In this example, we will choose 2 V as the minimum voltage, V_{Rmin}, across R_B. Now the value of R_B can be calculated from the following relationship.

$$R_B = \frac{V_{Rmin}}{I_{Dmin} + I_{Lmax}/(h_{FE} + 1)} \tag{16.11}$$

For this example, $R_B = 2/(5 \text{ mA} + 20 \text{ mA}) = 80 \; \Omega$.

The minimum voltage input to the regulator is $20.6 + 2 = 22.6$ V. This is the voltage v_{Omin} shown in Fig. 16.3 as v_{Lmin}.

The required value of filter capacitance and the voltage rating of the transformer can be determined from Fig. 16.3 after we have chosen the current rating of the transformer secondary and determined R_{eq}. If the load on the power supply is expected to draw near 1.0 A dc for long periods of time, then the transformer secondary must have a current rating greater than 1.0 A because the high peak currents that flow in the transformer windings cause higher power dissipation than sinusoidal currents of the same average value. Therefore, a generous safety factor (perhaps 30 percent) must be allowed for the current rating of the transformer. On the other hand, a 1.0 A rating is adequate if the load draws 1.0 A only occasionally. Audio-frequency amplifiers have this latter characteristic and we shall assume our load to be of that type. Then, from Table 16.1, assuming the secondary voltage rating to be about 25 V, $R_s \simeq 2.4 \; \Omega$, and from Eq. 16.3, $R_{eq} = 20 \; \Omega/2.4 \; \Omega \simeq 8$. We can now move to the right along the $R_{eq} - 8$ line on Fig. 16.3. Observe that as we move to the right, the ripple decreases and the capacitance C_{eq} increases. Let us assume that a 10 percent difference between v_{Lmax} and v_{Lmin} is satisfactory. This difference is obtained approximately at $C_{eq} = 5,000 \; \mu F$ where $v_{Lmin} = 67$ percent and $v_{Lmax} = 76$ percent, and the actual difference is 9 percent. Thus, using Eq. 16.4, the actual filter capacitance $C_f = C_{eq}/R_s = 5,000/2.4 \simeq 2,000 \; \mu F$. The maximum *open-circuit* voltage V_{peak} out of the rectifier is therefore $v_{Lmin}/0.67 = 22.6/0.67 = 33.7$ V, and we must add approximately 1.0 V for the drop across the two rectifiers giving 34.7 as the peak open-circuit voltage across the transformer secondary. However, transformers are normally rated at full-load voltage, which is about 10 percent less than the open-circuit voltage. But the open-circuit voltage we have determined is at *minimum* line voltage, which is about 10 percent lower than the nominal line voltage. Therefore, 34.7 V is also the peak secondary voltage at full load and nominal input voltage. The secondary rms voltage rating is, therefore, $34.7 \times 0.707 = 24.5$ V. A 25 V transformer would be suitable.

The maximum power dissipation of the reference diode must yet be determined. This dissipation will occur when the primary voltage is maximum and the load current is minimum. The maximum full load rms secondary voltage of 27 V occurs when the primary voltage is 130 V. The corresponding no-load voltage is 10 percent above this value, or about 30 V. The peak voltage into the rectifiers is therefore $(2)^{1/2}(30) = 42$ V and the maximum voltage applied to the regulator is approximately $42 - 1 = 41$ V, allowing 0.5 V for each rectifier at minimum current. Let us assume that the minimum load current is zero. Then the ripple is approximately zero and the dc voltage applied to the regulator is nearly equal to the 41 V maximum value. The current through R_B is $I = (41 - 20.6)\,V/80\,\Omega = 255$ mA, which all flows through the reference diode since $I_B = 0$ when $I_L = 0$. Thus, the maximum power dissipation of the reference diode is approximately $20.6\,V(0.255\,A) = 5.25$ W. A 10 W diode would probably be purchased.

Notice that the high dissipation requirement of the reference diode in the preceding circuit resulted from the low value of R_B, which in turn resulted from the low value of minimum voltage (2 V) that we allowed across R_B. The large change in reference-diode current also causes a wider voltage variation than may be desired. On the other hand, an increase of the minimum voltage across R_B requires higher input voltages and lower efficiency. Of course, R_B is larger and ΔI_D is smaller if the maximum load current is smaller.

The characteristics of the emitter-follower regulator can be improved and the reference diode dissipation can be reduced greatly in high-current regulators if a Darlington-connected amplifier is used in the circuit (Fig. 16.8). The maximum base current I_{B1} is then $I_L/(h_{FE2} + 1)(h_{FE1} + 1)$, which is usually less than 1 mA. Then a low-power reference diode with minimum currents of the order of 1 mA can be used. In this circuit, R_B can be much larger than in Example 16.3.

Example 16.4

Let us add a silicon transistor T_1 to the circuit of Example 16.3 so it has the form given in Fig. 16.8. Assume that T_1 has $h_{FE1} = 100$ and $V_{BE} = 0.6$ V at $I_E = 20$ mA. Then $I_{B1max} \approx 20/100 = 0.2$ mA. Then, if $I_{Dmin} = 1.0$ mA and we allow a 2.0 V minimum across R_B, the value of R_B is $2.0\,V/1.2\,mA = 1.67$ kΩ. Any standard value between 1.5 kΩ and 1.8 kΩ would be suitable. The design would then proceed as before, except the additional 0.6 V must be added to 20.6 V to yield a 21.2 V reference-diode voltage. The voltages all along the line back to the primary voltage must therefore be appropriately increased. The important result is that $I_{Dmax} = 19.8\,V/1.67\,k\Omega = 12$ mA and the maximum power dissipation of the reference diode is $(21.2\,V)(12\,mA) =$

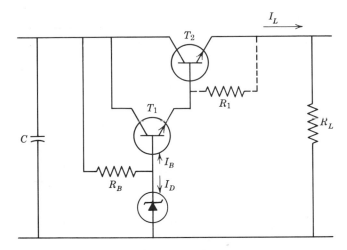

Fig. 16.8. Emitter-follower regulator using a Darlington-connected amplifier.

254 mW, approximately. In addition, the change of I_D is $\Delta I_D - 11$ mA. The maximum dissipation of the power transistor (T_2) must also be considered. This dissipation occurs at maximum line voltage and maximum load current. The minimum input voltage to the Darlington regulator is $22.6 + 0.6 = 23.2$ V at full load and 105 V input to the transformer primary. Then, if the input to the transformer rises to 130 V, $V_{min} = 23.2 \times 130/105 = 28.8$ V. Since the ripple is about 10 percent, the average input voltage to the regulator is about 5 percent above 28.8 V or 30.3 V. The average voltage across the regulating transistor under these conditions is $30.3 - 20 = 10.3$ V and the maximum power dissipation is 10.3 V $\times 1$ A $= 10.3$ W. If germanium transistors are used, the dashed resistor R_1 (Fig. 16.8) may be needed to reduce the thermal currents, as was discussed in Chapter 10.

The reduction of the ripple voltage by the emitter-follower regulator can be determined from the equivalent circuit of Fig. 16.9. This equivalent circuit was initially discussed in Chapter 3. The ripple voltage is the ac component in either the input or output of the regulator. The ripple voltage in the output is essentially the same as the ripple voltage, v_r, across r_d. Thus,

$$v_r = \frac{v_r' \, r_d}{R_B + r_d} \qquad (16.12)$$

where v_r' is the ripple voltage across the filter capacitor. The ripple voltage is maximum at full load and should be calculated for this condition. For example, if $r_d = 30 \ \Omega$ (a typical value) at $I_D = 1$ mA and the

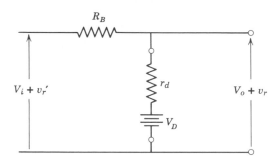

Fig. 16.9. Equivalent circuit used to determine the ripple voltage at the regular output.

peak-to-peak ripple voltage across the capacitor is about 3.5 V, as considered in the preceding examples, the peak-to-peak ripple voltage in the output of the Darlington-connected circuit example is $v_r = 3.5 \times 30 / 1.7$ kΩ $= 0.06$ V.

Sometimes the good regulation of a regulated supply is not needed but low ripple voltage is required. Then the reference diode in the circuit of Fig. 16.8 may be replaced by a capacitor, as shown in Fig. 16.10. This circuit can be designed in the same manner as the emitter-follower regulator except there is no reference diode current, and jX_C replaces r_d in determining the output ripple from Eq. 16.12.

Example 16.5

Let us assume that the active filter circuit of Fig. 16.10 will be used to replace the Darlington-connected emitter-follower regulator of Fig. 16.8, the power supply requirements and transistors being the same as in Example 16.3. Thus, $V_{omin} = 20$ V, $I_{Lmax} = 1$ A, $h_{FE2} = 50$, $h_{FE1} = 100$, $C = 2,000$ μF. Now I_{Bmax} is ≈ 1 A/(50)(100) or 0.2 mA. If we

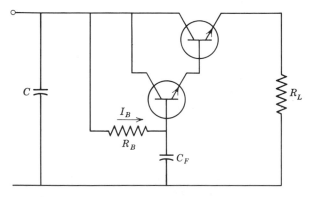

Fig. 16.10. Active filter for reducing ripple.

allow a 2 V maximum drop across R_B, the value of R_B is 2 V/0.2 mA = 10 kΩ. The value of v_r can be calculated from Eq. 16.10 for a given value of C_F. However, a preferable procedure is to specify the desired value of v_r and calculate the required value of C_F. If jX_C is used to replace r_d in Eq. 16.12, this equation becomes $v_r = jv_r' X_C/(R_B + jX_C)$. However, if $R_B \gg X_C$, the denominator of this relationship is $\simeq R_B$ and the magnitude of v_r is

$$|v_r| = \frac{v_r' X_{CF}}{R_B} \tag{16.13}$$

Solving for X_{CF},

$$X_{CF} = \frac{1}{2\omega C_F} = \frac{|v_r| R_B}{v_r'} \tag{16.14}$$

Solving for C_F,

$$C_F = \frac{v_r'}{2\omega R_B |v_r|} \tag{16.15}$$

where ω is the angular frequency $(2\pi f)$ of the primary power frequency. The factor 2 appears because of the full-wave rectification. In this example, let us specify the peak-to-peak ripple voltage in the output to be 0.01 V. Then with $v_r' = 3.5$ V and $f = 60$ Hz, $C_F = 3.5/(4\pi \times 60 \times 0.01 \times 10^4) = 47$ μF.

The filter capacitor C_F can also be used in parallel with a reference diode to further reduce the ripple voltage in the output of an emitter-follower regulator.

**16.5
CLOSED-LOOP
REGULATORS**

Although the emitter-follower regulators provide satisfactory performance for many applications, their output resistance cannot be reduced below the value given by Eq. 16.10. Also, large values of C_F (Fig. 16.10) are required to provide very low values of ripple. On the other hand, regulators that employ the principle of negative feedback can provide almost any desired value of output resistance and ripple quite easily. The basic philosophy of the closed-loop regulator is illustrated by the block diagram of Fig. 16.11. A fraction of the output voltage ηV_o is compared with a reference voltage V_{REF} and their difference is amplified and used to control the series regulator, which in turn controls the output voltage. A typical circuit diagram that will perform this basic function is shown in Fig. 16.12. In low-current regulators, the series regulator may be a single transistor, but a Darlington-connected ampli-

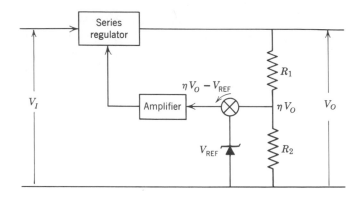

Fig. 16.11. Block diagram of a closed-loop regulator.

fier is usually used as shown in Fig. 16.12 and as previously used in the emitter-follower regulator (Fig. 16.8). The differential amplifier consisting of transistors T_1 and T_2 provides both the voltage comparison and the amplification functions. The resistors R_1 and R_2 are chosen so that their ratio provides the desired ratio between V_{REF} and V_O, while their sum provides a bleeder current through the resistors that is large compared to the base current of transistor T_1. If these conditions are met,

$$V_{REF} \simeq \frac{R_2}{R_1 + R_2} V_O \qquad (16.16)$$

and the current I_R through R_1 and R_2 is

$$I_R = \frac{V_O}{R_1 + R_2} \qquad (16.17)$$

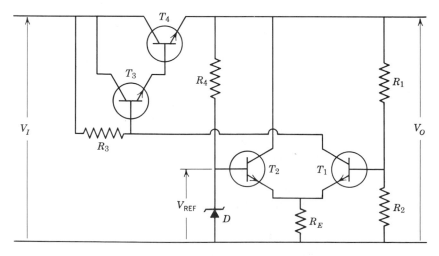

Fig. 16.12. Typical closed-loop regulator circuit.

These equations can be solved simultaneously to find R_1 and R_2. By substituting Eq. 16.17 into Eq. 16.16,

$$V_{REF} = R_2 I_R \qquad (16.18)$$

and

$$R_2 = \frac{V_{REF}}{I_R} \qquad (16.19)$$

Substituting this expression for R_2 into Eq. 16.17 and solving for R_1

$$R_1 = \frac{V_O}{I_R} - R_2 \qquad (16.20)$$

The resistor R_4 is chosen so that the current through the reference diode D is large in comparison with the base current of transistor T_2. Then, the current through D is essentially constant, and hence the reference voltage V_{REF} is very constant. The diode D should have essentially zero temperature coefficient if V_{REF}, and hence V_O, are to be independent of temperature.

The operating principles of the closed-loop regulator will be illustrated by assuming that the output voltage V_O becomes more positive because of either a reduction in load current or an increase of input voltage, or both. Then, the base of transistor T_1 becomes more positive than V_{REF} and the current through T_1 increases. The increased current through T_1 causes the drop across resistor R_3 to increase and the forward bias of the series regulator transistors T_3 and T_4 to decrease. This decreased forward bias increases V_{CE} and therefore reduces the output voltage or, in other words, tends to cancel the assumed rise in output voltage. Note that a polarity reversal must be provided in the amplifier so that negative feedback is obtained in the closed loop.

The effectiveness of the amplifier and feedback system in improving the characteristics of the power supply will now be investigated. A semiblock-diagram similar to the one in Fig. 16.11 is shown in Fig. 16.13, as an aid in the investigation. Observe that the output voltage v_{CE} of the series-pass transistor T_4 is the difference between the output voltage V_O and the input voltage V_I or $(V_I - V_O)$. Let K_v be the voltage amplification between the input of the amplifier and the output (V_{CE}) of the series-pass transistor T_4. Then,

$$V_I - V_O = (\eta V_O - V_{REF}) K_v \qquad (16.21)$$

But we are primarily interested in the ratio of the change in output voltage to the change in the input voltage in order to determine the reduction in ripple or input voltage variations. Therefore, we write Eq. 16.21 in terms of voltage variations instead of total voltages.

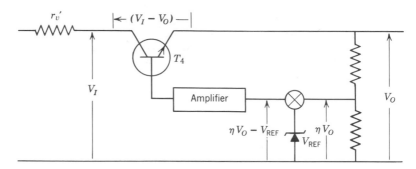

Fig. 16.13. Block diagram of a closed-loop regulator.

$$\Delta(V_I - V_O) = \Delta(\eta V_O - V_{\text{REF}})\,K_v \qquad (16.22)$$

But $\Delta V_{\text{REF}} = 0$ if the reference voltage is constant. Then,

$$\Delta V_I - \Delta V_O = \Delta \eta V_O K_v \qquad (16.23)$$

Solving for ΔV_O, we have

$$\Delta V_O (1 + \eta K_v) = \Delta V_I \qquad (16.24)$$

or

$$\Delta V_o = \frac{\Delta V_I}{1 + \eta K_v} \qquad (16.25)$$

Observe from Eq. 16.25 that variations in the output voltage are reduced by the factor $(1 + \eta K_v)$ compared with variations of V_I that are caused by ripple, primary voltage variations, and so forth. Note that Eq. 16.25 is a standard feedback equation with $\mathcal{F}$ replaced by η.

If both sides of Eq. 16.25 are divided by the change in load current ΔI_L, the effectiveness of the regulator in reducing the power supply output resistance may be obtained. Thus

$$\frac{\Delta V_O}{\Delta I_L} = \frac{\Delta V_I / \Delta I_L}{1 + \eta K_v} \qquad (16.26)$$

But $\Delta V_O / \Delta I_L$ is the input resistance of the power supply after regulation and $\Delta V_I / \Delta I_L$ is the output resistance r_o' of the filter before regulation, where we should think of the change in voltage ΔV_I as being caused by the change of current ΔI_L. Then,

$$r_{of} = \frac{r_o'}{1 + \eta K_v} \qquad (16.27)$$

where r_{of} is the output resistance of the regulated supply. Observe that, as in other feedback circuits, the output resistance is reduced by the factor $(1 + \eta K_v)$. As noted in Chapter 13, low power-supply impedance is helpful in maintaining stability in a multistage amplifier.

Example 16.6

Let us consider the design of the type regulator shown in Fig. 16.12 with the following data and specifications given: $V_O = 20$ V, $I_{Lmax} = 1.0$ A, $h_{FE4} = 80$, $h_{FE3} = h_{FE2} = h_{FE1} = 200$, pprr = 0.15. Notice in Example 16.3 that 0.10 V_{peak} is about equal to 0.15 V_{AVE} since $V_{AVE} = 0.7\ V_{peak}$. All transistors are silicon. R_3 (which was R_B in the preceding examples) can be determined as before. At maximum load $I_{B3} = 1$ A $/ (80 \times 100) = 0.12$ mA. If we assume that V_{Imin} occurs at I_{Lmax}, this is the worst case. Under these conditions, the minimum collector current flows in T_1. We shall let this minimum current be 0.2 mA. Then, if we allow 2.0 V minimum across R_3, $R_3 = 2 / 0.32$ mA $= 6$ kΩ. The rectifier circuit and filter capacitor could now be determined as in Example 16.3. We shall select $V_{REF} = 6.8$ V since diodes in this voltage area have minimum dynamic resistance and minimum thermal coefficient. It is preferable to calculate the remaining resistance values in the circuit using nominal input voltage, rather than minimum. This minimum V_I is approximately $20 + (2 \times 0.6) + 2 = 23.2$ V. The peak-to-peak ripple at full load is about $24 \times 0.15 = 3.6$ V so that the average V_I at 105 primary volts (assuming 117 V, 60 Hz nominal input power) is $23.2 + 3.6 / 2 = 25$ V. (We guessed 24 V in calculating this average above.) Then the *nominal* average $V_I = 25 \times 117 / 105 = 27.8$ V. Under this condition, the average current through $R_3 = (27.8 - 21.2) / 6$ k$\Omega = 1.1$ mA. Since $I_{B3max} = 0.12$ mA, the nominal collector current in transistor T_1 is $1.1 - 0.12 \approx 1.0$ mA. Thus $I_{B1} = 1.0$ mA $/ 100 - 10$ µA. Since the current through the voltage divider circuit should be very large in comparison with I_{B1}, we choose this bleeder current $I_R = 10$ mA, which is only 1 percent of the full load current. Then, by using Eqs. 16.23 and 16.24, $R_2 = 6.8$ V $/ 10$ mA $= 680\ \Omega$, $R_1 = 2$ k$\Omega - 680 = 1.32$ kΩ. An adjustable resistance that will span the 1.32 kΩ value is usually used for R_2 so that the output voltage can be adjusted.

At nominal conditions, the base current $I_{B2} = I_{B1} = 10$ µA. The current through r_4 should be large compared with this value so that the current through D varies only a few percent. We choose this current to be approximately 2 mA. Then $R_4 = (20 - 6.8)$ V $/ 2$ mA $= 6.6$ kΩ. A standard 6.8 kΩ resistor will suffice.

The amplifier voltage gain K_v must be calculated to determine the characteristics of the regulated supply. The easiest method of calculating voltage gain is probably using the relationship $K_v = K_i R_L / R_{in}$. The following approximations will be used: $K_i = h_{FE}$, $R_{in} = 2h_{ie}$ for the differential amplifier, $R_{in} = (h_{ie3} + h_{FE3} h_{ie4})$ for the Darlington amplifier, and $h_{ie} \approx \beta_o / g_m$. Then $h_{ie} \approx 100 / 0.04 = 2.5$ kΩ and $R_{in} \approx 5.0$ kΩ

for the differential amplifier. In addition, $h_{ie} \simeq 80/40 = 2.0 \, \Omega$ for transistor T_4 at $I_L = 1$ A and $h_{ie} = 100/0.48 = 220 \, \Omega$ for transistor T_3. Then, $R_{in} = 220 + 200 = 420 \, \Omega$ for the Darlington connection. This resistance, in parallel with R_3 is the load resistance for the differential amplifier. This fact points up the need for the Darlington amplifier to provide a suitable (400 Ω) load resistance for the differential amplifier. The load resistance for the series regulator is the output resistance r_o' as seen in Fig. 16.13. Then, for the differential amplifier, $K_v \simeq 100 \, (400)/5 \, k\Omega = 8$. The value of r_o' can be accurately obtained from a plot of V_I as a function of I_L, as given in Fig. 16.5. the value of r_o' can also be obtained by calculating the increase of the average value of V_I as I_L decreases from its full load value to essenitally zero, and then taking $\Delta V_I/\Delta I_L$. This voltage rise is approximately the peak ripple voltage plus about a 2 V reduced drop in the rectifier diodes and the transformer windings. The total voltage rise in this example is, therefore, about $(0.15 \times 25/2) + 2 \simeq 4$ V and since ΔI_L is 1 A, $r_o' \simeq 4 \, \Omega$. Then the voltage gain of the series regulator is $K_i r_o'/R_{in} = 100 \times 80 \times 4/420 = 76$. Thus the total gain $K_v = 8 \times 76 = 608$ and $\eta = V_{REF}/V_o = 6.8/20 = 0.34$, so $\eta K_v = 207$ and $1 + \eta K_v = 208$. Therefore, the peak-to-peak ripple voltage is $3.7/208 = 0.018$ V and $r_{of} = 4/208 = 0.02 \, \Omega$ for the regulated power supply.

**16.6
INTEGRATED
VOLTAGE
REGULATORS**

A wide variety of monolithic integrated regulator chips have been developed by various manufacturers of solid-state devices. The schematic diagram of a typical IC regulator is shown in Fig. 16.14. This circuit operates in the same basic manner as the closed-loop regulator in the preceding section (16.5). A number of refinements and special features have been added to improve the performance of the basic regulator previously discussed. These features and circuits are briefly examined in the paragraphs which follow.

You may recall that the output voltage of the basic closed-loop regulator discussed in Section 16.5 could be adjusted by the voltage divider having voltage ratio η in the output of the regulator. However, the minimum voltage available was equal to the reference voltage V_R, which is typically about 6.5 volts because of the good reference diode characteristics available in that voltage area. The circuit of Fig. 16.14 provides a stable, adjustable, temperature-compensated reference voltage that may be adjusted down to 3.46 volts. This reference voltage circuit is shown in Fig. 16.15. The part of the circuit enclosed by the

*Fig. 16.14. Schematic diagram of the MC1561 monolithic IC regulator.
(Courtesy of the Motorola Semiconductor Company, Phoenix, Arizona)*

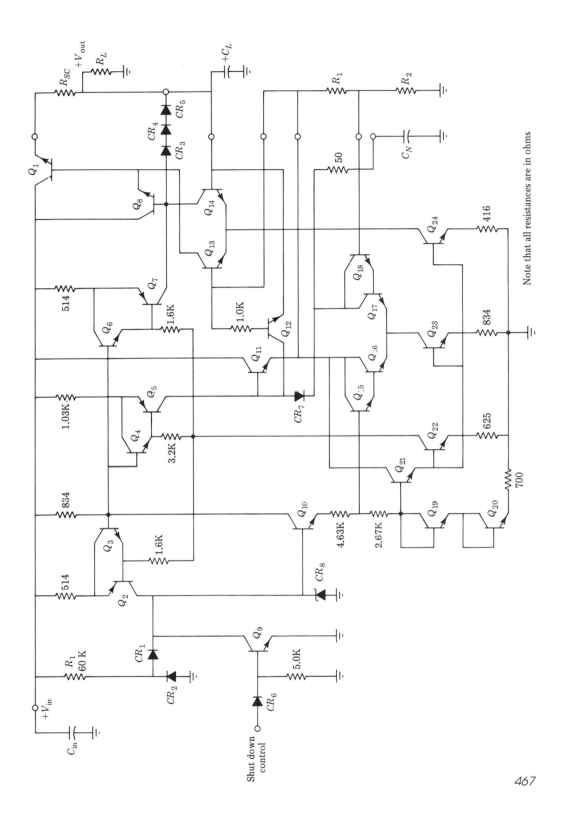

Note that all resistances are in ohms.

467

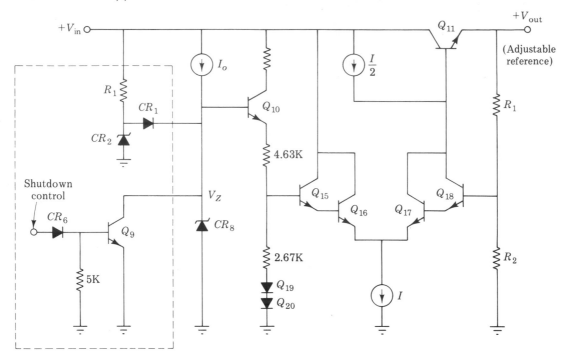

Fig. 16.15. Schematic diagram of the temperature-compensated, adjustable reference voltage section of the IC regulator shown in Fig. 16.14.

dashed line is not part of the reference voltage circuitry but includes a start-up and a shut-down circuit. Current from the unregulated input voltage V_{in} flows through R_1 and produces a voltage across the reference diode $CR2$. This voltage is coupled through the diode $CR1$ into the regulator circuitry which turns on the circuit by activating the current sources, which will be discussed later. Conversely, when a positive voltage greater than about 1.2 V is applied to the shut-down terminal, transistor $Q9$ is driven into saturation, thus reducing the reference voltage across $CR8$ to near 0, which in turn reduces the output voltage to approximately zero.

As the reference circuit (Fig. 16.15) turns on, the current source I_O provides an essentially constant current to the 7.0 volt main reference diode $CR8$. Then diode $CR1$ decouples the starting circuit because there is no forward voltage across $CR1$. The base current of $Q10$ is negligible compared with I_O. The constant current through $CR8$ maintains a stable reference voltage V_Z across $CR8$. However, $CR8$ has a positive temperature coefficient and the V_{BE} of transistor $Q10$ has a negative temperature coefficient. These coefficients are additive in producing a positive temperature coefficient at the emitter terminal of $Q10$. On the other hand, the diode-connected transistors Q19 and Q20 have negative temperature coefficients which are additive in producing a negative temperature coefficient at the anode of Q19. Therefore, there must be a point along the resistive path between the emitter of Q10 and the anode of Q19 at which the temperature coefficient is zero. This occurs, by design, at the junction of the 2.67 kΩ and 4.63 kΩ resistors. The voltage at this junction, with reference to ground, is 3.46 V. This well regulated, zero-temperature-coefficient reference voltage is used in the conventional closed-loop regulator circuit consisting of the series-pass transistor Q11 and the Darlington-connected differential amplifier using transistors Q15 through Q18. The Darlington connection essentially eliminates the loading on both the 3.46 V reference and the voltage divider consisting of R_1 and R_2. The output voltage of this regulator is controlled by the ratio of R_1 and R_2, as discussed in Section 16.5. However, the output voltage of this regulator *does not* provide power to the load but only serves as an adjustable reference voltage for the main regulator consisting of the Darlington connected series-pass transistor Q1 (Fig. 16.14) and the differential amplifier Q13 and Q14. The output voltage of this main regulator is connected directly to the base of the differential transistor Q14 without the use of a voltage divider. Thus the output voltage is the same as the adjustable reference voltage, and the loop gain is increased by eliminating the voltage divider.

Figure 16.15 shows a current source in the base circuit of the series-pass transistor Q11 in the adjustable reference regulator. This current source replaces the resistor previously used in this location to provide bias current to the series-pass transistor and collector current for the differential amplifier. The current source provides higher voltage gain than the resistor because of the increased load resistance on the differential amplifier. The main regulator uses a similar current source. These current sources and the current source used in the voltage-reference circuit each use a p-n-p–n-p-n transistor combination, one of which is shown in Fig. 16.16a.

The current source I_C in Fig. 16.16a uses a p-n-p transistor which has a very high output resistance and low h_{FE}. The essentially constant current that flows through Q10 into the voltage regulator circuit (shown in Fig. 16.16b) also flows through R_B in the base circuit of the n-p-n transistor Q4. Therefore, the voltage drop across R_B is essentially constant, since the base current of the high β transistor Q4 is negligible. But, assuming the V_{BE} of Q4 and Q5 to be equal, the voltage V_E at the emitter of Q5 is equal to V_B at the base of Q4. Then the constant voltage drop across R_B also appears across R_E and the current through R_E must be constant. This current must flow through either Q5 or Q4 and must be equal to the sum of the currents $I + I_C$, neglecting the base current of Q4. Then since I is a constant current, I_C must also be a constant current.

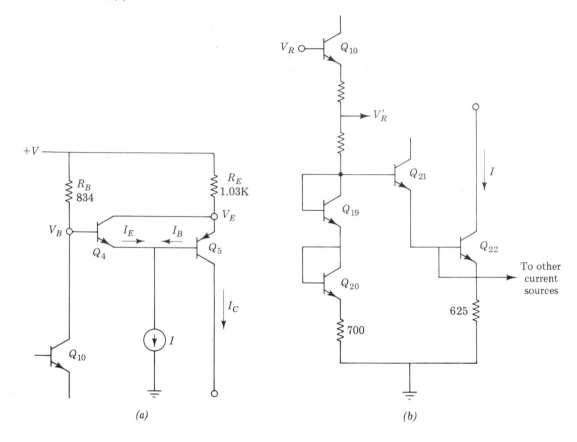

Fig. 16.16. (a) Schematic diagram of a pnp-npn current source used in the
voltage regulator circuit of Fig. 16.14. (b) Schematic diagram of the
current source I shown in (a).

This relationship holds, regardless of the h_{FE} of Q5 so long as I exceeds
I_B so Q4 remains turned on.

The current source I is shown in detail in Fig. 16.16b. The
essentially-constant current through Q10 also flows through the 700 Ω
resistor to ground, producing a constant voltage across this resistor.
This constant voltage also appears across the 625 Ω emitter resistor of
Q22, since the base to emitter voltages of Q19 and Q20 are essentially
equal to those of Q21 and Q22. Thus I, controlled by the 625 Ω resistor,
is constant. Actually, the current I and the collector current of Q10 both
have a positive temperature coefficient, but since I_C (Fig. 16.16) is proportional to the difference between these two currents, I_C is essentially
independent of temperature. The other current sources shown in the
lower part of Fig. 16.14 operate in the same manner as the one shown
in Fig. 16.16b.

Some additional features of the circuit of Fig. 16.14 are:

1. C_N is a filter capacitor used to reduce the noise in the regulator, and diode $CR7$ prevents C_N from discharging into the regulator circuit during shutdown.

2. Transistor $Q13$ draws its collector current through $Q8$ in the Darlington series-pass device instead of from $+V_{in}$. This arrangement increases the quiescent current through $Q8$ and thus increases both h_{FE} and f_T in this transistor, especially under light load conditions.

3. Transistor $Q12$ prevents excessive voltage from appearing between the inputs of the differential amplifier $Q13$ and $Q14$ during an output short-circuit condition. $Q12$ is normally off. However, the input voltage to $Q13$ and $Q14$ is applied between the base and emitter of $Q12$. Therefore, a differential input voltage in excess of about 0.5 V causes $Q12$ to divert the drive current of the series-pass transistor in the adjustable reference-voltage supply to the output terminal of the regulator. Therefore, the reference voltage is reduced to approximately zero whenever the output is shorted.

4. The maximum output current, or short-circuit current of the regulator, is controlled by the external resistor R_{SC}. The load current flows through this resistor. When the voltage drop across R_{SC} exceeds about 0.5 volt, the diodes $CR3$, $CR4$, and $CR5$ are brought into conduction, thus diverting the drive current from transistor $Q8$ in the Darlington series-pass amplifier to the output terminal.

The typical performance characteristics given by the manufacturer for the MC1561 chip are shown in Table 16.2. The thermal resistance between the chip and the 9-pin TO-66 case is about 7°C/W.

Table 16.2. Typical performance characteristics of the MC1561.

Temperature Drift of V_{out}	±30 ppm/°C
Z_o (Independent of V_{out})	0.020 ohms
Input Regulation	0.003%/V
Transient Recovery Time	0.3 μs
(I_{load} = 150 mA; ΔI_{load} = 50 mA)	
Regulator Bias Current	5.0 mA
Maximum Load Current	600 mA
Output Noise	
(With C_N = 0.1 μF)	0.15 mVrms
Minimum Voltage Differential	2.1 volts
($V_{in} - V_{out}$)	

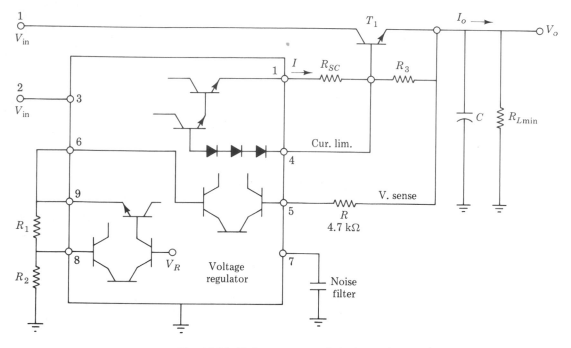

Fig. 16.17. High-power regulated supply.

A high-power regulated supply may be constructed by using a regulator IC to drive a series-pass high-power transistor, as shown in Fig. 16.17. The available current is then h_{FE} of the power transistor times the maximum current available from the regulator IC. The Darlington arrangement may be used, if needed, for high current gain. The unregulated input voltages $V_{in}1$ and $V_{in}2$ may be the same, but in some applications, $V_{in}1$ may be from a separate, lower voltage supply in order to minimize the power dissipation in the high-power transistor.

The capacitor C (Fig. 16.17) serves two purposes. First, it maintains a low output impedance at frequencies above the pass band of the regulator where the impedance would otherwise rise because of the decreased open-loop gain. Second, it provides a low-frequency pole in the series-pass stage of the regulator and thus provides phase-lag compensation for the closed-loop system which has unity voltage gain.

The resistor R (Fig. 16.17) in the voltage-sense lead to the regulator chip is used to suppress parasitic oscillations caused by induced voltages in the external sense lead, particularly when this lead is quite long. This resistor should be located within 0.25 inch of the regulator chip terminal.

The output terminals of the regulator shown in Fig. 16.17 correspond to the MC1560 regulator given in Fig. 16.14, so these figures may be correlated. However, a number of suitable regulators are available from the various manufacturers. The data sheets of these regulators give recommended external circuit component values and special circuits for protecting the regulator and enhancing its capabilities. The data sheets should be studied before selecting or using a regulator IC.

Inexpensive high-power regulators are available in specific output voltage ratings. The output voltage of these regulators may be adjusted upwards by using the simple voltage-divider circuit shown in Fig. 16.18. The current I must be large in comparison with the small current that flows out of the common terminal 3 of the regulator. Then, since $V_R = IR_2$ and $V_O = I (R_1 + R_2)$, the output voltage is

$$V_o = V_R \frac{(R_1 + R_2)}{R_2} \tag{16.28}$$

**16.7
SWITCHING
REGULATORS**

The efficiency η of a conventional power-supply regulator may be expressed by

$$\eta = \left(\frac{I_L}{I_L + I_R}\right)\left(\frac{V_O}{V_{in}}\right) \tag{16.29}$$

The load current I_L is normally large in comparison with the current I_R (which is used by the regulator) except for situations where the load is very light and the efficiency is then unimportant. Therefore, the first term in Eq. 16.29 usually approaches unity at normal power supply loads. However, the available primary power source V_{in} is sometimes fixed by the voltage requirements of other parts of the system and may be much higher than the regulated load voltage V_O. For example, the regulator for a 5.0 V digital system in a satellite having a 30 V battery as a primary power source would have an efficiency less than 16 per-

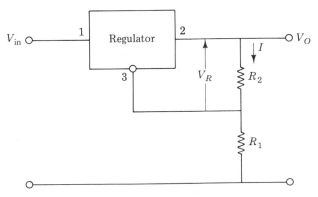

Fig. 16.18. Voltage-adjusting circuit for a fixed-voltage regulator.

cent. In addition to being wasteful of the precious battery power, the large power dissipation required by a high-power inefficient regulator greatly increases the size and the cost of the regulator. A *switching regulator* may overcome these difficulties.

The basic operating principle of a switching regulator is illustrated in Fig. 16.19. The base voltage of the series-pass transistor T_1 consists of a series of rectangular pulses that drive the transistor into either the saturation region or the cutoff region. Thus, the transistor behaves as a switch. While the transistor is *on*, current flows through transistor T_1 and inductor L to the load R_L and the filter capacitor C. When transistor T_1 is switched *off*, the current through L tends to decrease so the voltage polarity across L reverses and diode D becomes forward biased. Thus, diode D provides current to the load while the transistor is *off*. If the time constant L/R is long in comparison with the switching period, the current I_L cannot change appreciably during either the on time or the *off* time of the switch, and the output voltage V_O is fairly well filtered by the inductance. The capacitor C further reduces the ripple. The output voltage V_O is controlled by the ratio of t_{on} to t_{off}. If both the transistor saturation voltage and the forward bias voltage of the diode are negligible compared with V_{in} and V_O, it may be seen from Fig. 16.19 that

$$V_O t_{off} = (V_{in} - V_O) t_{on} \tag{16.30}$$

This relationship results from the fact that the incremental increase in $V_O (\Delta V_O$ in Fig. 16.19b) must equal the incremental decrease in V_O during a cycle in the steady state, and the rate of change of inductor current, and hence V_O, is proportional to the voltage across the inductor.

A conventional voltage regulator circuit or an op amp may be used to

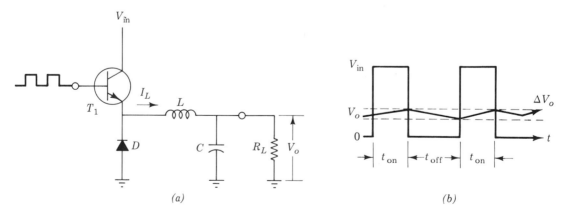

(a)   (b)

Fig. 16.19. Illustration of the basic operating principle of a switching regulator: (a) switching circuit; (b) voltage waveforms.

regulate the output voltage of the switching circuit of Fig. 16.19, as shown in Fig. 16.20. As the power supply is turned on, the output voltage V_O rises until it becomes slightly positive with respect to the reference voltage V_R plus a small voltage produced across R_1 due to the current through R_2. Then, since the output voltage is applied to the-inverting amplifier terminal, the amplifier output goes negative and cuts off the series-pass transistor Q. This action causes the diode D to conduct and the voltage at the upper end of R_2 drops to a few tenths of a volt above ground. Then the current through R_1 and R_2 reverses direction and the noninverting input terminal of the amplifier becomes slightly negative with respect to V_R. This condition remains until the decreasing current through R_L causes V_O to become slightly negative with respect to V_R minus the small voltage across R_1. Then the output voltage of the amplifier becomes positive and the cycle repeats. You may have observed that the amplifier acts as a voltage comparator with hysteresis or, in other words, a *Schmitt Trigger*.

The peak-to-peak ripple voltage in the output is somewhat larger than the change in reference voltage ΔV_R at the noninverting amplifier input terminal. But ΔV_R is the difference between the voltage drops across R_1 when the series-pass transistor is switched *on* and when it is switched *off*. Therefore, neglecting both the forward diode drop and the transistor Q saturation voltage (see Fig. 16.20),

$$\Delta V_R = \frac{(V_{in} - V_R)\,R_1}{R_1 + R_2} - \frac{-V_R\,R_1}{R_1 + R_2} = \frac{V_{in}\,R_1}{R_1 + R_2} \qquad (16.31)$$

Thus ΔV_R, and consequently the peak-to-peak output ripple, may be restricted to a few millivolts if R_2 is a few thousand times as large as R_1. For example, if $V_{in} = 30$ V, $R_1 = 1$ kΩ and $R_2 = 3$ MΩ, $\Delta V_R \approx 10$ mV.

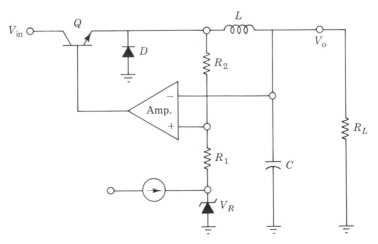

Fig. 16.20. Closed-loop regulator for the switching circuit of Fig. 16.19.

The switching frequency f may be determined from the basic relationships

$$V_{in} - V_O = L\frac{di}{dt} \simeq L\frac{\Delta i_L}{t_{on}} \tag{16.32}$$

and

$$\Delta i_L = C\frac{\Delta V_O}{t_{on}} \tag{16.33}$$

where Δi_L is the change in the inductor current during t_{on}. Substituting Eq. 16.33 into Eq. 16.32,

$$V_{in} - V_O = \frac{LC\,\Delta V_O}{t_{on}^2} \tag{16.34}$$

Solving explicitly for t_{on},

$$t_{on} = \sqrt{\frac{LC\,\Delta V_0}{V_{in} - V_O}} \tag{16.35}$$

The total period $T = t_{on} + t_{off}$ and using Eq. 16.30 to obtain t_{off} in terms of t_{on},

$$T = t_{on}\left(1 + \frac{V_{in} - V_O}{V_O}\right) = \frac{V_{in}}{V_O}t_{on} = \frac{V_{in}}{V_O}\sqrt{\frac{LC\,\Delta V_O}{V_{in} - V_O}} \tag{16.36}$$

Assuming that $\Delta V_O \simeq \Delta V_R$ and using Eq. 16.31 with $R_2 \gg R_1$,

$$f = \frac{1}{T} = \frac{V_O}{V_{in}}\sqrt{\frac{(V_{in} - V_O)\,R_2}{LC\,V_i\,R_1}} \tag{16.37}$$

The actual frequency is somewhat lower than the value given by Eq. 16.37 because ΔV_O is greater than ΔV_R. Since the inductor usually produces audible sounds at the switching frequency, this frequency should preferably be above the audio range. A minimum frequency of 20 kHz is desired. Also, as seen by Eq. 16.37, the required values of L and C decrease as the switching frequency increases. However, the power loss in the series-pass switching transistor increases with frequency because most of the loss occurs during the rise and fall times of this transistor. Therefore, a high-speed switching transistor is required in order to maintain high efficiency at the desirable switching frequencies. The rise time of the switching transistor should not exceed about 5 percent of the switching period.

Once the switching frequency has been established, the value of inductance L (Fig. 16.20) may be determined from the ratio of the peak current we will allow through the inductor to the full-load average current through the load (or inductor). Since the peak current $i_p = I_{Omax} + \Delta i_L/2$, we may write

$$\Delta i_L = 2\,(i_p - I_{Omax}) \tag{16.38}$$

Solving Eq. 16.32 explicitly for Δi_L,

$$\Delta i_L = \frac{(V_{in} - V_O)\,t_{on}}{L} \tag{16.39}$$

Combining Eqs. 16.38 and 16.39,

$$L = \frac{(V_{in} - V_O)\,t_{on}}{2\,(i_p - I_{Omax})} \tag{16.40}$$

and, from Eq. 16.36, $t_{on} = V_O T/V_{in} = V_O/f V_{in}$. Therefore,

$$L = \frac{(V_{in} - V_O)\,V_O}{2\,(i_p - I_{Omax})\,f\,V_{in}} \tag{16.41}$$

A typical value of peak current i_p is $1.2\,I_{Omax}$. Using this value,

$$L = \frac{(V_{in} - V_O)\,V_O}{0.4\,I_{Omax}\,f\,V_{in}} = \frac{2.5\,(V_O - V_O^2/V_{in})}{f\,I_{Omax}} \tag{16.42}$$

Both the recovery time and overshoot of the output voltage due to transient load curent changes are directly proportional to the inductance L. Therefore, L should not be larger than necessary.

The value of the filter capacitor C may now be determined from a rearrangment of Eq. 16.34 and 16.36.

$$C = \frac{(V_{in} - V_O)\,t_{on}^2}{L\,\Delta V_O} = \frac{(V_{in} - V_O)}{L\,\Delta V_O}\left(\frac{V_O}{f\,V_{in}}\right)^2 \tag{16.43}$$

You may observe from Fig. 16.14 that the MC1561 regulator may be used as a switching regulator, since both inputs of the differential amplifier are available at external terminals. The only additional external components required, as compared to the conventional regulator, are the inductor, the diode (sometimes called a *catch* diode or *freewheeling* diode), and the resistors R_1 and R_2. Most other regulator ICs will also function as switching regulators.

Switching regulator efficiencies typically range from about 70 percent to 90 percent. The main disadvantages of switching regulators compared with conventional regulators are larger output ripple, slower transient response, and noise coupled back into the primary power source. The input power lines may require filtering in order to reduce this noise to acceptable levels.

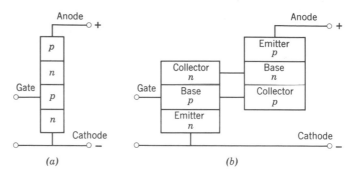

*Fig. 16.21. Silicon-controlled rectifier: (a) actual construction;
(b) equivalent electrical circuit.*

**16.8
SILICON-
CONTROLLED
RECTIFIERS
AND TRIACS**

Voltage or current regulation may also be accomplished through the use of special semiconductor devices known as thyrodes, silicon-controlled switches (SCSs), silicon-controlled rectifiers (SCRs), or triacs. We shall first consider how these devices operate and then study several circuits that employ them.

The thyrode, or SCR, is constructed as shown in Fig. 16.21a. The action of this device can be explained by the equivalent circuit, which is shown in Fig. 16.21b. This equivalent circuit shows an n-p-n transistor and a p-n-p transistor interconnected. In fact, the SCR action can be produced by connecting two transistors as shown in Fig. 16.21b. If the gate is negative (or near zero volts), the n-p-n transistor will be in the cutoff condition with essentially no collector current. The collector current of the n-p-n transistor furnishes the base current for the p-n-p transistor and vice versa. Thus, if one transistor is cut off, the other transistor is also cut off, and the impedance from anode to cathode is very high.

If the bias on the gate is made positive until collector current flows in the n-p-n transistor, this collector current becomes the base current or the p-n-p transistor, which begins to conduct. The collector current of each transistor becomes the base current for the other transistor. A cumulative action is therefore initiated since an increase of current in one unit causes an increase of current in the other unit. This cumulative action culminates when both transistors are driven into saturation. When both transistors are saturated, the impedance from anode to cathode is very low. Thus, the gate is said to "trigger" the SCR "ON."

As soon as the self-regeneration action commences, the gate loses control over the action. Generally, the collector current from the p-n-p unit is much larger than the external gate current. As a result, the external gate circuit can turn the SCR ON but has difficulty turning the switch OFF. (The action is much the same as the action in the gas triode or thyratron.) To turn the SCR OFF, the gate bias must be in the reverse direction and the anode voltage must be reduced essentially to zero. In this condition, no current flows through the p-n-p unit, and the gate regains control of the current.

If the gate is maintained at cutoff (negative potential) and the voltage of the anode is varied, the characteristics are as indicated in Fig. 16.22. As the negative anode voltage is increased, avalanche breakdown of the SCR occurs. On the other hand, as the anode voltage is made more positive, the center junction is reverse biased while the other two junctions are forward biased. As the center junction approaches avalanche breakdown, the avalanche current across this junction has the same polarity as a positive gate current. Thus, as avalanche breakdown approaches, the SCR turns itself ON. This anode "turn-on" potential (with reverse-biased gate) is known as the *breakover voltage* of the SCR.

In normal operation, the SCR is operated with an anode potential below the breakover voltage. Then the device is turned ON at the appropriate time by the gate. However, remember the device can be turned ON by high anode potentials. In addition, if the anode potential *changes* at a sufficiently high rate, current from the junction capacitances in the SCR may be large enough to supply the gate with sufficient current to turn the SCR ON. The symbol for the SCR is shown in Fig. 16.23a.

Another p-n-p-n device is known as a *Schockley diode*. The construction of this device is similar to an SCR without the gate lead. The Schockley diode can be turned ON by a high potential (which exceeds the breakover voltage) and turned OFF by a zero or reverse potential. The symbol for a Schockley diode is shown in Fig. 16.23b.

A more versatile p-n-p-n arrangement has a lead brought out from each area. A device with this configuration is known as a *silicon-*

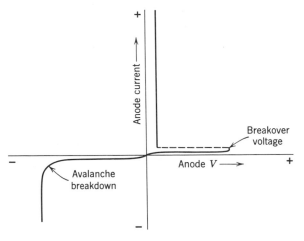

Fig. 16.22. Current-voltage characteristics of the SCR with gate reverse biased.

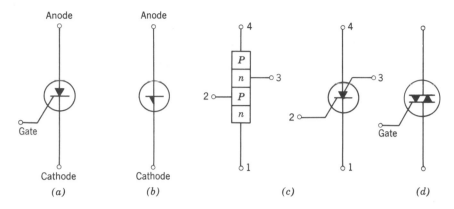

Fig. 16.23. Symbolic representation of p-n-p-n devices: (a) SCR;
(b) Schockley diode; (c) SCS; (d) triac.

controlled switch (SCS). The configuration of an SCS and its symbolic representation is shown in Fig. 16.23c.

Two SCRs may be mounted in the same case for full wave applications. These devices are known as *triacs* and have the anode for each device connected internally to the cathode of the other device. A single gate lead (connected internally to the two gates) is provided. These devices can be switched from a blocking to a conducting state for either polarity of applied anode potential with positive or negative gate triggering. The symbol for a triac is given in Fig. 16.23d.

16.9
APPLICATIONS
OF SCRs

As the name implies, SCR devices can often be used as controlled rectifiers. For example, the SCR is widely used in a light-dimmer control or for the speed control of simple series motors (such as electric saws and drills). A simple control of this type is shown in Fig. 16.24. To

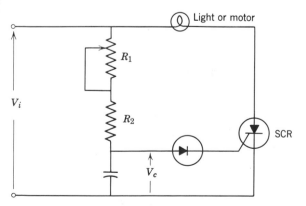

Fig. 16.24. Light dimmer or motor speed control.

visualize the effect of this circuit, an analysis of the RC circuit must be considered. If the current to the gate of the SCR is assumed to be quite small, the current through the capacitor is

$$I_c = \frac{V_i}{(R_1 + R_2) + (1/j\omega C)} \tag{16.44}$$

where V_i is the input voltage. The voltage across the capacitor is $I_c Z_c$ or

$$V_c = \frac{V_i}{(R_1 + R_2) + (1/j\omega C)} \cdot \frac{1}{j\omega C} = \frac{V_i}{j\omega C(R_1 + R_2) + 1} \tag{16.45}$$

Note that if $j\omega C(R_1 + R_2)$ is much smaller than 1, V_c is essentially equal to V_i. In contrast, if $j\omega C(R_1 + R_2)$ is much larger than 1, V_c becomes approximately equal to $V_i/[j\omega C(R_1 + R_2)]$. This later value V_c will lag V_i by approximately 90° and will have a magnitude much less than V_i. Thus, for a given value of ωC, the *magnitude* and the *phase angle* of the gate voltage V_c in this analysis can both be adjusted by changing $R_1 + R_2$ (or really just R_1) in Fig. 16.24.

By adjusting the magnitude and the phase angle of the gate voltage, the time when the SCR fires can be controlled. This effect is illustrated in Fig. 16.25. In Fig. 16.25a, the value of $\omega C(R_1 + R_2)$ is much smaller than 1, so the gate voltage has almost the same magnitude and the same phase as the applied voltage. Since the gate potential exceeds the firing potential very early in the cycle, current flows as if the SCR were a conventional diode. At the end of the positive half cycle, the anode voltage and current are reduced to zero so the gate regains control of the SCR. The voltages then reverse on the anode and gate. (The diode in Fig. 16.24 protects the gate against the large reverse-bias voltage.) With reversed potentials, almost no current flows through the SCR. As the gate and anode voltages again become positive, the cycle repeats.

In Fig. 16.25b, the value of $\omega C(R_1 + R_2)$ is approximately equal to 1. Then from Eq. 16.29, the gate voltage lags the applied voltage by 45°. In this case, the gate will not permit the SCR to fire until the voltage across the load and the SCR has progressed 45° through the positive half-cycle. Thus, the current waveform has the shape shown in Fig. 16.25b. Notice that the average value of this current waveform is less than the average value of the current waveform shown in Fig. 16.25a. Since the power delivered to the load is $I^2 R$, the power to the load is reduced as the conduction angle (the period when the SCR is conducting) is reduced.

In Fig. 16.25c, the value of $\omega C(R_1 + R_2)$ is much greater than 1. The gate voltage now lags the applied voltage by almost 90°. In addition, the amplitude of the gate potential is reduced to such a small magnitude that the gate will not fire until the gate voltage is almost at its peak value. Thus, the gate fires when the applied voltage to the load and the

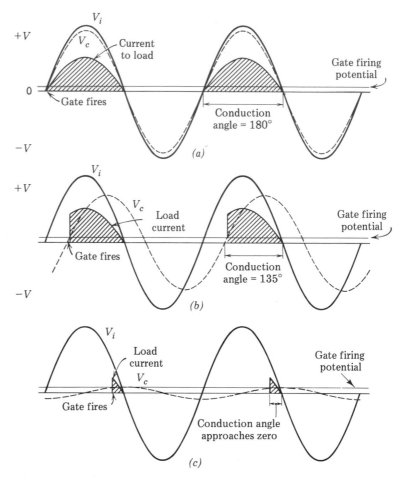

Fig. 16.25. Effect of adjusting R_1 in Fig. 16.24 on the voltage and current waveforms: (a) $\omega C\,R_1 + R_2 \ll 1$; (b) $\omega C\,R_1 + R_2 \simeq 1$; (c) $\omega C\,R_1 + R_2 \gg 1$.

SCR has almost reached the end of its positive half cycle. A short pulse of current flows and then ceases as the applied voltage to the load and SCR drops to zero.

The design considerations for this type of circuit will be illustrated through the use of an example.

Example 16.7

Let us design a light-dimmer circuit to handle a 100 W light globe. We shall use the circuit shown in Fig. 16.26. The SCR must be able to handle about 1 A (current of the 100 W light) and block at least 185 V (the 120 V line may increase to 130 V with 185 V peak). Since the 2N3562, which has data given in Table 16.3, has a blocking voltage

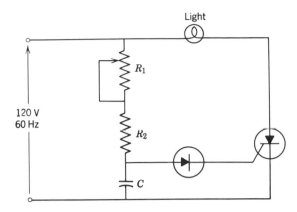

Fig. 16.26. Light-dimmer circuit used in Example 16.7.

of 200 V and an average anode current of 1A, it will be used. The gate will have current pulses longer than 8 ms, so the average gate power dissipation of 100 mW must be used to determine the maximum allowable gate current in this circuit. Curves of gate-cathode voltage versus gate current (present in the Texas Instrument Manual) indicate typical 2N3562 devices have 1.3 V from the gate to cathode when the gate current is 150 mA. If the rms values of 1.3 V and 0.15 A are used, the average power dissipation would seem to be 195 mW. However, since this circuit is a half-wave circuit, the actual average dissipated power is really only 98 mW, which is below the allowable gate dissipation. Maximum gate current flows when R_1 is zero. Then the value of R_2 should limit the gate current to 0.15 A rms. The value of R_2 is 120 V / 0.15 = 800 Ω. To allow an extra safety margin, let us choose R_2 to be 1,000 Ω.

From Eq. 16.45 we note that if R_1 is zero, $\omega C R_2 \ll 1$. Since ω is $2\pi \times 60 = 377$ rad / s, we have $C \ll 1 / 377 \times 10^3$ or $C \ll 2.66$ μF. Let us pick a value approximately one-fifth of this value. Then $C = 0.5$ μF will be used. As a final calculation, $\omega C(R_1 + R_2) \gg 1$ when all of R_1 is in the circuit. Then, $377 \times 5.0 \times 10^{-7}(R_1 + R_2) \gg 1$

Table 16.3. Data for the 2N3562 SCR.

Silicon-controlled rectifier
Forward blocking voltage = 200 V
Continuous forward anode current = 1.6 A
Surge current = 18 A
Minimum gate voltage to trigger = 0.3 V at 75°C
Gate trigger voltage at 25°C = 0.35 to 0.8 V
Maximum gate surge current = 250 mA for 8 ms

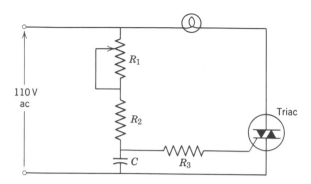

Fig. 16.27. Full-wave control circuit using a triac.

or $(R_1 + R_2) \gg 5{,}200 \ \Omega$. Again, let us choose $(R_1 + R_2)$ about five times as large as this value or $R_1 = 25{,}000 \ \Omega$.

The maximum reverse voltage permitted on the gate is 5 V, so the diode shown in Fig. 16.26 will be required. This diode must be able to withstand the total reverse voltage applied to the circuit as noted by Eq. 16.45. Thus, the diode must have a reverse voltage rating greater than 185 V and must be able to pass a current equal to the maximum gate current (150 mA in this example) when forward biased.

The circuit shown in Fig. 16.26 has one serious limitation. The maximum current this circuit will pass is a half-wave signal. Since a regular 120 V light globe is intended for use in a full-wave circuit, the light from the lamp in Example 16.9 will vary from essentially zero to about one-third of its brilliance in a half-wave circuit. Full-wave circuits have been developed and are available in the literature.[1] These circuits are very useful for controlling dc motors or other devices which obtain their dc power by rectification from an ac source. However, a much simpler way to control power to a load that will operate on ac power is to use a *triac*, as shown in Fig. 16.27. In this circuit, the SCR is replaced by a triac. Since the triac can switch from a blocking to a conducting state for either polarity of applied anode voltage and with either positive or negative gate triggering, the current through the load will have the waveform shown in Fig. 16.28. In this circuit, the current is controlled from zero to a full conduction value by adjusting R_1 through the proper range.

Another application for the triac is the *solid-state* relay shown diagrammatically in Fig. 16.29. In this circuit, a small control current, of

---

[1] *SCR Manual*, General Electric Company.

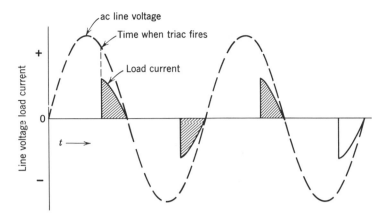

Fig. 16.28. Voltage and current waveform for the circuit in Fig. 16.27.

the order of 1 mA, will cause the light-emitting diode (LED) to turn on the photo transistor and thus cause sufficient current flow through R_2 to fire the triac. The resistance of R_2 must be small enough to allow the triac to conduct through about 170° of each half cycle and yet large enough to protect the photo transistor and the triac gate. The resistor R_1 is chosen to provide the proper LED current from the available control voltage source. The light-coupled amplifier is used to fire the triac because it provides complete isolation between the control circuit and the switching circuit.

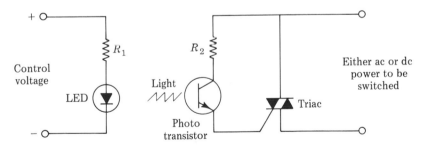

Fig. 16.29. Circuit diagram of a solid-state relay.

Problems   *Section 16.3*

16.1   Determine the average value of output resistance for the power supply with the characteristics shown in Fig. 16.5 with $C = 2,000$ F.

16.2   Design a power supply to provide 25 V and 1.5 A dc output with a 3.75 V peak-to-peak ripple voltage. Use a center-tapped transformer with 120 V 60 Hz primary power. List the component ratings.

16.3   Design a power supply that will provide 40 V at 2 A dc output, using a bridge rectifier and 120 V, 60 Hz primary power. Use pprr = 0.1.

16.4   A given high-voltage power supply must produce 200 V at 200 mA. We desire the peak-to-peak ripple to be no greater than 2 V. Design the power supply and list the ratings of each component used.

Section 16.4

16.5   Design an emitter-follower regulator that will provide 25 V at 2.0 A max to a load. Use the circuit of Fig. 16.8 with $h_{FE1} = 100$ and $h_{FE2} = 50$. Allow for a primary voltage range from 105 V to 130 V at 60 Hz. Specify the component ratings and determine the ripple voltage in the output, if $r_d = 25$ ohms, $I_{Dmin} = 1$ mA and the input pprr = 0.15.

16.6   Use an active filter instead of the regulator in the power supply of Problem 16.5 and determine the value of capacitance C_f required to provide a 5 mV peak-to-peak maximum ripple in the output.

Section 16.5

16.7   A given rectifier-filter has $V = 30$ V with a peak-to-peak ripple voltage = 4 V at $I_L = 0.5$ A and $r_o = 6$ ohms. What must be the open-loop voltage gain of a regulator that will provide 1 mV peak ripple in the output? What will be the output resistance of the regulator?

Section 16.6

16.8   The regulator of Fig. 16.17 is required to deliver 10 A max at 20 V. What is the minimum h_{FE} that I_1 may have if the regulator is an MC1560? What will be the power dissipation of both the regulator chip and T_1 at $I_O = 10$ A if T_1 has the minimum required h_{FE} and $V_{in}1 = V_{in}2 = 30$ V? Determine suitable values for R_{SC}, R_1, and R_2.

Section 16.7

16.9   A regulator is needed to supply 1.0 A at 5.0 V to a given load. The primary power source provides approximately 25 V, unregulated. The acceptable ripple level in the output is 10 mV peak-to-peak. Design a switching regulator for this application, using the circuit of Fig. 16.20. Assume that the voltage amplifier has a very high gain and the series-pass switching transistor has a rise time of 2 μs.

16.10   Use the MC1561 IC regulator to design the power supply of Problem 16.9. Assume that the series-pass switching transistor has a rise time of 2 μs. Compare the efficiency of your switching regulator to that of a conventional regulator using the MC1561 for the same application.

Section 16.8

16.11   A circuit is connected as shown in Fig. 16.26. The circuit elements are as given in Example 16.9. Sketch the input voltage V_1; the capacitor voltage V_C, assuming the gate loading to be negligible; and the current through the lamp i_L if $R_1 = 10$ k ohms. What is the conduction angle in this case?

16.12   A circuit is connected as shown in Fig. 16.27. The triac is an MAC2 and has the following specifications:

     Peak blocking voltage $-$ 200 V

     RMS conduction current $=$ 8 A

     Peak gate power $=$ 10 W

     Peak gate current $=$ 2 A

     Typical ON voltage $=$ 1 V

     Typical gate trigger voltage $=$ 0.9 V (2 V maximum)

Find the value of each circuit element in Fig. 16.27 for control over the current range from zero to essentially full wave. What maximum wattage rating may the light bulb have? What is the maximum power dissipation of the triac?

17

Oscillator Circuits

In this chapter, an electronic oscillator will be defined as a device that generates a sinusoidal voltage or current waveform. As with most electronic devices, a source of dc power is required for operation. In general, either one-terminal pair devices such as klystrons, tunnel diodes, and so forth, or two-terminal pair devices such as conventional transistors, tubes, etc. can be used as the active elements. Only the two-terminal pair elements will be discussed here. The active elements must work in conjunction with passive (R, L, and C) networks. In some microwave devices, the values of R, L, and C are distributed over the circuit rather than being separate lumped elements. However, even in these instances equivalent circuits of R, L, and C elements can be developed to help visualize the action of the device.

We have encountered oscillating circuits in the previous work. The feedback circuits of Chapter 12 were found to oscillate under some conditions. In these circuits the oscillations were undesirable and the amplifier circuits were said to be unstable when the oscillations occurred. Therefore, we learned how to avoid oscillations. However, we will soon discover that all practical communications systems require signal generating circuits known as *oscillators*; so we now need to reevaluate our previous assumption that all oscillations are to be avoided. The aforementioned circuits could be, and sometimes are, used to fill this need. But in general, special techniques are required to enable the oscillator to meet the following requirements:

1. The ability to maintain a constant frequency of oscillation. This characteristic is known as *frequency stability*.

2. The ability to maintain a constant amplitude output. This charac-
    teristic is known as *amplitude stability*.

3. Single-frequency output which requires a good sinusoidal wave-
    form.

The techniques will be the subject of this chapter.

**17.1
RC
OSCILLATORS**

A commonly used oscillator type is the RC oscillator illustrated in Fig.
17.1a. Positive feedback is applied to the amplifier through the circuit
consisting of C_b, R_b, R_a, and C_a. This circuit provides the in-phase
feedback voltage at only one frequency and is therefore the frequency-
determining network. At the operating frequency, the voltage V_f at the
noninverting input terminal lags the feedback current I_f by the same
phase angle that the current I_f leads the output voltage V_o. The negative
feedback circuit consisting of R_f and R_1 limits the voltage gain of the
amplifier to the value required to give a sinusoidal output voltage.
Increased understanding of the circuit operation may be obtained from
an elementary circuit analysis.

The basic feedback equation (12.7) shows that the amplifier gain
becomes infinite, and hence oscillation occurs when $K_{vn} \mathscr{F}_{vp} = 1$ and the
feedback is positive, where K_{vn} is the voltage gain of the amplifier with
negative feedback and $\mathscr{F}_{vp}$ is the positive voltage feedback factor.

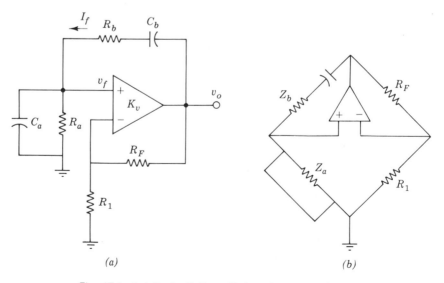

(a)   (b)

*Fig. 17.1. (a) Basic R-C oscillator circuit. (b) Circuit of (a)
drawn in a bridge configuration.*

But

$$\mathcal{F}_{vp} = \frac{Z_a}{Z_a + Z_b} = \frac{\dfrac{R_a / j\omega C_a}{R_a + 1/j\omega C_a}}{\dfrac{R_a / j\omega C_a}{R_a + 1/j\omega C_a} + R_b + 1/j\omega C_b} \tag{17.1}$$

Multiplying both numerator and denominator by $(R_a + 1/j\omega C_a)$

$$\mathcal{F}_{vp} = \frac{R_a / j\omega C_a}{(R_a / j\omega C_a) + (R_a + 1/j\omega C_a)(R_b + 1/j\omega C_b)} \tag{17.2}$$

Equation 17.2 is simplified greatly if we let $C_a = C_b = C$ and $R_a = R_b = R$. Then

$$\mathcal{F}_{vp} = \frac{R}{R + j\omega CR^2 + 2R + \dfrac{1}{j\omega C}} \tag{17.3}$$

and, since $K_{vn} \mathcal{F}_{vp} = 1$,

$$K_{vn} = \frac{1}{\mathcal{F}_{vp}} = 1 + j\omega CR + 2 + \frac{1}{j\omega CR} \tag{17.4}$$

However, K_{vn} is real over the mid-frequency range of the amplifier. Therefore, the sum of the j-terms must be 0 and

$$j\omega CR = \frac{-1}{j\omega CR} \tag{17.5}$$

Solving for ω, the required frequency of oscillation,

$$\omega = \frac{1}{RC} \tag{17.6}$$

Substituting this value of ω into Eq. 17.4,

$$K_{vn} = 3 \tag{17.7}$$

Thus the amplifier with its negative feedback must have a voltage gain of 3. However,

$$K_{vn} \simeq \frac{1}{\mathcal{F}_{vn}} = \frac{R_F + R_1}{R_1} = 3 \tag{17.8}$$

Therefore,

$$R_F = 2R_1 \tag{17.9}$$

The parallel combination of R_1 and R_F should equal R_a in order to minimize the output offset voltage.

The circuit of Fig. 17.1a is redrawn in the form of a bridge circuit in Fig. 17.1b to show why it is commonly known as a *wein bridge oscillator*. The differential input voltage of the amplifier is so very small in comparison with the output voltage that the circuit essentially appears as a balanced bridge.

The major problem with the circuit of Fig. 17.1 is that the output voltage will be distorted if the negative feedback factor is too small. On the other hand, the circuit will not oscillate if the negative feedback factor is too large. In other words, the circuit operation is much too sensitive to the feedback resistance values. This problem may be overcome by several different techniques, two of which are shown in Fig. 17.2.

In the circuit of Fig. 17.2a, R_F is slightly greater than $2R_1$, so the s-plane poles are in the right half plane when the oscillator is switched on. Therefore, the oscillations increase in amplitude until the peak voltage across R_F exceeds the breakdown voltage of the *double anode clipper* D_{Z2} (two reference diodes back-to-back). Due to the shunting effect of the clipper on R_F, the amplifier gain then decreases slightly, moving the poles to the $j\omega$ axis, and the peak ac voltage across R_F is held constant at approximately the breakdown voltage of the double anode

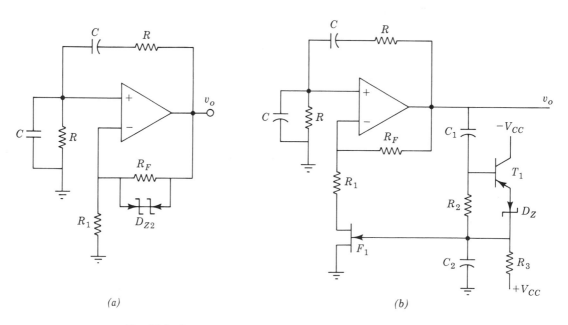

(a)   (b)

Fig. 17.2. Automatic gain control and limiting circuits for
Wein-Bridge oscillators.

clipper. The output voltage V_o must also remain essentially constant since it is 1.5 times as large as the voltage across R_F.

In the circuit of Fig. 17.2b the FET F_1 acts as a variable resistance in the negative feedback loop to control the gain of the amplifier. You may recall that the drain resistance of an FET is essentially the ohmic channel resistance when the magnitude of the drain-to-source voltage is less than the pinch-off voltage and this channel resistance is controlled by the gate voltage. Thus, the feedback factor and hence the amplifier gain can be controlled by the output voltage magnitude. When power is first applied to the circuit, the gate of F_1 is forward biased by the positive supply and the channel resistance is very low. Then the negative feedback factor is essentially $R_1 / (R_F + R_1)$ which is chosen to be slightly less than 1/3. The initial turn-on and noise cause oscillations to build up at the frequency ω_o until the output voltage negative peak exceeds the Zener diode voltage plus the forward V_{BE} of T_1. Then T_1 conducts and the emitter current of T_1 increases the voltage across R_3 and thus negatively charges capacitor C_2. Therefore, F_1 becames reverse biased and the channel resistance increases until the negative feedback factor becomes 1/3 and the amplifier gain becomes precisely 3. The supply voltages $+V_{CC}$ and $-V_{CC}$ must be large enough so that the peak output voltage (about 7 V if $V_E = 6.5$ V) is well within the linear range of the amplifier.

The circuit shown in Fig. 17.2a is simpler than the circuit of Fig. 17.2b, but the double anode clipper in the feedback network causes some distortion by flattening the peaks of the output wave. This flattening may be slight and the distortion may be less than 1 percent if precision resistors are used and ratio $R_1 / (R_F + R_1)$ is only slightly less than 1/3.

The Wein-Bridge oscillator has been popularly used in commercial signal generators which cover the frequency range from approximately 1 Hz to several MHz. This wide range is usually accomplished by switching the capacitors C to change capacitor values in decade steps and by using *ganged* potentiometers to simultaneously change the values of R to cover each decade range.

17.2
FUNCTION
GENERATORS

While sinusoidal signals are widely used, often it is useful or necessary to generate other waveforms. For example, digital circuits require square waves or pulse waveforms for proper operation. As a result, many of the laboratory signal generators are capable of generating a variety of signal waveforms. These signal sources are normally called *function generators*.

One type of function generator (suggested by National Semiconductor Corporation[1]) is shown in Fig. 17.3. The op amp A_1, and its

---

[1] *Linear Applications Handbook*, National Semiconductor Corporation, 2900 Semiconductor Drive, Santa Clara, California.

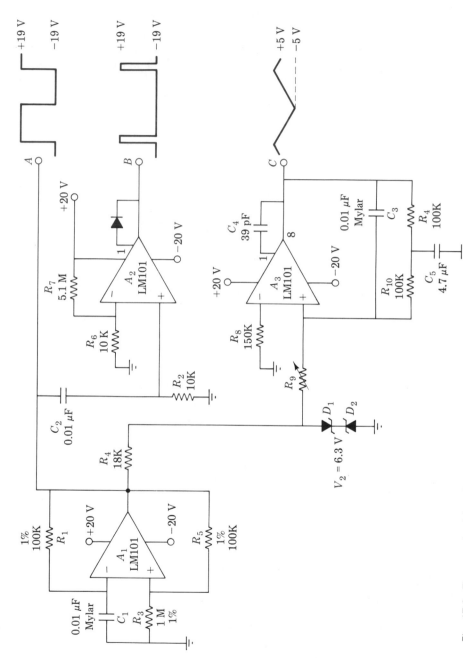

Fig. 17.3. Three-output function generator.

associated circuitry, is a square wave generator. To see how this circuit operates, let us assume the − input terminal of A_1 is a volt or so more negative than the + input terminal. Then the output voltage of A_1 will be saturated just about one volt below $+V_{CC}$. The voltage on the + input terminal (V_{i+}) is determined by the ratio of resistors R_3 and R_5.

$$V_{i+} = V_o \frac{R_3}{R_3 + R_5} \tag{17.10}$$

In Fig. 17.3, the value of V_{i+} will be 17.27 V when V_o is 19 V. The capacitor C_1 charges through R_1 toward V_o. When V_{i-} is a few millivolts below V_{i+}, the voltage of V_o will begin to decrease. V_{i+} will also begin to decrease, but V_{i-} is held essentially constant by C_1, and V_o will decrease even more. The effect is accumulative; so V_o will drop abruptly from $+19$ V to about a volt above $-V_{CC}$ or -19 V. The voltage V_{i+} will then drop to -17.27 V. The voltage on C_1 will decrease from 17.27 V toward -19 V with a time constant of R_1C_1. The equation for V_{i-} during this time will be

$$V_c = V_{i-} = V_f + (V_{in} - V_f) e^{-t/\tau} \tag{17.11}$$

where V_f is the final value of V_c (-19 V in our circuit) and V_{in} is the initial value of V_c (17.27 V in our circuit). Of course, when V_c reaches -17.27 V, the op amp switches and V_o will become 19 V. If these known values of voltages are substituted into Eq. 17.11, we have

$$-17.27 = -19 + (17.27 + 19) e^{-t/R_1C_1} \tag{17.12}$$

or

$$t = \left(\ell n \frac{36.27}{1.73} \right) 10^{-2} = 0.03 \text{ second} \tag{17.13}$$

Then,

$$f = \frac{1}{2t} = 16.4 \text{ Hz} \tag{17.14}$$

As noted when V_{i-} reaches -17.27 V, the output will shift to $+19$ V and the positive half cycle begins again. The operating frequency of this square wave generator is determined by the ratio of R_5 and R_3, which sets the value of V_i where switching occurs, and by the time constant R_1C_1.

The op amp A_2 and its associated circuitry in Fig. 17.3 is a *monostable multivibrator*. This circuit has one stable state, with V_{i-} held at about 0.4 V and V_{i+} at less than 0.4 V and usually at about 0 V. When the output signal of A_1 flips to $+19$ V, this signal change of 38 V (from

−19 V to +19 V) is coupled to the V_{i+} terminal of A_2 and V_o of A_2 flips to +19 V. The charge on C_2 is drained through R_2 to ground. Since the time constant of R_2C_2 is much less than the time constant of R_1C_1, the voltage on V_{i+} of A_2 drops below 0.4 V long before one-half period of the square wave. The pulse output of A_2 drops back to −19 V whenever V_{i+} is less than 0.4 V. The duration of the output pulse of A_2 can be found from Eq. 17.11, when V_c is 0.4 V, V_i is +38 V, V_f is 0 V, and τ is R_2C_2. In general, the pulse width of A_2 is determined by the ratio of R_6 and R_7, since these resistors determine the switching voltage for V_i of A_2. In addition, the time constant R_2C_2 determines how fast V_{i+} drops below the switching voltage. Of course, the pulse width of A_2 cannot exceed the time for one-half cycle of the square wave.

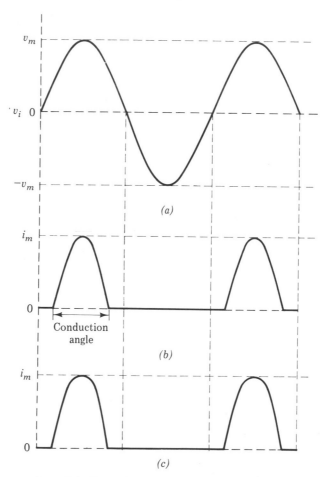

Fig. 17.4. Current waveforms in class C amplifiers:
(a) input voltage waveform;
(b) conduction in linear region;
(c) conduction into saturation region.

The op amp A_3 and its associated circuitry is a dc stabilized integrator circuit. Its output is a ± 5 V triangular wave. The resistor R_4 and diodes D_1 and D_2 reduce the ± 19 V output signal to a ± 6.3 V square wave. The amplitude of the triangular wave is determined by the V_z of the diodes D_1 and D_2 and by the time constant of $R_9 C_3$. The dc stabilization is accomplished by the T circuit R_{10}, R_{11}, and C_5.

Function generators may take a variety of forms. Usually the basic circuit is a sine wave generator and then other waveforms are derived to give the desired output. The manufacturers of the different ICs provide a large array of suggested waveform generators and function generators.

**17.3
CLASSES OF
OPERATION**

In previous chapters, we have talked about class A and class B amplifiers. When one desires to amplify a single frequency or a narrow band of frequencies, a class C amplifier may be used. In the class C amplifier, current flows for less than 180° in each cycle. In fact, current usually flows in pulses, as shown in Fig. 17.4c. As this figure shows, the current waveforms are greatly distorted. This distortion takes the form of harmonics added to the fundamental signal components. This distortion may be eliminated by passing the signal through a band pass filter that will pass the fundamental components, but reject the higher frequency components. Simple L-C filters in the form of high Q resonant circuits work well in the radio-frequency spectrum. At these frequencies, the inductors are relatively small and light and may have Qs of 100 or more.

Since the class B amplifiers have much higher efficiencies than the class A amplifier, the reader may suspect the class C amplifiers have even higher efficiencies. Such is indeed the case. For example, it can be shown[2] that if the current waveform is a part of a sinusoidal wave as in Fig. 17.4b, the efficiency is

$$\eta_p = \frac{\theta - \sin \theta}{4 \sin \theta / 2 - 2 \cos \theta / 2} \tag{17.15}$$

where θ is the conduction angle or the number of degrees in the cycle during which conduction occurs. Unfortunately, as the conduction angle decreases, so does the power output for a given peak collector current. Thus, a compromise between power output and efficiency must be achieved.

If the input signal to a class C amplifier increases in magnitude, the collector may be driven into saturation. Then, the sinusoidal collector current becomes flattened on the top, as shown in Fig. 17.4c. In this case, a Fourier analysis of this waveform must be used to find the

---

[2]C. L. Alley and K. W. Atwood, *Electronic Engineering*, 3rd edition, John Wiley and Sons, New York, New York, 1973, pp. 590–593.

magnitude of the fundamental signal component and the dc component in order to determine the efficiency. The flattened waveform suggests a method for achieving greater efficiency in the amplifier. When the transistor is cutoff, there is no collector current, so the power dissipated in the collector circuit is zero. When the transistor is in saturation, the collector current is high, but the collector voltage is only a few tenths of a volt, so the power dissipated in the collector is also very low. In fact, the time when the collector power dissipation becomes significant occurs only while the transistor is in the linear region. Therefore, let us quickly switch between the cutoff region and the saturation region and the collector power dissipation can be drastically reduced. Operation in this mode is referred to as *class D operation*.

One form of a class D amplifier is shown in Fig. 17.5a. The input signal is a square voltage wave that alternately saturates one transistor and cuts off the other transistor. If the transistor saturation voltage is ignored, the effect is the same as if the two transistors are replaced by the single-pole double-throw switch shown in Fig. 17.5b. The input voltage v_I to the tuned circuit is a square wave with a peak-to-peak value of V_{CC}, as shown in Fig. 17.5c. A Fourier analysis of this square wave yields

$$v_I = V_{cc} \left(\frac{1}{2} + \frac{2}{\pi} \sin \omega_o t + \frac{2}{3\pi} \sin 3\omega_o t + \frac{2}{5\pi} \sin 5\omega_o t + \cdots \right)$$

where ω_o is the fundamental frequency and $1/f_o$ equals the period of one cycle of the square wave. The resonant frequency of the series tuned circuit is ω_o, so essentially only the fundamental component of the signal current flows through the tuned circuit. The voltage across R_L is therefore

$$v_{RL} = \frac{2V_{CC}}{\pi} \sin \omega_o t \qquad (17.16)$$

The current through the resonant circuit has the waveform shown in Fig. 17.5d.

$$i_{RL} = \frac{2V_{CC}}{\pi R_L} \sin \omega_o t \qquad (17.17)$$

The positive half of this current waveform flows through Q_1 when it is saturated, and the negative half of this current waveform flows through Q_2 when it is saturated. The capacitor C_b must be included in the circuit to prevent signal currents from circulating through the power supply. This bypass capacitor must have negligible reactance at the resonant frequency of the tuned circuit.

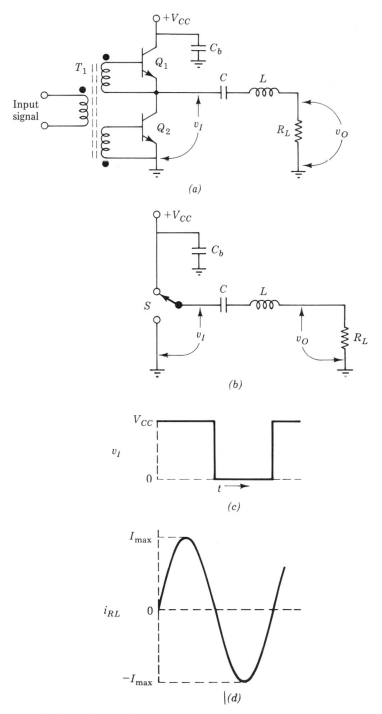

Fig. 17.5. Class D amplifier: (a) actual circuit; (b) an equivalent circuit;
(c) input voltage waveform to the tuned circuit;
(d) resulting current in the tuned circuit.

For proper filtering action, the series resonant circuit should have a Q of at least 5. The maximum allowable third harmonic component determines the acceptable minimum value of Q. Of course, additional filter elements could be used if required. The power output of this amplifier is

$$P_o = \frac{2V_{CC}}{\pi\sqrt{2}} \times \frac{2V_{CC}}{\pi R_L \sqrt{2}} = \frac{2V_{CC}^2}{\pi^2 R_L} \tag{17.18}$$

The average dc current drawn from the power supply is

$$I_{DC} = \frac{I_{max}}{\pi} = \frac{2V_{CC}}{\pi^2 R_L} \tag{17.19}$$

Then, the power drawn from the power supply is

$$P_I = I_{DC}V_{CC} = \frac{2V_{CC}^2}{\pi^2 R_L} \tag{17.20}$$

The efficiency would be 100 percent with the assumptions we have made. Actually, both v_{RL} and the V_{CC} term in i_{RL} are reduced by the collector saturation voltage (V_{Csat}) across the two transistors. Hence, the actual power output is

$$P_o = \frac{2(V_{CC} - 2V_{Csat})^2}{\pi^2 R_L} \tag{17.21}$$

In addition, we have assumed the turn-on and turn-off times of the transistors to be negligible. However, note that the current through each transistor is essentially zero near the switching times. Therefore, if the switching times are finite, but still less than 5 degrees of an input signal cycle, this loss may be neglected. If this effect must be included, the current and voltage values must be multiplied as we integrate over the switching times. Other effects such as shunt capacitances and lead inductance may have an effect, but in general the efficiencies of these amplifiers are very high if suitable transistors are available. Several other high-efficiency circuits are listed in the literature.[3] An example may help to further clarify these concepts.

Example 17.1

Design a class D amplifier which will deliver maximum output to a 50 Ω load (perhaps a transmission line) at a frequency of 1.8 MHz from a 12 V power supply.

First, let us assume the maximum collector saturation voltage of

[3] H. L. Krauss, C. W. Bostian, F. H. Raab, *Solid-State Radio Engineering*, John Wiley and Sons, New York, New York, 1980. pp. 432–471.

the transistors is 0.25 V. Then, the peak-to-peak voltage at v_I in Fig. 17.5 is $12 - 2(0.25) = 11.5$ V. The peak value of collector current is (Eq. 17.17) $I_{max} = 2 \times 11.5 / (\pi \times 50) = 146$ mA. The power output is (Eq. 17.21) $P_o = 2(11.5)^2 / \pi^2 50 = 536$ milliwatts. The average dc current drawn from the power supply is (Eq. 17.19) $I_{DC} = 2 \times 11.5 / (\pi^2 \times 50) = 46.6$ mA. The power input to the collector circuit of the amplifier is $P_I = I_{DC} V_{CC} = 0.0466 \times 12 = 559$ milliwatts. If the saturation voltage is the main loss of power in these transistors, the efficiency of the collector circuit is $\eta = (536 / 559)100 = 95.9$ percent. The two transistors must dissipate $559 - 536 = 23$ milliwatts between the two of them. Our transistors must have the following characteristics:

$$V_{CE} \text{ max} = 12 \text{ V or more}$$

$$I_C \text{ max} = 150 \text{ mA or more}$$

$$P_D \text{ max} = 12 \text{ mW or more}$$

The remaining circuit elements must now be found. Let us assume the ohmic resistance in the inductor is small compared to 50 Ω. In order to obtain a good sinusoidal output waveform, let us design for a circuit Q of 10. Since $Q = \omega_o L / R_L$, the value of L is $Q R_L / \omega_o = 10 \times 50 / (2\pi \times 1.8 \times 10^6) = 44.2$ μH. At resonance, $\omega_o L = 1 / \omega_o C$, so $C = 1 / \omega_o^2 L = 1 / (2\pi \times 1.8 \times 10^6)^2 \times 44.2 \times 10^{-6}) = 176$ pF. Let $(1 / \omega_o C_b) = 1\Omega$ or $C_b = 1 / (2\pi \times 1.8 \times 10^6) = 8.84 \times 10^{-8}) = 0.0884$ μF. Let us use a value of 0.1 μF for the bypass capacitor C_b.

In the foregoing example, the power output of the given amplifier was quite low. In order to increase the power output, either the supply voltage V_{CC} must be increased or the load resistance R_L must be decreased. An easy way to reduce the effective load resistance is to use an impedance transformer that has been designed to operate at the desired frequency. At the higher frequencies, these transformers can be physically small and light in weight.

17.4 TUNED AMPLIFIERS An important class of oscillators known as LC oscillators are basically tuned amplifiers with positive feedback. Therefore, we need to consider the fundamental principles of tuned amplifiers before proceeding to the various oscillator circuits that are identified primarily by their tuning and feedback techniques.

A tuned FET amplifier is shown in Fig. 17.6. Observe that the basic difference between this amplifier and the RC-coupled amplifier is the tuned circuit which replaces the resistor in the drain circuit. In order to

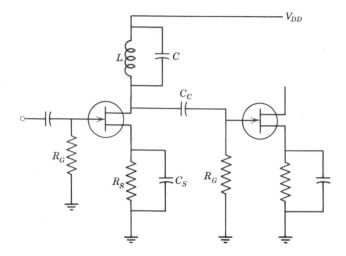

Fig. 17.6. Tuned FET amplifier.

determine the voltage gain and the bandwidth of the amplifier, the equivalent circuit of Fig. 17.7 is given. The capacitor C includes the shunt capacitance of the amplifiers and the distributed wiring capacitance. This absorption of the shunt capacitance into the tuned circuit makes possible the amplification of very high frequencies, approaching the f_T of the transistor.

The equivalent circuit of Fig. 17.7 would have only parallel elements and, hence, the analysis would be comparatively easy except for the series resistance of the coil R_{ser}. Therefore, we will determine a parallel combination of resistance and inductance which will have the same impedance over the pass band as the series combination. This will be accomplished by finding the admittance of the series combination.

$$Y = \frac{1}{Z} = \frac{1}{R_{ser} + j\omega L} \tag{17.22}$$

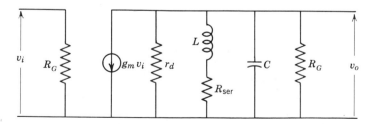

Fig. 17.7. Equivalent circuit for the tuned amplifier of Fig. 17.6.

Rationalizing,

$$Y = \frac{R_{\text{ser}} - j\omega L}{(R_{\text{ser}} + j\omega L)(R_{\text{ser}} - j\omega L)} = \frac{R_{\text{ser}} - j\omega L}{R_{\text{ser}}^2 + (\omega L)^2} \qquad (17.23)$$

But the ratio of ωL to R_{ser} is the Q of the coil, known as Q_o. Therefore, if Q_o is ten or higher, which is usually the case, $(\omega L)^2$ is at least 100 times R_{ser}^2, so this latter term can be neglected in the denominator of Eq. 17.23. Using this simplification,

$$Y \simeq \frac{R_{\text{ser}} - j\omega L}{(\omega L)^2} = \frac{R_{\text{ser}}}{(\omega L)^2} - j\frac{1}{\omega L} \qquad (17.24)$$

This admittance is of the form $G + jB$ and represents a conductance $G = R_{\text{ser}}/(\omega L)^2$ in parallel with an inductive susceptance $B = 1/(\omega L)$, as shown in the dashed enclosure in the equivalent circuit of Fig. 17.8. Since the Q of a coil can be written as $Q_o = \omega L/R_{\text{ser}}$, the conductance of the parallel combination representing the coil can be written

$$G_P = \frac{R_{\text{ser}}}{(\omega L)^2} = \frac{1}{Q_o \omega L} \qquad (17.25)$$

This conductance may also be expressed as a resistance

$$R_{\text{par}} = \frac{1}{G_p} = \frac{(\omega L)^2}{R_{\text{ser}}} = Q_o \omega L \qquad (17.26)$$

Observe that this equivalent parallel resistance is a function of frequency. However, the tuned amplifier which uses a high Q circuit ($Q \geq 10$) amplifies only a narrow band of frequencies near the resonant frequency ω_o. Therefore, the effective parallel resistance of the coil may be assumed constant over the pass band with the value

$$R_{\text{par}} = Q_o \omega_o L \qquad (17.27a)$$

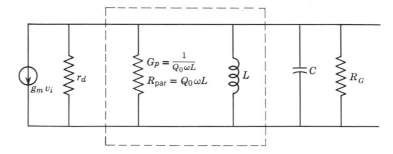

Fig. 17.8. Equivalent circuit containing only parallel elements.

Note that the Q_o of the coil is the ratio R_{par}/ω_oL and that small values of series resistance R_{ser} which provide high Q_o give large values of effective parallel resistance R_{par} because R_{par} is proportional to the Q_o of the coil. In fact, rearranging Eq. 23.6 yields

$$Q_o = R_{par}/\omega_oL. \tag{17.27b}$$

The effective parallel resistance of the coil can be combined with the other parallel resistance elements in the equivalent circuit of Fig. 17.8 to produce the simplified equivalent circuit of Fig. 17.9, where R represents the parallel combination of r_d, R_{par}, and R_G of Fig. 17.8. In calculating Q_o (the coil Q), we have used R_{par}/ω_oL where R_{par} accounts for the energy loss in the coil. However, the characteristics of a tuned circuit depend upon the energy loss of the *entire* circuit. Thus, a *circuit* Q will be defined as

$$Q = \frac{R}{\omega_oL} = R\omega_oC \tag{17.28}$$

where R is the effective total shunt resistance in parallel with L and C, as shown in Fig. 17.8.

The symbol Q, given without subscripts, will ways represent the *circuit* Q in this discussion. The circuit Q is a very important parameter in a tuned circuit because it determines the bandwidth and affects the amplifier gain. Since the impedance of a lossless parallel tuned circuit is infinite at the resonant frequency, Fig. 17.9 shows that, at resonance, the total load impedance is R and the output voltage V_o of the FET amplifier is

$$V_o = -g_mV_iR \tag{17.29}$$

Then, the voltage gain at resonance is

$$G_v = \frac{V_o}{V_i} = -g_mR \tag{17.30}$$

Observe that the voltage gain is proportional to R, which is equal to $Q\omega_oL$ at the resonant frequency. Thus, the gain is proportional to both

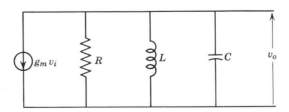

Fig. 17.9. Simplfied equivalent circuit for a tuned amplifier.

the circuit Q and the inductive reactance of the coil. Let us consider an example.

Example 17.2

A circuit is connected as shown in Fig. 17.6. The inductance L has a value of 1 mH and a Q_o of 100. The capacitance is 1,000 pF and $R_G = 1$ MΩ. The FET has $r_d = 2.5 \times 10^5$ Ω and $g_m = 2 \times 10^{-3}$ S. Determine the voltage gain of this circuit.

The value of ω_o is 10^6 rad/s and $R_{par} = 100,000$ Ω. The parallel combination of R_{par} (100,000 Ω), r_d (2.5×10^5 Ω), and R_G (10^6 Ω) is $R = 67$ kΩ. Then, the circuit $Q = R/\omega_o L = 6.7 \times 10^4/10^6 \times 10^{-3} = 67$. Finally, the voltage gain at resonance (Eq. 17.30) is

$$G_v = -g_m R = -2 \times 10^{-3} \times 6.7 \times 10^4 = -134$$

17.5 TUNED-AMPLIFIER BANDWIDTH

The desired circuit Q is determined by the bandwidth requirement of the amplifier, however, and not by the desired voltage gain. The relationship between the bandwidth and the circuit Q may be obtained by writing a general expression for the output voltage V_o. From Fig. 17.9,

$$V_o = -g_m V_i Z = \frac{-g_m V_i}{G + j\omega C + \dfrac{1}{j\omega L}} \tag{17.31}$$

where $G = 1/R$. If we now multiply both numerator and denominator of the right-hand side of Eq. 17.31 by R and then divide both sides of the equation by V_i,

$$G_v = \frac{-g_m R}{1 + j\omega CR + \dfrac{R}{j\omega L}} \tag{17.32}$$

But $R/\omega_o L = Q$ and similarly $R\omega_o C = Q$, since $\omega_o L = 1/\omega_o C$ at the resonant frequency ω_o. Making these substitutions into Eq. 23.11,

$$G_v = \frac{-g_m R}{1 + jQ\left(\dfrac{\omega}{\omega_o} - \dfrac{\omega_o}{\omega}\right)} \tag{17.33}$$

Since the numerator in Eq. 17.33 is equal to the voltage gain at resonance, when $\omega = \omega_o$, the gain decreases to the half-power value when the magnitude of the j part of the denominator is equal to unity. But there are two frequencies at which half-power gain occurs, as shown in Fig. 17.10. These frequencies are designated as ω_L and ω_H. Thus, when

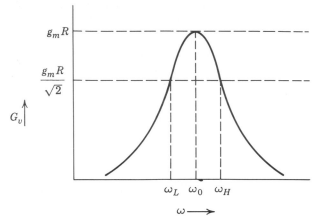

Fig. 17.10. Frequency response of the single-tuned amplifier.

ω is equal to ω_L (or ω_H), the magnitude of the j part of the denominator of Eq. 17.33 is equal to one.

$$Q\left(\frac{\omega_H}{\omega_o} - \frac{\omega_o}{\omega_H}\right) = 1 \tag{17.34}$$

$$Q\left(\frac{\omega_L}{\omega_o} - \frac{\omega_o}{\omega_L}\right) = -1 \tag{17.35}$$

and

$$\frac{\omega_H}{\omega_o} - \frac{\omega_o}{\omega_H} = \frac{\omega_o}{\omega_L} - \frac{\omega_L}{\omega_o} \tag{17.36}$$

Rearranging terms, we have

$$\frac{\omega_H}{\omega_o} + \frac{\omega_L}{\omega_o} = \frac{\omega_o}{\omega_L} + \frac{\omega_o}{\omega_H} \tag{17.37}$$

$$\frac{\omega_H + \omega_L}{\omega_o} = \omega_o \frac{\omega_H + \omega_L}{\omega_H \omega_L} \tag{17.38}$$

Therefore,

$$\omega_o^2 = \omega_H \omega_L \tag{17.39}$$

Again, using Eq. 17.34, we find

$$Q\left(\frac{\omega_H^2 - \omega_o^2}{\omega_o \omega_H}\right) = 1 \tag{17.40}$$

and substituting the value of $\omega_o^2 = \omega_H \omega_L$ from Eq. 17.39,

$$Q\left(\frac{\omega_H^2 - \omega_L \omega_H}{\omega_o \omega_H}\right) = Q\left(\frac{\omega_H - \omega_L}{\omega_o}\right) = 1 \qquad (17.41)$$

But $\omega_H - \omega_L$ is the bandwidth B, in radians per second, as seen from Fig. 17.10. Then,

$$B = \frac{\omega_o}{Q} \qquad (17.42)$$

Note that both sides of Eq. 17.42 may be divided by 2π to obtain the bandwidth in Hertz.

$$B \text{ (Hertz)} = \frac{f_o}{Q} \qquad (17.43)$$

17.6 TUNED COUPLING CIRCUITS The capacitively-coupled circuit of Fig. 17.6 is adequate for coupling FET stages because both the input and output impedances of an FET are high and reasonably good impedance matching is achieved. However, a coupling circuit that provides impedance matching as well as tuning may yield much higher power gain when used to couple bipolar transistors. One such circuit is the inductively-coupled circuit shown in Fig. 17.11.

A simplified y-parameter circuit, in which y_{re} is neglected, is given in Fig. 17.12. The output and input conductances are assumed to be approximately g_{oe} and g_{ie}, respectively. The $jb_{oe} = j\omega_o C_{oe}$ and the trans-

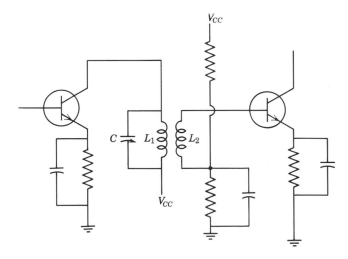

Fig. 17.11. Inductively-coupled tuned transistor amplfier.

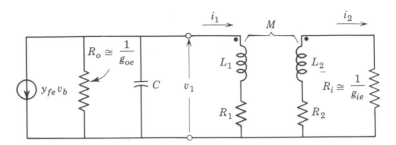

Fig. 17.12. Simplified y-parameter equivalent circuit.

formed jb_{ie} are absorbed in the tuning capacitance C. The mutual impedance of the transformer $j\omega M$, where M is the mutual inductance, is
defined as the ratio of the voltage induced in one winding to the current
flowing in the other winding. Therefore, referring to Fig. 17.12,

$$V_1 = (j\omega L_1 + R_1)I_1 - j\omega MI_2 \qquad (17.44)$$

where the currents and voltages are assumed to be sinusoidal. Since the
voltage induced in the secondary is $j\omega MI_1$,

$$I_2 = \frac{j\omega MI_1}{j\omega L_2 + R_i + R_2} \qquad (17.45)$$

where $R_i \simeq 1/g_{ie}$. Substituting this value of I_2 into Eq. 17.44,

$$V_1 = (j\omega L_1 + R_1)I_1 + \frac{(\omega M)^2 I_1}{j\omega L_2 + R_i + R_2} \qquad (17.46)$$

Therefore, the impedance seen at the primary terminals is

$$Z_1 = \frac{V_1}{I_1} = j\omega L_1 + R_1 + \frac{(\omega M)^2}{j\omega L_2 + R_i + R_2} \qquad (17.47)$$

The third term on the right-hand side of Eq. 17.47 is the impedance
coupled into the primary (in series) as a result of the secondary current
I_2. Note that this term is complex, and thus reactance as well as resistance is coupled into the primary. This reactance, which will affect
the tuning of the primary, is a function of g_{ie}. Thus, the coupling circuit
will be detuned by any change which affects g_{ie}. This undesirable detuning effect can be essentially eliminated if R_i is large in comparison
with ωL_2. This condition can be met in most circuits. In addition, ωL_2
is usually much larger than R_2. Then the impedance seen at the primary
terminals is

$$Z_1 \simeq j\omega L_1 + R_1 + \frac{(\omega M)^2}{R_i} \qquad (17.48)$$

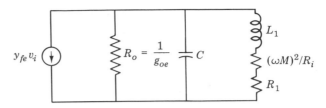

Fig. 17.13. Simplified equivalent circuit with $R_i \gg \omega L_2$.

This transformer primary impedance can replace the transformer in the equivalent circuit of Fig. 17.12 to produce the simplified primary circuit shown in Fig. 17.13.

We now need a basis for choosing the primary inductance L_1. One sensible basis is the obtaining of maximum power transfer, and thus maximum gain. However, maximum power into the primary may not yield maximum power in the load because part of the primary power is lost in the primary resistance R_1. This problem can be eliminated if we transform this loss resistance to its effective parallel value and combine it with R_o to obtain a modified output resistance R_o', as shown in Fig. 17.14. Using the relationship $R_{par} = Q_o \omega_o L_1$ at the resonant frequency (Eq. 17.27) and $R_o' = R_{par} R_o / (R_{par} + R_o)$,

$$R_o' = \frac{R_o Q_o \omega_o L_1}{R_o + Q_o \omega_o L_1} \tag{17.49}$$

If we now match the impedance of the tuned primary, Fig. 17.14, to R_o', maximum power will be transferred to the load, since the power loss in the transformer has been removed. Therefore, we will transform the series resistance $(\omega M)^2 / R_i$ to its equivalent parallel value and equate it to R_o'.

$$R_o' = \frac{(\omega L_1)^2}{(\omega M)^2 / R_i} = \frac{L_1^2}{M^2} R_i \tag{17.50}$$

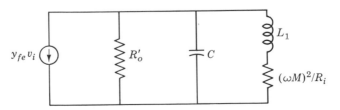

Fig. 17.14. Equivalent circuit showing the modified output resistance R_o', which includes the primary winding loss.

But from basic coupled-circuit theory, we know that the mutual inductance is

$$M = k\sqrt{L_1 L_2} \tag{17.51}$$

where k is the coefficient of coupling. Substituting this value for M in Eq. 17.50 we have

$$R_o' = \frac{L_1^2 R_i}{k^2 L_1 L_2} = \frac{L_1 R_i}{k^2 L_2} \tag{17.52}$$

We will now determine L_1 from the bandwidth requirement. Since the circuit Q is ω_o/B, which is known, and the total shunt resistance R is $R_o'/2$ in the matched amplifier, we can write

$$\frac{R_o'}{2} = Q\omega_o L_1 \tag{17.53}$$

and

$$R_o' = 2Q\omega_o L_1 \tag{17.54}$$

Substituting the value of R_o' given in Eq. 17.49 into Eq. 17.54, we have

$$\frac{R_o Q_o \omega_o L_1}{R_o + Q_o \omega_o L_1} = 2Q\omega_o L_1 \tag{17.55}$$

simplifying,

$$\frac{R_o Q_o}{R_o + Q_o \omega_o L_1} = 2Q \tag{17.56}$$

Then

$$R_o Q_o = 2QR_o + 2QQ_o \omega_o L_1 \tag{17.57}$$

Solving for L_1, Eq. 17.57 becomes

$$L_1 = \frac{R_o(Q_o - 2Q)}{2QQ_o\omega_o} = \frac{R_o}{\omega_o}\left(\frac{1}{2Q} - \frac{1}{Q_o}\right) \tag{17.58}$$

Note that an impedance match cannot be obtained unless the primary Q_o is greater than two times the required circuit Q.

From the y-parameter circuit (Fig. 17.13), the signal voltage at the collector terminal is

$$V_o = \frac{-y_{fe}V_i}{y_{oe} + Y_L} \tag{17.59}$$

However, at the resonant frequency ω_o, the total susceptance $j(b_{oe} + B_L)$ is zero, leaving only $g_{oe} + G_L$ as the total admittance through which the current $y_{fe}V_i$ flows. Then the magnitude of the voltage gain from the base to the collector at resonance is

$$K_{vc} = \frac{V_o}{V_i} = \frac{y_{fe}}{g_{oe} + G_L} \tag{17.60}$$

If maximum power transfer is achieved, $g_{oe} + G_L = 2g'_{oe}$ where g'_{oe} includes the parallel resistance $Q_o\omega_oL$ of the tuned primary.

The voltage gain of the entire stage, from base to base, is usually desired. In a lossless coupling circuit, the voltage ratio must be equal to the square root of the impedance transformation ratio. Therefore, using Eq. 17.60,

$$K_v = \frac{y_{fe}}{g_{oe} + G_L} \sqrt{\frac{R_i}{R'_o}} \tag{17.61}$$

17.7
TAPPED-TUNED
CIRCUITS

Impedance transformation can be accomplished in a tuned circuit by tapping a single coil, as an auto transformer, as shown in Fig. 17.15a. In the equivalent circuit, Fig. 17.15b, R_o is the output reistance of the amplifying device ($1/g_{oe}$, approximately, for a transistor and r_d for a MOSFET or FET). The series resistance in the coil is R_1, and R_i is the input resistance of the following stage. There are n_2 turns between the tap and the RF ground terminal of the coil.

The equivalent circuit of Fig. 17.15b can be simplified somewhat by transforming the coil resistance R_1 into its equivalent parallel value $R_{par} = Q_o\omega_oL$ and then combining this resistance with the output resistance R_o to provide a modified output resistance R'_o, as illustrated in Fig. 17.16.

Since there is no energy loss in the coupling circuit of Fig. 17.16b, the power delivered to the tuned circuit is equal to the power consumed by the load. Let us assume that we want the tuned coupling circuit to provide a resistance R'_o as a load for the driving amplifier so maximum power transfer will be achieved. Then the power delivered to the coupling circuit is V_1^2/R'_o where V_1 is the rms voltage applied to the tuned coupling circuit at resonance. Also, the power delivered to the actual load, which we have assumed is the input resistance R_i of the following amplifier, is V_2^2/R_i where V_2 is the rms voltage across R_i. Then

$$\frac{V_1^2}{R'_o} = \frac{V_2^2}{R_i} \tag{17.62}$$

We will assume that the voltage $d\phi/dt$ generated in each turn is equal to the voltage generated in every other turn. This is rigorously true if

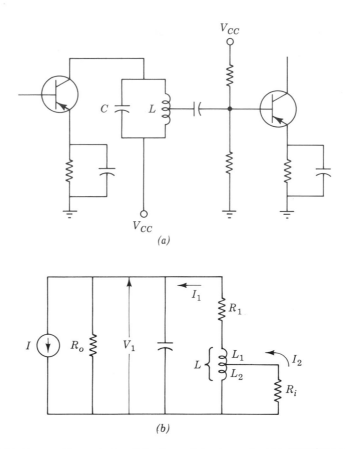

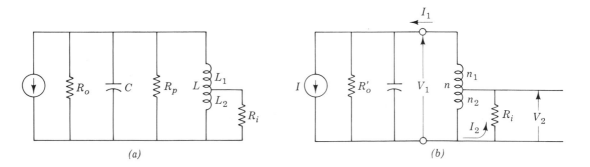

Fig. 17.15. Tapped-tuned circuit that provides impedance transformation:
(a) actual circuit; (b) an equivalent circuit.

Fig. 17.16. (a) Equivalent parallel resistance R_p, which represents the power
loss in coil L, is combined with the transistor output resistance R_o to produce
the modified output resistance in R_o' in (b).

equal flux cuts every turn, as it does in either tightly-coupled coils or long slender coils with small mutual coupling, and is a good approximation for the coil configurations normally used. Then the volts per turn in the coil are V_1/n and

$$V_2 = \frac{V_1 n_2}{n} \tag{17.63}$$

Substituting Eq. 17.63 into Eq. 17.62,

$$\frac{V_1^2}{R_o'} = \frac{n_2^2 V_1^2}{n^2 R_i} \tag{17.64}$$

Solving for n_2 explicitly,

$$n_2 = n \left(\frac{R_i}{R_o'} \right)^{1/2} \tag{17.65}$$

In the event a driving point impedance R_L other than R_o' is desired, that value may be obtained by merely replacing R_o' by R_L in Eq. 17.65.

The required value of inductance L for the total coil can be determined from the bandwidth requirement and the total shunt resistance across the tuned circuit, in precisely the same manner as for the inductively-coupled circuit. Thus, if maximum power transfer is desired, Eq. 17.58 is applicable. If maximum power transfer is not desired, then L can be determined from the relationship $\omega_o L = R/Q$ where R is the parallel combination of R_o' and R_L and $Q = \omega_o/B$.

The impedance transformation could have been accomplished by "tapping the capacitor" instead of the coil, as shown in Fig. 17.17. The relationship between C_1 and C_2 required to provide the desired impedance transformation can be determined from the energy relationship given by Eq. 17.62 for the tapped coil. Let us assume that the current circulating around the tuned circuit is large compared with the external currents. Actually, the circulating current is Q_o times the terminal current, where Q_o is determined at the terminal concerned. Then the current I through C_2 is approximately equal to the current through C_1. $V_2 = -jX_{C2}i$ and $V_1 \simeq -jX_C I$, where C is the capacitance of C_1 and C_2 in series and is the total capacitance that tunes L. These expressions for V_1 and V_2 may be substituted into Eq. 17.62 to yield

$$\frac{X_C^2}{R_o'} = \frac{X_{C2}^2}{R_i} \tag{17.66}$$

Replacing X_C with $1/\omega_o C$ and X_{C_2} with $1/\omega_o C_2$, Eq. 17.66 becomes

$$\frac{1}{C^2 R_o'} = \frac{1}{C_2^2 R_i} \tag{17.67}$$

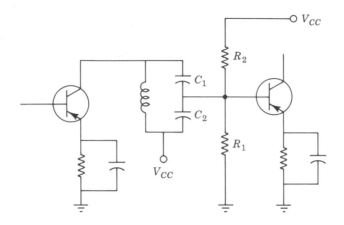

Fig. 17.17. Alternate impedance matching system.

When Eq. 17.67 is solved for C_2, we have

$$C_2 = C \left(\frac{R_o'}{R_i}\right)^{1/2} \tag{17.68}$$

The tuning capacitance $C = 1/\omega_o^2 L$ and since capacitors in series add like resistors in parallel, $C = C_1 C_2/(C_1 + C_2)$. Therefore,

$$C_1 = \frac{C_2 C}{C_2 - C} \tag{17.69}$$

The tapped-tuned circuit has an advantage over the inductively-coupled circuit because of the simplicity and availability of the single

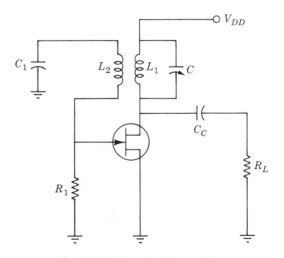

Fig. 17.18. Tuned-drain oscillator.

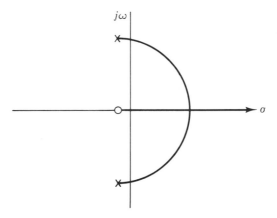

Fig. 17.19. Pole-zero and root-locus for the tuned-drain oscillator.

coil. However, part of the signal is shunted through the bias resistors R_1 and R_2 (Fig. 17.17).

**17.8
LC
OSCILLATORS**
The single-turned inductively-coupled amplifier can be used as an oscillator if some of the energy in the output circuit is coupled back to the input circuit, as shown in Fig. 17.18.

    The pole-zero plot for the voltage gain of the tuned-drain amplifier is given in Fig. 17.19. The root-locus plot for positive feedback is also given in this figure. Observe that the frequency of oscillation is very nearly the resonant frequency of the tuned circuit if the circuit Q is high because the amplifier (open-loop) poles are very near the $j\omega$ axis. The actual frequency of oscillation and the mutual inductance requirements may be determined from an analysis of the equivalent circuit of Fig. 17.20. Small-signal operation is assumed initially because the equivalent circuit is valid only for small signals. Since the transformer primary impedance is $j\omega L_1 + (\omega M)^2/R_1$ (Eq. 17.48), assuming $R_1 \gg \omega L_2$, the nodal equation for the drain node may be written

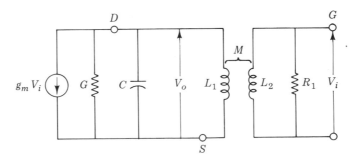

Fig. 17.20. Equivalent circuit for the tuned-drain oscillator.

$$g_m V_i = \left[G + j\omega C + \frac{1}{j\omega L_1 + (\omega M)^2 / R_1} \right] V_o \qquad (17.70)$$

where G includes the conductance of the load resistance R_L. If $R_1 \gg \omega L_2$, as previously assumed, the gate voltage V_i is essentially $j\omega M I_d$, where I_d is the signal current through L_1. Then, since $I_d = V_o / (j\omega L_1 + \omega^2 M^2 / R_1)$,

$$V_i = \frac{j\omega M V_o}{j\omega L_1 + [(\omega M)^2 / R_1]} \qquad (17.71)$$

The voltage gain with positive feedback may now be written using $G_a = V_o / V_i$, which is obtainable from Eq. 17.70, and $\mathscr{F} = V_i / V_o$ from Eq. 17.71.

$$G_{af} = \frac{G_a}{1 - G_a \mathscr{F}} = \frac{g_m / [G + j\omega C + R_1 / (j\omega L_1 R_1 + \omega^2 M^2)]}{1 - \dfrac{j\omega M g_m}{(j\omega L_1 + \omega^2 M^2 / R_1)(G + j\omega C) + 1}} \qquad (17.72)$$

Oscillation occurs when the denominator of Eq. 17.72 is zero. Then

$$j\omega L_1 G - \omega^2 L_1 C + \frac{\omega^2 M^2 G}{R_1} + \frac{j\omega^3 M^2 C}{R_1} + 1 - j\omega M g_m = 0 \quad (17.73)$$

Equating the real terms to zero

$$\omega^2 L_1 C = \frac{\omega^2 M^2 G}{R_1} + 1 \qquad (17.74)$$

or

$$\omega^2 = \frac{1}{L_1 C - (M^2 G / R_1)} \approx \frac{1}{L_1 C} + \frac{G M^2}{L_1^2 C^2 R_1} \qquad (17.75)$$

Since $1 / L_1 C = \omega_n^2$, where ω_n is the undamped resonant frequency of the tuned circuit, the frequency of oscillation is

$$\omega = \omega_n \sqrt{1 + \frac{(\omega_n M)^2 G}{R_1}} \qquad (17.76)$$

Equation 17.76 shows that the frequency of oscillation depends on the load conductance G, the input resistance R_1, and the coupling impedance $\omega_n M$. The oscillation frequency will be very nearly ω_n if $(\omega_n M)^2 G / R_1$ is small in comparison with unity, which is true if the circuit Q is high so G is small, or R_1 is large, as it may be when the amplifier is an FET.

The required value of mutual inductance may be obtained by equating the imaginary terms of Eq. 17.73 to zero. Then

$$\frac{\omega^3 M^2 C}{R_1} - \omega M g_m + \omega L_1 G = 0 \qquad (17.77)$$

or

$$M^2 - \frac{g_m R_1 M}{\omega^2 C} + \frac{L_1 G R_1}{\omega^2 C} = 0 \qquad (17.78)$$

Using the quadratic formula,

$$M = \frac{g_m R_1}{2\omega^2 C} \pm \sqrt{\left(\frac{g_m R_1}{2\omega^2 C}\right)^2 - \frac{L_1 G R_1}{\omega^2 C}} \qquad (17.79)$$

$$M = \frac{g_m R_1}{2\omega^2 C} \left(1 \pm \sqrt{1 - \frac{4\omega^2 L_1 C G}{g_m^2 R_1}}\right) \qquad (17.80)$$

If the circuit Q is high, $\omega^2 \simeq 1/L_1 C$. Then

$$M \simeq \frac{g_m R_1 L_1}{2} \left(1 \pm \sqrt{1 - \frac{4G}{g_m^2 R_1}}\right) \qquad (17.81)$$

The solution which results from the positive sign preceding the radical in Eq. 17.81 may be discarded because it yields unrealizable values of M. Normally, $4G/g_m^2 R_1 \ll 1$; therefore, the approximation $(1-x)^{1/2} \simeq 1 - x/2$ can be used to simplify Eq. 17.81. Then,

$$M \sim \frac{L_1 G}{g_m} \qquad (17.82)$$

Note that the required value of M is independent of frequency, provided that tuning is achieved by using a variable capacitor and L_1 remains fixed. The value of M obtained from Eq. 17.82 results in class A operation. Larger values of M cause increased gate drive and result in class B or class C operation, depending on M. The resistor R_1 and capacitor C_1 (Fig. 17.18) automatically provide the increased reverse bias required for class B or class C operation as the mutual inductance M is increased and the gate junction draws current on the positive peaks of the input voltage.

A bipolar transistor can be used instead of the FET in Fig. 17.18. Then, g_m is replaced by the y_{fe} of the transistor and y_{fe} becomes complex at frequencies above f_β. Thus the analysis becomes more complicated at these higher frequencies. Also, the lower input resistance R_1 of the bipolar transistor causes the frequency to be more dependent upon the transistor parameters.

The Hartley oscillator circuit shown in Fig. 17.21 is similar to the tuned-collector circuit except a single, tapped coil is used and the tuning capacitor tunes the entire coil. In this arrangement, the coupling between the collector and base circuits does not depend on the mutual inductance between L_1 and L_2 because an ac voltage across L_1 is also applied across the series combination of C and L_2. Therefore, signal current flows through L_2, and the voltage across L_2 is the feedback voltage to the base-emitter junction. The signal currents through L_1 and L_2 are essentially equal (because the tuned circuit current is large compared with the collector current), so the ratio of collector voltage to base voltage is essentially L_1/L_2. This ratio is the voltage gain, and the reciprocal L_1/L_2 is the feedback ratio. Thus $\mathscr{F}G = 1$, provided that L_2/L_1 is large enough to cause oscillation. Note that one end of the tuned circuit is at the same signal potential as the collector, and the other end of the tuned circuit is the same signal potential as the base. The coil tap is at the same signal potential as the emitter. Since the base and collector are at opposite potentials with respect to the emitter, the feedback is positive. This signal arrangement always holds for the Hartley oscillator. Since the input signal is not referenced to ground, any one of the three coil or electrode terminals may be at signal-ground potential. The oscillator operation is unaltered by the choice of ground point except, of course, that the output terminal must *not* be at the signal-ground point. An example will be used to illustrate one method of designing a Hartley oscillator.

Example 17.3

A 2N4957 transistor with y-parameters given in Fig. 17.22 is to be used in the Hartley oscillator circuit of Fig. 17.21. Let us assume the load resistance R_L to be 10 kΩ and the Q_o of the coil to be 100. The oscillator frequency is to be 1.0 MHz. The circuit Q should be high to ensure good frequency stability. In this example, we will design for $Q = 50$. The base driving power is very small in comparison with the power furnished to the load or dissipated in the tuned circuit and, therefore, its effect on the circuit Q will be neglected. Figure 17.22 shows that $g_{oe} \simeq 0.1$ mS at $I_C = 2$ mA, so the parallel combination of R_L and R_o is $R_X = 5$ kΩ.

The inductance L_1 can be determined by the following method. The total shunt collector circuit resistance is

$$R_{sh} = Q\omega_o L_1 \qquad (17.83)$$

The portion of this shunt resistance contributed by the coil resistance in the tuned circuit is

$$R_{par} = Q_o\omega_o L_1 \qquad (17.84)$$

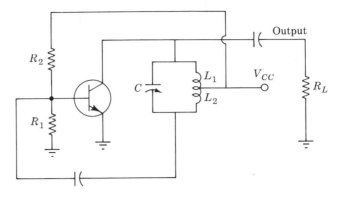

Fig. 17.21. Hartley oscillator circuit.

But R_{sh} is the parallel combination of R_{par} and R_X. Then, using Eqs. 17.84 and 17.83

$$R_X = \frac{R_{par}R_{sh}}{R_{par} - R_{sh}} = \frac{Q_o Q}{Q_o - Q}\,\omega_o L_1 \tag{17.85}$$

and

$$L_1 = \frac{R_X (Q_o - Q)}{\omega_o (Q_o Q)} = \frac{R_X}{\omega_o}\left(\frac{1}{Q} - \frac{1}{Q_o}\right) \tag{17.86}$$

In this example

$$L_1 = \frac{5 \times 10^3\,(50)}{6.28 \times 10^6\,(100)\,(50)} = 8.0 \text{ } \mu\text{H}$$

The class A voltage gain of the amplifier is $y_{fe}R_{sh}$. Also, R_{sh} is 5 kΩ in parallel with $Q_o\omega_o L_1 = 5$ kΩ, or 2.5 kΩ. The Q-point collector current will be chosen as 2.0 mA. Then y_{fe} (Fig. 17.21) is 58 mS and

$$G_v = y_{fe}R_{sh} = 58 \times 10^{-3}\,(2.5 \times 10^3) = 145$$

The oscillator will operate class A if

$$L_2 = L_1 / G_v = 8 / 145 = 0.055 \text{ } \mu\text{H}$$

However, a change in parameters or loading might stop the oscillation in this class A mode. The oscillator will be much more dependable if the inductance L_2 is increased by a factor of at least 4 or 5. The oscillator will then have much better amplitude stability and greater power output. Then

$$L_2 \simeq 5\,(0.055) \text{ } \mu\text{H} = 0.27 \text{ } \mu\text{H}$$

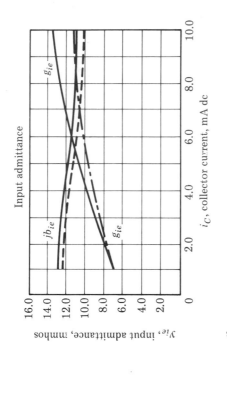

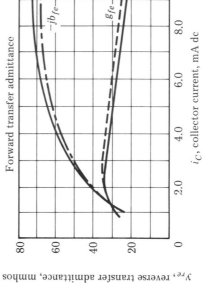

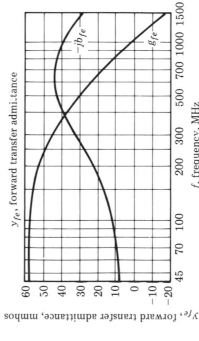

(a) Y parameters versus frequency
$I_C = -2$ mA, $V_{CE} = -10$ V

(b) Y parameters versus current
$f = 450$ MHz, $V_{CE} = -10$ V

Fig. 17.22. Common-emitter y parameters for 2N4957, 2N4958, and 2N4959 transistors. (Courtesy of Motorola Semiconductor Products, Inc)

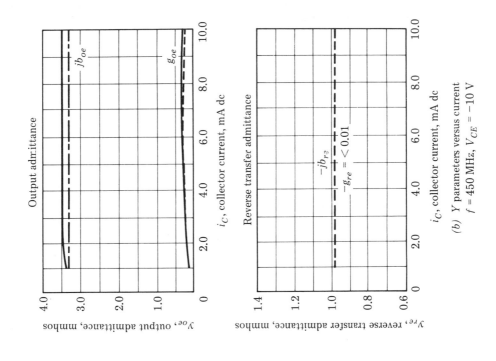

Output admittance

y_{oe}, output admittance

Reverse transfer admittance

y_{re}, reverse transfer admittance

(b) Y parameters versus current
$f = 450$ MHz, $V_{CE} = -10$ V

y_{oe}, output admittance

y_{re}, reverse transfer admittance

(a) Y parameters versus frequency
$I_C = -2$ mA, $V_{CE} = -10$ V

Figure 17.22. (Continued)

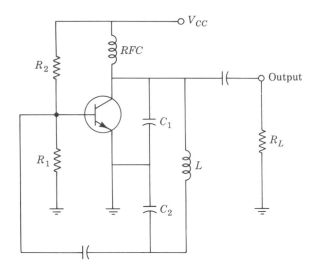

Fig. 17.23. Colpitts oscillator.

The tuning capacitance can be found from the relationship

$$C = \frac{1}{\omega_o^2 L} = \frac{1}{(6.28 \times 10^6)^2 (8.27 \times 10^{-6})} = 3023 \text{ pF}$$

The bias components are chosen to provide about 2.0 mA quiescent collector current and the blocking capacitor C should have reactance equal to approximately one-tenth of the bias resistance.

The Colpitts oscillator shown in Fig. 17.23 is almost identical to the Hartley except that the tuned circuit capacitance, instead of the inductance, is tapped. Also an RF choke has been added to permit the application of direct current to the collector and to present a very high impedance at the oscillation frequency. The design of a Colpitts oscillator may follow the pattern given for the Hartley oscillator in Example 17.3 but with $j\omega L_1$ and $j\omega L_2$ replaced by $1/j\omega C_1$ and $1/j\omega C_2$, respectively.

**17.9
CRYSTAL-
CONTROLLED
OSCILLATORS**

A general class of oscillators which achieve very good frequency stability because of the exploitation of a high Q circuit is the *crystal-controlled* oscillator. In the "crystal" oscillator, the conventional LC circuit is replaced by a quartz crystal. The crystal has the property of producing a potential difference between two parallel faces on one axis when the crystal is strained or deformed along a second axis. Conversely, when a potential difference is applied across the parallel faces along the first axis, it will either compress or expand along the second axis. This property, which is known as the *piezoelectric effect* after its discoverer, causes the crystal to behave as a very high Q resonant circuit. The crystal will vibrate readily at its *mechanical resonant frequency*, but because of its associated electrical properties the crystal behaves as though it were an LC circuit with extremely high Q (of the order of tens of thousands). The crystal is cut into very thin slices and then carefully ground to the desired resonant frequency. The orientation of the slice, with reference to the crystal axes, determines the properties of the crystal, such as vigor of oscillation and variation of frequency with temperature.

The equivalent electrical circuit of a crystal is given in Fig. 17.24. The crystal itself behaves as a series RLC circuit. However, the electrical connections must be made to the crystal faces by conducting electrodes or plates, known as a crystal holder. The crystal holder provides a capacitance, shown as C_h in Fig. 17.24, which is in parallel with the crystal circuit. Thus the crystal behaves as a series resonant circuit at its natural resonant frequency. But at a slightly higher frequency the net inductive reactance of the crystal resonants with the crystal holder capacitance to produce parallel resonance. The parallel resonant frequency is only slightly higher than the series resonant frequency because the equivalent inductance of the crystal may be of the order of henries while the effective capacitance C is a small fraction of a pF. This extremely high equivalent inductance, small capacitance, and low mechanical loss account for the extremely high Q of the crystal and provide a very impressive rate of change of reactance with frequency.

The reactance of a typical crystal in a holder is sketched as a function of frequency in Fig. 17.25. Note that the reactance is inductive only between the series resonant frequency ω_s and the parallel resonant frequency ω_p. These frequencies differ by a very small percentage (a few hundred Hz per MHz); therefore, the effective inductance changes very rapidly with frequency in this region.

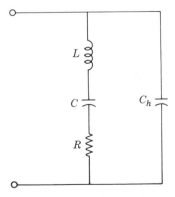

Fig. 17.24. Equivalent circuit of a crystal mounted in a holder.

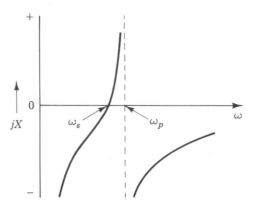

Fig. 17.25. Sketch of reactance as a function of frequency for a crystal in a holder.

Example 17.4

A 1 MHz crystal may typically have $L = 1.20$ H, $C = .02113$ pF, $R = 250 \ \Omega$ and $Q = 30,000$. Crystal holders usually have capacitance values in the range of 5 to 10 pF. Let us assume that the crystal holder for the typical 1 MHz crystal has $C_h = 7$ pF. Then the total capacitance is $7(.02113)/(7 + .02113) = .020924$ pF and the parallel resonant frequency is approximately 1.005 MHz, which is only ½ of 1 percent higher than the series resonant frequency. Thus, the effective inductance of the crystal in its holder changes from essentially 0 to ∞, while the frequency changes 0.5 percent.

Crystal oscillator circuits may be very much like the conventional RC or LC circuits previously discussed except that crystals are used as the frequency-determining elements. For example, the oscillator circuit shown in Fig. 17.26 is known as a *Pierce* oscillator. However, close observation reveals that the Pierce oscillator is actually a Colpitts oscillator with the inductive element replaced by a crystal. This oscillator may be designed as a conventional Colpitts and the crystal will provide the proper inductance for operation at a frequency between the series and parallel resonant frequencies of the crystal. The capacitors C_1 and C_2 appear as tuning capacitors but a change in their capacitance has little influence on the frequency of oscillation because of the compensating change of effective inductance provided by the crystal. The magnitudes of C_1 and C_2 do control the load impedance seen by the transistor and the ratio C_2/C_1 determines the voltage gain and hence the class of operation of the oscillator, as previously discussed.

The crystal may be used in a Hartley-type circuit by either of the techniques shown in Fig. 17.27. In Fig. 17.27*a* the crystal provides the

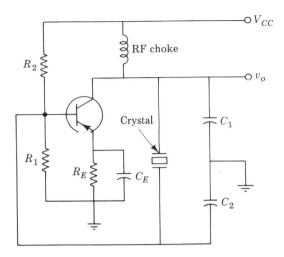

Fig. 17.26. Pierce oscillator.

proper reactance, in series with the coil, to cause the oscillator to operate near the series resonant mode of the crystal. In the circuit of Fig.
17.27b, the crystal acts as a bypass for R_E and permits the circuit to
operate as a Hartley oscillator at frequencies very near the series mode
of the crystal.

The crystal may be used with ICs to produce oscillators, as shown in
Fig. 17.28. The circuit of Fig. 17.28a is the Pierce, or Colpitts, type
where the crystal acts as an inductance and the capacitor ratio C_2/C_1
determines the voltage gain, or class of operation, of the IC. The circuit
of Fig. 17.28b operates as a Wein-Bridge oscillator in which the crystal

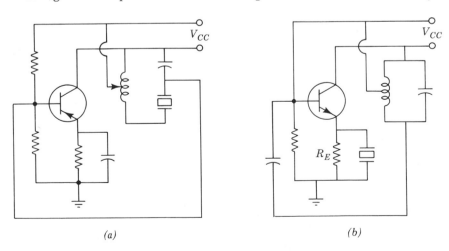

(a)   (b)

Fig. 17.27. Hartley-type crystal oscillators.

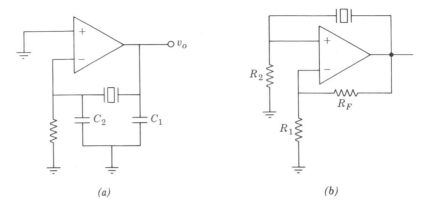

(a)   (b)

Fig. 17.28. (a) Pierce type IC crystal oscillator.
(b) Wein-Bridge type IC crystal oscillator.

provides the proper feedback phase at its series-mode frequency and the resistors R_F and R_1 control the voltage gain. Gain limiting could be used in this circuit, as previously discussed, to provide low distortion. However, the required voltage gain in this circuit, and hence the required ratio of R_F to R_1, is determined by the ratio of the series crystal resistance to the resistance of R_2. If no negative feedback is used the oscillator will provide essentially square waves in its output.

Crystals are commercially available in the fundamental range at any specified frequency between 70 kHz and 20 MHz. One of the authors constructed a crystal oscillator which operated at 6 kHz in the flexural mode of the crystal. Other crystal oscillators may be achieved for frequencies below 70 kHz by the use of divide, or countdown, circuits such as multivibrators. Fundamental mode crystals are not available at frequencies above about 20 MHz because the frequency of oscillation is inversely proportional to the thickness of the crystal slice, and the crystal becomes so fragile at frequencies above 20 MHz that the stress caused by typical output voltages may break the crystal. However, crys-

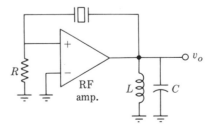

Fig. 17.29. Harmonic operation of a crystal oscillator obtained
by tuning the LC circuit in the output to an odd harmonic
of the fundamental crystal frequency.

tals are distributed elements and will, therefore, vibrate in harmonic modes if encouraged to do so. Crystals will satisfactorily operate on the odd harmonics up to the ninth. Therefore, they may be used up to about 180 MHz. The encouragement needed is the use of a tuned circuit as a load for the oscillator as shown in Fig. 17.29. The LC circuit is tuned to an odd harmonic of the fundamental crystal frequency and the resistance R is made small enough that the closed-loop gain is sufficiently high to produce oscillation only in the neighborhood of the LC circuit resonant frequency.

Problems *Section 17.1*

17.1 Design a Wein-Bridge oscillator that will cover the frequency range from 10 Hz to 1 MHz. Choose the amplifier type, select a method of amplitude stabilization, and determine suitable values for the component parts.

Section 17.2

17.2 The stability of the function generator in Fig. 17.3 can be increased by connecting the right-hand side terminals of R_1 and R_5 to the zener diode side of R_4. Then, loading on terminal A will not effect the frequency of A_1. Make this change and determine the frequency of the square wave signal and the pulse width of the pulse at terminal B. How will this change effect the signal on terminal C?

17.3 Modify the function generator in Fig. 17.3 so that the square wave has a frequency of 400 Hz. The pulse width of the pulse generator is 15 μs, and the amplitude of the signal on terminal G is still ±5 V.

Section 17.3

17.4 Design a class D amplifier which will deliver the maximum power output to a 10 Ω load at a frequency of 3 MHz. Use a 12 V power supply and assume the maximum collector saturation voltage is 0.2 V.

17.5 Design a class D amplifier which will deliver 50 W to a 50 Ω load at a frequency of 2 MHz. Assume the maximum collector saturation voltage is 0.25 V. Use a circuit Q of 10.

17.6   Design a class D amplifier which will provide 0.5 W output power to a 50 Ω load at 27 MHz (CB use). Use a circuit Q of 10 and assume the inductance has a Q_o of 100. Assume $V_{Csat} = 0.25$ V.

Section 17.4

17.7   A 1 mH coil having $Q_o = 100$ is connected in parallel with a 1,000 pF capacitor. Determine the resonant frequency of the combination and the effective parallel resistance.

17.8   In the tuned FET amplifier of Fig. 17.6, $L = 1$ mH, $Q_o = 100$, $C = 500$ pF and $R_g = 1$ megohm. The FET has $g_m = 2$ mS and $r_d = 2.5 \times 10^5$ ohms. Determine the voltage gain and the resonant frequency of the amplifier.

Section 17.5

17.9   An RF amplifier is needed for a standard broadcast receiver. It is to have 20 kHz bandwidth at 1.0 MHz center frequency. Design a single-tuned capacitively-coupled amplifier for this frequency and bandwidth using a FET with $g_m = 2$ mS, $r_d = 250$ kohm, R_g of the following transistor = 1 megohm, $C_o = 5$ pF and $C_i = 25$ pF for the transistors used, at the recommended q-points. Determine the values for L, C, and the voltage gain if the Q_o of the coil = 150.

Section 17.6

17.10   A transistor with $r_o = 10$ kohm is to be coupled to a transistor having $r_i = 1$ kohm. The resonant frequency is 5 MHz and the desired bandwidth is 200 kHz. Use an inductively-coupled circuit to provide maximum power gain and determine L_1, L_2, C and the voltage gain if $y_{fe} = .1$ S. What is the ratio of ωL_2 to r_i? Assume coil $Q_o = 100$.

Section 17.7

17.11   The inductance of a cylindrical coil may be obtained from the following empirical formula:

$$L = \frac{n^2 r^2}{9r + 10l} \times 10^{-6} \text{ H}$$

where n is the number of turns, r is the coil radius and l the length of the coil winding space in inches. Use a tapped coil to couple the amplifiers of Problem 17.10. Determine the total turns n of the coil and the turns n_2 to the tap point if $r = 0.1$ inch and $l = 0.25$ inch.

17.12   Use a "tapped-capacitor" circuit to couple the amplifiers of Problem 17.10. Determine the values of C_1 and C_2.

Section 17.8

17.13   Design a Hartley oscillator to operate at 1 MHz using an FET having $g_m = 2$ mS and $r_d = 100$ kohm at $I_D = 2$ mA.

17.14   Design a Colpitts oscillator which uses a 2N4957 transistor and has the same specifications and load resistance as the Hartley oscillator of Example 17.2.

Section 17.9

17.15   Use the crystal of Example 17.4 to design a Colpitts-type crystal oscillator. How far off the series resonant frequency of the crystal will the oscillator operate?

17.16   Use a crystal in the harmonic mode and a high-frequency op amp to design a 150 MHz crystal-controlled oscillator. Assume that the effective series resistance of the crystal is 50 Ω and the actual load on the oscillator is 2 kΩ. Specify components that will prevent oscillation at the wrong harmonic. What should be the fundamental crystal frequency?

Appendix

Transistor
Characteristics

MAXIMUM RATINGS

Collector-to-base voltage	−75 max V
Collector-to-emitter voltage.........................	−50 max V
Emitter-to-base voltage.............................	−1.5 max V
Collector current	−5 max A
Base current	−1 max A
Emitter current...................................	5 max A

Transistor dissipation:

At mounting-flange temperatures up to 81°C.........	12.5 max W
At mounting-flange temperatures above 81°C	Derate 0.66 W / °C

Temperature range:

Operating (junction) and storage...............	−65 to 100 °C
Lead temperature (for 10 seconds maximum)	255 max °C

CHARACTERISTICS

Collector-to-base breakdown voltage (with collector mA = −10 and emitter current = 0)	−75 min V
Collector-to-emitter breakdown voltage (with collector mA = −100 and base current = 0)	−50 min V
Base-to-emitter voltage (with collector-to-emitter V = −10 and collector mA = −50)	−0.24 V
Collector-cutoff current (with collector-to-base V = −40 and emitter current = 0)	−1 max mA
Collector-cutoff saturation current (with collector-to-base V = −0.5 and emitter current = 0).......	−70 max μA
Emitter-cutoff current (with emitter-to-base V = −1.5 and collector current = 0)	−2.5 max mA

Theremal resistance:

Junction-to-case	1.5 max °C / W

In Common-Emitter Circuit

dc forward current-transfer ratio (with collector-to-emitter V = −1 and collector mA = −1000)	150
Gain-bandwidth product (with collector-to-emitter V = −5 and collector mA = −500)	4 MHz

(Continued)

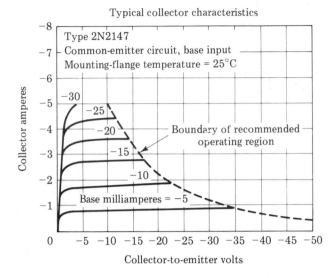

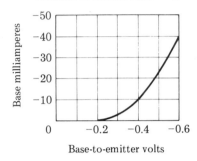

Typical collector characteristics

Type 2N2147
Common-emitter circuit, base input
Mounting-flange temperature = 25°C

Boundary of recommended operating region

Base milliamperes = −5

Collector amperes

Collector-to-emitter volts

Typical base characteristic
Type 2N2147
Common-emitter circuit, base input
Mounting-flange temperature = 25°C
Collector-to-emitter volts = −2

Base milliamperes

Base-to-emitter volts

Fig. A.1. Characteristics of the 2N2147 transistor. (Courtesy of Radio Corporation of America.)

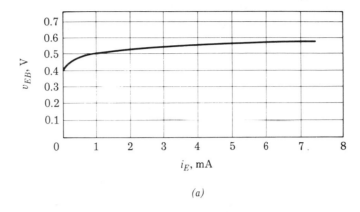

(a)

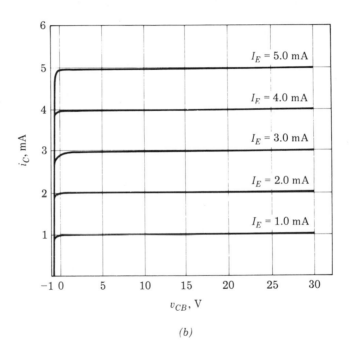

(b)

Fig. A.2a. Common-base characteristics of the 2N3903 transistor: (a) input characteristics; (b) collector characteristics.

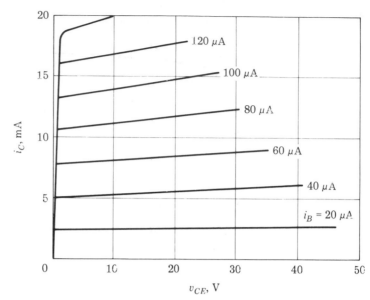

Fig. A.2b. Collector characteristics of the 2N3903 transistor. Maximum ratings: $T = 25$ C, $V_{CEO} = 40$ V, $I_{CO}=0.05$ μA, $I_C=100$ mA, $P_{diss} =310$ mW, derate 2.81 mw/°C. Small signal characteristics: $I_C=5$ mA, $v_{CE}=10$ V, $f = 1$ kHz, $f_t=250$ MHz, $C_{ob}=2$ pF, $h_{fe}=150$, $h_{ie}=850$ Ω, $h_{oe}=30$ μS, $h_{re}=2 \times 10^{-4}$.

Index

A Acceptor level, 32
Active filters, 377–403
   double-pole high pass, 391–392
   high-order, 386–387
   multiple-feedback, 392–398
   stabiliy, 364–365
   twin-T, 398–400
   universal, 400–404
Amplifier, 79
   cascode, 252
   characteristics, 151
   common-base, 81, 139
   common-collector, 145
   common emitter, 89, 132
   common-gate FET, 229
   common-source FET, 224
   complementary-symmetry, 258
   Darlington, 240
   design, 159
   differential, 245
   direct-coupled, 239
   emitter-follower complementary-symmetry, 260
   FET, 194
   FET–BJT, 232
   frequency response, 176
   input compensation, 362–364
   multistage, 267
   n–p–n and p–n–p, 243
   noise, 279–281
   operational (op amp), 3
   quasi-complementary symmetry, 420–424
   source follower FET, 226
   time response, 176
   tuned, 501, 515
Audio amplifier, 1
Automatic gain control, 235
Avalanche breakdown, 62

B   Backward diode, 67
Base, 77
Base width modulation, 93
Beta cutoff frequency, 127
Bias circuit design for FET, 217
Bias for MOSFET, 204
Bipolar transistors, 77
Block diagram, 1
Bode plots, 273–279
Boltzmann constant, 25
Boltzmann equation, 24
Boot strapping circuit, 172
Bridge rectifier circuit, 59
Bypass capacitors, 173

C   Capacitance
    diffusion, 102
    junction, 101
Capacitor input filters, 448
Carrier mobilities, 23
Cascaded stages, 267
Cascode amplifier, 252
Catch diode, 477
Cathode-ray tube, 9
Characteristic curves, transistor, 85
Charge carrier, 20
    density, 23
Class A amplifier, 410
Class B amplifier, 411–416
Class C operation, 497
Class D operation 498–501
Closed-loop regulators, 461-466
Collector, 77
Color organ, 379–383
Colpitts oscillator, 522
Common-base amplifier, 81, 139
Common-collector amplifier, 145
Common-emitter amplifier, 89, 132
Common-gate FET amplifier, 229
Common-source FET amplifier, 224
Comparator, voltage, 367–373
Compensation for fast rise time, 356–364
Complementary-symmetry amplifier 258, 416–420
Conductance, 21
Conduction band, 19
Conduction in semiconductor, 20
Conductor, 18

Coupling capacitors, 173
Coupling circuits, tuned, 507–511
Covalent bonds, 18
Crossover distortion, 412–413
Crystal-controlled oscillators, 522, 526
Current density, 21
Current notation, standard, 82
Current stability factor, 240

D Damping ratio ζ, 327
Darlington connection, 240
dB gain, 272
Decibels, 153
Deflection plates, 9
Density, charge carrier, 23
Depletion-mode MOSFET, 197
Depletion region, 43
Derating curve, 96
Derating factor, 96, 168
Differential amplifiers, 245
Diffusion capacitance, 52, 72, 102
Diffusion of charges, 41
Diffusion constant, 42
Diffusion current, 42
Diode equation, 49, 51
Diode model, piecewise linear, 55
Diode noise, 283
Diode-stabilized Darlington amplifier, 243
Diodes, 39
Direct-coupled amplifiers, 239
Donor level, 28
Doped semiconductors, 27
Drift current, 42, 43
Dual gate MOSFET, 206

E Ebers-Moll model, 83
Electric field intensity, 21
Elmore's rise time, 271
Emitter, 77
Emitter-coupled amplifier, 249
Emitter-follower complementary-symmetry amplifier, 260
Emitter-follower regulators, 455
Enhancement-mode MOSFET, 200
Excess carrier lifetime, 46
Extrinsic material, 27

F   Feedback circuits, 307–312
Feedback effect on band width, 301, 302
Feedback effect on distortion, 300–301
Feedback effect on gain, 299
Feedback factor, 297
Feedback stability, 312–332
FET amplifier, 194, 217
FET–BJT amplifier, 232
FET model, 192
FET noise, 292
Field effect transistors (FET), 187
Fixed bias, 100
Fletcher–Munson curves, 441–442
Forbidden band, 19
Free-wheeling diode, 477
Full-wave bridge rectifier, 58
Function generators, 493–496

G   Gain margin, 314
Gallium-arsenide material, 32
Gap energy, 25, 26
Graded-base transistor, 123

H   h-parameter model, 118
h-parameters, 119
Half-power frequencies, 268
Half-wave rectifier, 58
Hartley oscillator, 518–522
Heat sinks, 435–438
High-frequency model FET amplifier, 224
Hole, 20
Hybrid-π model, 113
Hybrid-π parameters, 121
Hysteresis, 369–371

I   IC oscillators, 525–527
IC power supplies, 466–472
IGFET, 197
Indium phosphide, 34
Injection current, 43
Input offset voltage, 348–349
Integrated circuits, 337–351
Integrator, 351
Intrinsic material, 20
Inverting op amps, 347

J Junction capacitance, 68, 101
Junction FET (JFET), 187
Junction transistors, 77

L Lateral p–n–p transistors, 341
LC oscillators, 515–522
Lifetime, excess carrier, 46
Light dimmer, 480–485
Linear integrated circuits, 337–351
Linear small-signal model, 108
Log-gain plots, 323–325
Loudness control, 441–442

M Majority current, 43
Materials, properties of, 34
Minority carrier diffusion, 46
Minority current, 43
Mobility of carriers, 21, 23
Model, Ebers-Moll, 83
 h-parameter, 118
 hybrid π, 113
 linear small signal, 108
 π, 113
 small-signal, 107
 y-parameter, 116
 z-parameter, 117
Modulation, base-width, 93
Monolithic circuits, 337–341
MOSFET, 197
Multistage amplifier bandwidth, 269–270
Multistage amplifier gain, 267

N n–p–n to p–n–p amplifiers, 243
n-type material, 27
Negative feedback, 297–302
 effect on input impedance, 303–305
 effect on output impedance, 305–306
Noise bandwidth, 282
Noise contours, 291
Noise figure, 285
Noise-figure contours, 160
Noise power, 281–282
Noise temperature, 289
Non-inverting op amps, 346

O   Op-amp compensation, 351–353
Operation, regions of, 91
Operational amplifiers, 3, 345–351
Oscillator amplitude stability, 490
Oscillator frequency stability, 489
Oscillators, 489
   Colpitts, 522
   crystal-controlled, 522–526
   Hartley, 518–532
   IC, 525–527
   LC, 515, 522
   Pierce, 525
Oscilloscope, 10

P   p-type material, 30
Passivation, 125
Peak-to-peak ripple, 60
Peak-to-peak ripple ratio, 448
Peaking of the frequency response, 325–329
Perceived-loudness contours, 441–442
Phase-lag compensation, 330–332
Phase-lead compensation, 318–322
Phase margin, 318
π model, 113
Pierce oscillator, 525
Potential hill, 43
Power amplifiers
   control circuits, 441–443
   IC, 425–428
   output and efficiency, 407–415
   VMOS, 428–434
Power supplies, 57, 447–485
   active filter, 460–461
   integrated circuit, 466–472
   output resistance, 454
   regulation, 448
   requirements, 447
   ripple, 448
Projected cutoff bias, 260
Properties of materials, 34
Push-pull amplifiers, 410–416

R   Radio receiver, 2
Ratings, transistor, 95
RC oscillators, 490
Rectifier circuit, 58

Rectifiers, full-wave, 449–450
Reference diodes, 62, 63, 66
Regulation, 61
Resistance, 21
Resistor noise, 281
Reverse current, 45
Ripple factor, 61
Rise time, 271
Root-locus plots, 314, 317

S S-domain, 272
Saturation current, 43
Sawtooth waveform, 11
Schmitt triggers, 371
Schockley diode, 479
Semiconductors, 17
Silicon-controlled rectifiers, 478–484
Silicon-controlled switch, 479–480
Slew rate, 353–356
Small-signal model, 107
SNR calculations, 293
Solid-state relay, 484–485
Source follower FET amplifier, 226
Spot noise figure, 285
Stabilized-bias circuits, 344–345
Stabilized bias circuit for FET, 218
Stabilized bias design, 167
Stabilizing-bias circuits, 163
Standard voltage–current notation, 82
Static charge on MOSFET, 206
Straight-line approximations, 276
Substrate voltage effect, 205
Summing amplifier, 350
Switching regulators, 473–477
Switching regulator efficiency, 477

T Tapped-tuned circuits, 511–515
Thermal conduction, 435–438
Thermal resistance, 97
Thermal runaway, 439–441
Time response of amplifier, 176
Transition frequency, 128
Transistors
    bipolar, 77
    capacitances, 101
    characteristic curves, 85

Transistors (*Cont'd.*)
  junction, 77
  noise, 284–293
  noise optimization, 289–291
  ratings, 95
Triac, 484–485
Tuned amplifiers, 501–515
Tuned amplifier bandwidth, 505–507
Tuned-circuits, tapped, 511–515
Tuned coupling circuits, 507–511
Twin-T active filter, 398–400

U   Universal active filters, 400–404

V   Valence band, 19
Variations of transistor parameters, 161
VMOS power amplifiers, 428–434
Voltage breakdown characteristics, 99
Voltage comparator, 367–373
Voltage-controlled resistance, 233
Voltage follower, 364–367
Voltage notation, standard, 82
Voltage regulator, 66
Volume units (vu), 272

W   Weinbridge oscillator, 492

Y   y-parameter model, 116
y parameters, 518–521

Z   Zener breakdown, 62
Zener diodes, 63, 66
Z-parameter model, 117